Dynamics of Deflagrations and Reactive Systems: Flames

Edited by
A. L. Kuhl
Lawrence Livermore National Laboratory
Los Angeles, California

J.-C. Leyer
Université de Poitiers
Poitiers, France

A. A. Borisov
USSR Academy of Sciences
Moscow, USSR

W. A. Sirignano
University of California
Irvine, California

Volume 131
PROGRESS IN ASTRONAUTICS AND AERONAUTICS

A. Richard Seebass, Editor-in-Chief
University of Colorado at Boulder
Boulder, Colorado

Technical papers presented from the Twelfth International Colloquium on Dynamics of Explosions and Reactive Systems, Ann Arbor, Michigan, July 1989, and subsequently revised for this volume.

Published by the American Institute of Aeronautics and Astronautics, Inc., 370 L'Enfant Promenade, SW, Washington, DC 20024-2518.

American Institute of Aeronautics and Astronautics
Washington, D.C.

Library of Congress Cataloging-in-Publication data

International Colloquium on Dynamics of Explosions and Reactive
Systems (12th : 1989 : Ann Arbor, Mich.)
Dynamics of deflagrations and reactive systems—flames / edited by A.L. Kuhl ...
[et al.].
p. cm.—(Progress in astronautics and aeronautics; v. 131)
"Technical papers presented from the Twelfth International Colloquium on
Dynamics of Explosions and Reactive Systems, Ann Arbor, Michigan, July 1989,
and subsequently revised for this volume."
Includes index.
ISBN 0-930403-95-9
1. Flame—Congresses. 2. Gas dynamics—Congresses.
I. Kuhl, A. L. II. American Institute of Aeronautics and Astronautics.
III. Title. IV. Series.
TL507.P75 vol. 131 [QD516] 629.1 s—dc20 [541.3'61] 91-7931

Table of Contents

Table of Contents for Companion Volume 132

Table of Contents for Companion Volume 133

Table of Contents for Companion Volume 134

Preface

The four companion volumes on *Dynamics of Deflagrations and Reactive Systems* and *Dynamics of Detonations and Explosions* present 91 of the 149 papers given at the Twelfth International Colloquium on the Dynamics of Explosions and Reactive Systems (ICDERS) held at the University of Michigan in Ann Arbor during July 1989.

These four volumes are included in the Progress in Astronautics and Aeronautics series published by the American Institute of Aeronautics and Astronautics, Inc. *Dynamics of Deflagrations and Reactive Systems: Flames* (Volume 131) and *Dynamics of Deflagrations and Reactive Systems: Heterogeneous Combustion* (Volume 132) span a broad area, encompassing the processes of coupling the exothermic energy release with the fluid dynamics occurring in any combustion process. *Dynamics of Detonations and Explosions: Detonations* (Volume 133) and *Dynamics of Detonations and Explosions: Explosion Phenomena* (Volume 134) principally address the rate processes of energy deposition in a compressible medium and the concurrent nonsteady flow as it typically occurs in explosion phenomena. The colloquium, in addition to embracing the usual topics of explosions, detonations, shock phenomena, and reactive flow, includes papers that deal primarily with the gasdynamic aspects of nonsteady flow in combustion systems, the fluid mechanic aspects of combustion (with particular emphasis on turbulence), and diagnostic techniques used to study combustion phenomena.

In this volume, *Dynamics of Deflagrations and Reactive Systems: Flames*, the papers have been arranged into chapters on ignition dynamics, diffusion flames with shear effects, dynamics of flames and shear layers, turbulent flames, and flame propagation in combustion engines. Although the brevity of this preface does not permit the editors to do justice to all papers, we offer the following highlights of some of the especially noteworthy contributions.

Chapter I, Ignition Dynamics, presents studies of the ignition of gaseous mixtures by a variety of mechanisms. *Kadowaki et al.* report on the results of the ignition of fresh mixtures of methane-air or propane-air by injection into high-temperature combustion products. They show that the ignition delay for methane was about the same as that for propane when high-temperature combustion products were used as the ignition source. This contrasts with traditional ignition studies using only fresh mixtures, where the ignition delay for methane is usually an order of magnitude larger than that for higher hydrocarbons. *Kono and coworkers* describe the results of two-dimensional numerical simulations of the ignition of propane-air mixtures by short-duration electrical sparks. The calculations show that the transformation of the hot gas kernel from an ellipsoidal shape to a torus is caused by the shock-induced inward flow along the spark electrodes. *García-Ybarra and Castillo* examine the ignition of a premixed combustible gas flowing along a hot flat plate.

Chapter II presents a number of articles on Diffusion Flames with Shear Effects. For example, *Tangirala et al.* describe numerical calculations of the structure of diffusion flames stabilized between counterflowing, laminar shear layers of hydrogen and air. For small rates of strain (30 sec^{-1}), a number of reactions are found to be in partial equilibrium over a major part of the flowfield, whereas for large strain rates (1440 sec^{-1}), nonequilibrium effects predominate. This is followed by a number of articles describing calculations of the structure of axisymmetric diffusion flames in hydrogen-air or methane-air mixtures (for example, by *Fukutani and coworkers*, and by *Behrendt and Warnatz*).

Chapter III addresses Dynamics of Flames and Shear Layers. *Ellzey and coworkers* report on numerical simulations of a 20-m/s hydrogen-nitrogen fuel jet surrounded by coflowing air. Conservation equations, including diffusion and conduction terms, are solved by means of an implicit, flux-corrected transport scheme developed at the Naval Research Laboratory. They find that the flame interface is unstable and examine the effects of heat release and viscosity on the formation of large-scale eddies and mixing processes in this diffusion flame. *Yamashita and coworkers* present the time-dependent behavior of a two-dimensional turbulent flame embedded in a shear layer. *Ghoniem and coworkers* describe calculations of flame dynamics based on their vortex element code. They present results for two cases: 1) an unstable, two-dimensional, two-stream, spatially developing and reacting shear layer; and 2) a reacting hot fuel jet emerging into a cold oxidizing environment. In the latter case, they show that density stratification and gravity play important roles in mixing and combustion, and that interactions between the various dynamical processes depend strongly on physical parameters such as the density ratio, the vorticity layer and flame thicknesses, the Froude number, and the form and amplitude of the perturbations.

Chapter IV contains a number of articles on Turbulent Flames. *Liu and coworkers* have investigated the influence of grid-generated turbulence (u_0') on the burning velocities of turbulent flames in methane-hydrogen-air mixtures. Using LDV techniques, they measure turbulent burning velocities (S_T) as large as 18 m/s. They found that this could be correlated with the laminar burning velocity (S_L) according to: $S_T = S_L + 5.3u_0' \sqrt{S_L}$. This chapter also contains articles on a multivariate pdf-closure model applied to the oxidation of CO in a turbulent flow by *Gutheil* and a numerical simulation of turbulent combustion in a boundary layer, among others.

Flame Propagation in Combustion Engines is the subject of Chapter V. *Bates* reports on a flame development in a four-stroke, spark-ignition, optical engine. Instantaneous flame images were used to study the dynamics of the flame, starting from its initiation at the spark plug, through the early kernel growth and mid-cycle burn, to completion at the combustion chamber wall. A range of flame development paths are identified and their implications of engine performance are explored. An experimental investigation of the influence of turbulent fluid motion on low-temperature autoignition (that is, the characteristic blue-flame appearance) caused by piston compression is described by *Ohta et al.* They conclude that swirl can be an effective tool for the elimination of knock in spark-ignition engines. *Gabano and coworkers* have measured autoignition delays for an n-butane and oxidizer mixture in a rapid compression machine. Pressure records and blue-light emission records clearly show a two-stage ignition process for temperatures above 800 K during the postcompression phase.

The companion volumes, *Dynamics of Deflagrations and Reactive Systems: Heterogeneous Combustion* (Volume 132), *Dynamics of Detonations and Explosions: Detonations* (Volume 133), and *Dynamics of Detonations and Explosions: Explosion Phenomena* (Volume 134), include papers on combustion of dust-air mixtures, droplet combustion, combustion at solid and liquid surfaces, combustion diagnostics, gaseous detonations, detonation initiation and transmission, nonideal detonations and boundary effects, multiphase detonations, vapor cloud explosions, blast wave reflections and interactions, and vapor explosions.

These four volumes will, we trust, help satisfy the need first articulated in 1966 and will continue the tradition of augmenting our understanding of the dynamics of explosions and reactive systems begun the following year in Brussels with the first colloquium. Subsequent colloquia have been held on a biennial basis: 1969 in Novosibirsk, 1971 in Marseilles, 1973 in La Jolla, 1975 in Bourges, 1977 in Stockholm, 1979 in Göttingen, 1981 in Minsk, 1983 in Poitiers, 1985 in Berkeley, 1987 in Warsaw, and 1989 in Ann Arbor. The colloquium has now achieved the status of a prime international meeting on these topics, and attracts contributions from scientists and engineers throughout the world.

To provide an enduring focal point for the administrative aspects of the ICDERS, the organization was formally incorporated in the state of Washington under the name Institute for Dynamics of Explosions and Reactive Systems (IDERS). Professor J. R. Bowen is serving as the current president. Communications may be sent to

Dean J. R. Bowen
President, IDERS
College of Engineering FH-10
University of Washington
Seattle, Washington, USA 98195

The proceedings of the first six colloquia have appeared as a part of the journal *Acta Astronautica*, or its predecessor, *Astronautica Acta*. With the publication of the Seventh Colloquium, the proceedings now appear as part of the Progress in Astronautics and Aeronautics series published by the American Institute of Aeronautics and Astronautics.

Acknowledgments

The Twelfth Colloquium was held under the auspices of the Department of Aerospace Engineering, University of Michigan at Ann Arbor, July 23–July 28, 1989. Local arrangements were organized by Professors C. W. Kauffman and M. Sichel. Publication of the Proceedings was made possible by grants from the National Science Foundation and the Defense Nuclear Agency of the United States.

Preparations for the Thirteenth Colloquium are now under way. The meeting will be held at Nagoya University, July 28–August 2, 1991. Professor T. Fujiwara of the University's Department of Aeronautical Engineering is chairman of the local organizing committee.

A. L. Kuhl
J.-C. Leyer
A. A. Borisov
W. A. Sirignano

January 1991

Chapter I. Ignition Dynamics

Detailed Numerical Modeling of H_2-O_2 Ignition by Hot Spots

U. Maas* and J. Warnatz *
Universität Stuttgart, Stuttgart, Germany

Abstract

Auto-Ignition in combustible mixtures is a phenomenon that occurs in many practical combustion processes (e.g., engine knock, ignition in rapid compression machines or shock tubes, etc). In many cases the combustible mixture can be treated as uniform in space, allowing zero-dimensional modelling. Somtimes, however, nonuniformities in temperature or pressure cause the formation of hot spots, that means locations in the mixture, that have an increased temperature with respect to the surrounding. Thus, if (as usual in combustion processes) the ignition delay is highly sensitive to the temperature, the hot spot will ignite much earlier than its surrounding, leading to space- and time-dependent processes that are governed by a superposition of various chemical, gas-dynamic, and transport processes.

This paper presents a mathematical model, which allows the simulation of ignition by hot spots in one-dimensional geometries. Mathematical modelling is performed by solving the conservation equations of the systems (i.e., conservation of mass, species mass, momentum, and energy) using detailed chemistry and a multispecies transport model. Spatial discretization of the resulting partial differential equation system on a statically adapted nonuniform grid leads to a system of ordinary differential and algebraic equations that is solved numerically using implicit methods. Results are presented for ignition processes in the hydrogen-oxygen system and show, that the

* Institut für Technische Verbrennung der Universität Stuttgart

model is able to describe the spatial and temporal development of explosions, as well as detonations, that are initiated by hot spots.

Introduction

Auto-Ignition in combustible mixtures is a phenomenon that occurs in many practical combustion processes (e.g., engine knock, ignition in rapid compression machines or shock tubes, etc). In many cases the combustible mixture can be treated as uniform in space, allowing zero-dimensional modelling.[1] Sometimes, however, nonuniformities in temperature or pressure cause the formation of hot spots, that means locations in the mixture, that have an increased temperature with respect to the surrounding. Thus, if (as usual in combustion processes) the ignition delay is highly sensitive to the temperature, the hot spot will ignite much earlier than its surrounding, leading to space- and time- dependent processes that are governed by a superposition of various chemical, gas-dynamic, and transport processes. These effects cause many auto-ignition processes to be multidimensional phenomena, that start at distinct flame kernels.[2] On the whole, ignition by hot spots may be characterized by three stages that occur more or less successively, namely, induction, excitation, and propagation.

During the induction phase, the increased temperature in the hot spot causes the formation of a radical pool, leading to ignition. This phase is mainly controled by the chemical kinetics and (if the induction time is sufficiently long) by diffusion and heat conduction. If the temperature in the hot spot is high enough, chemical processes will be faster than the various transport processes, and a zero-dimensional (i.e. a homogeneus) model can be used with sufficient accuracy to describe this induction phase (see, e.g., Ref. 3).

After the induction time (i.e., after the radical pool has grown), ignition of the hot spot follows. This phase (namely excitation phase) is mainly controlled by the chemical kinetics and gasdynamic processes. Simplified models, that consider the hot spot itself to be a homogeneous system surrounded by an inert mixture, allow an efficient modelling, as the governing equations (energy and species mass conservation and an equation for the motion of the interface) reduce to an ordinary differential equation system, that can be solved very efficiently.[4]

The last stage of the ignition is the propagation of the ignition. Here several processes may occur. In many cases the excitation phase is followed by the formation of a flame front and flame propagation. This flame propagation is mainly controlled by chemical kinetics and transport processes, and has been the object of many investigations (see, e.g., Refs. 3 and 5-11). However, the temperature in the unburned gas outside the hot kernel influences the flame propagation and, thus, the flame propagation can be influenced by the processes that have occurred in the hot kernel during the excitation phase (e.g., the formation of pressure waves [3,12,13]). If the initial pressure is high enough and the excitation time sufficiently short, high compression ratios in the hot kernel result.[4] These high compression ratios lead to the formation of strong shock waves, which may cause a detonation of the mixture.[14]

The previous examples show that the last stage of the ignition, namely the flame propagation, may take place in various ways, depending on the history of the ignition. Now, our aim is to present a method that allows the simulation of induction, excitation, and propagation with one common model. As previously shown, it is evident that the model has to account for detailed chemistry (especially for the description of the induction phase), detailed transport (important for the flame propagation) as well as for gasdynamic processes.

Our aproach, therefore, is a one-dimensional model, including detailed chemistry, multispecies transport, and allowing for spatial and temporal pressure fluctuations. Simulation is performed by solving the corresponding conservation equations (conservation of mass, species mass, momentum, and energy) that form a partial differential equation system which is solved numerically. In this way many phenomena occurring in hot spot ignitions can be investigated, especially processes like the transition from explosion to detonation, that do not allow simplifications.

Calculation Method

Equation System

Mathematical simulation of the ignition process is performed by solving the corresponding system of conservation equations[15] which may be written as

continuity

$$\frac{\partial \rho}{\partial t} + \text{div}\, \rho\, \vec{v} = 0 \qquad (1)$$

species mass

$$\rho\frac{\partial w_i}{\partial t} + \rho\,\vec{v}\text{grad}\ w_i + \text{div}\ \vec{j}_i = \dot{\omega}_i M_i \tag{2}$$

momentum

$$\frac{\partial \rho\,\vec{v}}{\partial t} + \text{grad}\ P + \text{div}\ \Pi + \text{div}\ (\rho\,\vec{v}\circ\vec{v}) = 0 \tag{3}$$

energy

$$\frac{\partial \rho h}{\partial t} - \frac{\partial P}{\partial t} + \text{div}\ (\rho\vec{v}h) - \vec{v}\text{grad}\ p + \text{div}\ \vec{j}_q + \Pi : \text{grad}\ \vec{v} = \dot{q} \tag{4}$$

P is the pressure, T the temperature, n_s the number of species, w_i the mass fraction of species i, M_i the molar mass of species i, $\dot{\omega}_i$ the molar scale rate of formation of species i, h the specific enthalpy, ρ the density, $\vec{v}$ the velocity, $\vec{j}_q$ the heat flux, $\vec{j}_i$ the diffusion flux of species i, Π the stretch tensor, $\dot{q}$ the source term for deposition of energy, and t the time.

In order to eliminate the convective terms, which cause much problems in the numerical solution, the equation system is transformed into Lagrangian coordinates. Restricting to one-dimensional geometries (infinite slab, infinite cylinder, or sphere), and using the ideal gas law, the following equation system results:

$$\frac{\partial r}{\partial \psi} - \frac{1}{\rho r^\alpha} = 0 \tag{5}$$

$$\frac{\partial \rho}{\partial t} + \rho^2 \frac{\partial}{\partial \psi}(v r^\alpha) = 0 \tag{6}$$

$$\frac{\partial v}{\partial t} + r^\alpha \frac{\partial p}{\partial \psi} - \frac{4}{3} r^\alpha \frac{\partial}{\partial \psi}\left(\rho\mu \frac{\partial}{\partial \psi}(v r^\alpha)\right) + 2\alpha r^\alpha \frac{v}{r}\frac{\partial \mu}{\partial \psi} = 0 \tag{7}$$

$$\begin{aligned}\frac{\partial T}{\partial t} - \frac{1}{\rho C_p}\frac{\partial P}{\partial t} - \frac{1}{C_p}\frac{\partial}{\partial \psi}\left(\rho r^{2\alpha}\lambda \frac{\partial T}{\partial \psi}\right) + \frac{r^\alpha}{C_p}\sum_{i=1}^{n_s} \rho w_i V_i C_{pi}\frac{\partial T}{\partial \psi} \\ + \frac{1}{\rho C_p}\sum_{i=1}^{n_s}\dot{\omega}_i h_i M_i - \frac{4\rho\mu}{3C_p}\left(\frac{\partial v r^\alpha}{\partial \psi}\right)^2 + \frac{2\alpha\mu}{C_p}\frac{\partial}{\partial \psi}\left(v^2 r^{\alpha-1}\right) = \frac{\dot{q}}{\rho C_p}\end{aligned} \tag{8}$$

$$\frac{\partial w_i}{\partial t} + \frac{\partial}{\partial \psi}\left(\rho r^\alpha w_i V_i\right) - \frac{\dot{\omega}_i M_i}{\rho} = 0 \tag{9}$$

$$P - \frac{\rho RT}{\overline{M}} = 0 \tag{10}$$

with $\alpha = 0$ for infinite slab, $\alpha = 1$ for infinite cylinder, $\alpha = 2$ for sphere, r is the radius, c_{pi} the constant pressure heat capacity of species i, λ the thermal conductivity of the mixture, M_i the molar mass of species i, V_i the diffusion velocity of species i, T the temperature, n_s the number of species, w_i the mass fraction of species k, $\dot{\omega}_i$ the molar scale rate of formation of species i, h_i the specific enthalpy of species i, and C_p the constant pressure heat capacity of the mixture.

The independent variables are t and ψ and the dependent variables r, T, P, w_i, v, and ρ.

Initial Values and Boundary Conditions

Pressure and composition of the mixture are assumed to be uniform in space at the beginning of the ignition and the system is considered to be in rest. Simulation of a hot spot is performed by using a space-dependent initial temperature profile.

$$P(\psi, t = t_o) = P_o \qquad w_i(\psi, t = t_o) = w_{i,o} \qquad v(\psi, t = t_o) = 0$$

$$T(\psi, t = t_o) = T_o + \Delta T \exp\left\{-\left(\frac{r - r_m}{r_s}\right)^{2\nu}\right\} \tag{11}$$

where r_s denotes the radius of the hot spot, r_m = location, ν = coefficient for the shape, and ΔT = maximum temperature difference. However, the mathematical model is not restricted to this particular choice. For spherical and cylindrical geometries at the center of the reaction vessel, i.e., at $\psi = 0$, symmetry conditions are used. Outer boundary conditions are simplified by assuming zero gradients of temperature and mass fractions at $\psi = \psi_0$: $\partial T/\partial\psi = 0$ and $\partial w_i/\partial\psi = 0$. Furthermore, $r = R_o$, $\partial P/\partial\psi = 0$, and $v = 0$ for a system at constant volume and $P = P_o$, $\frac{\partial v r^\alpha}{\partial\psi} = 0$ for a system with constant pressure at the outer boundary. R_o denotes the radius of the reaction vessel and P_o the pressure at the outer boundary.

For planar geometries, inner boundary conditions (i.e., at $\psi = 0$) are given by $\partial T/\partial\psi = 0$ and $\partial w_i/\partial\psi = 0$. $r = 0$, $\partial P/\partial\psi = 0$, and $v = 0$.

Solution Method

Spatial discretization using finite differences leads to a system of coupled ordinary differential and algebraic equations that can be solved numerically. Due to the large ratio of vessel diameter to flame front thickness and to the thickness of the hot spot, adaptive gridding has to be used. The procedure may be described as follows: After each time step a new grid point system is calculated. The grid point density is determined by equipartitioning the integral of a mesh function and inverse interpolation, the mesh function F given by a weighted norm of gradients and curvature of the dependent variables f_m

$$F(\psi) = \sum_{m=1}^{M} a_m \frac{\int_0^{\psi} \left| \frac{\partial f_m}{\partial \psi} \right| d\psi}{\int_0^{\psi_0} \left| \frac{\partial f_m}{\partial \psi} \right| d\psi} + b_m \frac{\int_0^{\psi} \left| \frac{\partial^2 f_m}{\partial \psi^2} \right| d\psi}{\int_0^{\psi_0} \left| \frac{\partial^2 f_m}{\partial \psi^2} \right| d\psi} \qquad (12)$$

a_m and b_m are the weighting factors. In the present computations, the integral mesh function distributes about 65 grid points over the physical coordinate. Additional grid points are inserted in order to improve stability, if the ratio of two neighboring grid point intervals is more than a factor of two

$$\frac{1}{2} \leq \frac{\psi_{i+1} - \psi_i}{\psi_i - \psi_{i-1}} \leq 2 \qquad (13)$$

The number of grid points that have to be inserted or deleted in the old grid point system is then determined. If there are only slight changes, the old grid point system is maintained and the integration continued. Otherwise, a new grid point system is generated by piecewise monotonic cubic hermite interpolation[16] and the integration is restarted.

A problem arising in the simulations is the resolution of very thin shock fronts. The thickness of shocks is about 1 μm. In order to avoid severe numerical instabilities, that result if the shock is not resolved, we apply an artificial viscosity term ("numerical diffusion") proposed by Richtmyer and Morton,[17] which spreads the shocks over some grid points.

The system of ordinary differential/algebraic equations is solved by the extrapolation code LIMEX,[18] because this one-step method turned out to need less computing time than backward differentiation formulation (BDF) codes if statical adaptive gridding is used.[3]

Table 1 Mechanism of the hydrogen–oxygen reaction

	Reaction	A	β	E_a
1	$O_2 + H \rightarrow OH + O$	2.00×10^{14}	0.00	70.30
2	$OH + O \rightarrow O_2 + H$	1.46×10^{13}	0.00	2.08
3	$H_2 + O \rightarrow OH + H$	5.06×10^{04}	2.67	26.30
4	$OH + H \rightarrow H_2 + O$	2.24×10^{04}	2.67	18.40
5	$H_2 + OH \rightarrow H_2O + H$	1.00×10^{08}	1.60	13.80
6	$H_2O + H \rightarrow H_2 + OH$	4.45×10^{08}	1.60	77.13
7	$OH + OH \rightarrow H_2O + O$	1.50×10^{09}	1.14	0.42
8	$H_2O + O \rightarrow OH + OH$	1.51×10^{10}	1.14	71.64
9	$H + H + M \rightarrow H_2 + M$	1.80×10^{18}	−1.00	0.00
10	$H_2 + M \rightarrow H + H + M$	6.99×10^{18}	−1.00	436.08
11	$H + OH + M \rightarrow H_2O + M$	2.20×10^{22}	−2.00	0.00
12	$H_2O + M \rightarrow H + OH + M$	3.80×10^{23}	−2.00	499.41
13	$O + O + M \rightarrow O_2 + M$	2.90×10^{17}	−1.00	0.00
14	$O_2 + M \rightarrow O + O + M$	6.81×10^{18}	−1.00	496.41
15	$H + O_2 + M \rightarrow HO_2 + M$	2.30×10^{18}	−0.80	0.00
16	$HO_2 + M \rightarrow H + O_2 + M$	3.26×10^{18}	−0.80	195.88
17	$HO_2 + H \rightarrow OH + OH$	1.50×10^{14}	0.00	4.20
18	$OH + OH \rightarrow HO_2 + H$	1.33×10^{13}	0.00	168.30
19	$HO_2 + H \rightarrow H_2 + O_2$	2.50×10^{13}	0.00	2.90
20	$H_2 + O_2 \rightarrow HO_2 + H$	6.84×10^{13}	0.00	243.10
21	$HO_2 + H \rightarrow H_2O + O$	3.00×10^{13}	0.00	7.20
22	$H_2O + O \rightarrow HO_2 + H$	2.67×10^{13}	0.00	242.52
23	$HO_2 + O \rightarrow OH + O_2$	1.80×10^{13}	0.00	−1.70
24	$OH + O_2 \rightarrow HO_2 + O$	2.18×10^{13}	0.00	230.61
25	$HO_2 + OH \rightarrow H_2O + O_2$	6.00×10^{13}	0.00	0.00
26	$H_2O + O_2 \rightarrow HO_2 + OH$	7.31×10^{14}	0.00	303.53
27	$HO_2 + HO_2 \rightarrow H_2O_2 + O_2$	2.50×10^{11}	0.00	−5.20
28	$OH + OH + M \rightarrow H_2O_2 + M$	3.25×10^{22}	−2.00	0.00
29	$H_2O_2 + M \rightarrow OH + OH + M$	2.10×10^{24}	−2.00	206.80
30	$H_2O_2 + H \rightarrow H_2 + HO_2$	1.70×10^{12}	0.00	15.70
31	$H_2 + HO_2 \rightarrow H_2O_2 + H$	1.15×10^{12}	0.00	80.88
32	$H_2O_2 + H \rightarrow H_2O + OH$	1.00×10^{13}	0.00	15.00
33	$H_2O + OH \rightarrow H_2O_2 + H$	2.67×10^{12}	0.00	307.51
34	$H_2O_2 + O \rightarrow OH + HO_2$	2.80×10^{13}	0.00	26.80
35	$OH + HO_2 \rightarrow H_2O_2 + O$	8.40×10^{12}	0.00	84.09
36	$H_2O_2 + OH \rightarrow H_2O + HO_2$	5.40×10^{12}	0.00	4.20
37	$H_2O + HO_2 \rightarrow H_2O_2 + OH$	1.63×10^{13}	0.00	132.71

A in cm,mole,s; E_A = kJ/mole; $k = AT^{\beta} \exp(-E_A/RT)$.
Third body collision efficiencies: $f_{H_2} = 1.0$, $f_{O_2} = 0.35$, $f_{H_2O} = 6.5$, $f_{N_2} = 0.4$

Efficiency of the method is improved by making use of the block-tridiagonal structure in the numerical evaluation of the Jacobian. The solution of the arising linear equation system is performed by LU (lower and upper triangular) decomposition of the block-tridiagonal iteration matrix. Typical computer times depend on the problem and range from 20 min to about 2 h on an IBM 3090 computer for the examples considered in this work.

Reaction Mechanism and Thermochemical Data

The calculation is performed using a detailed mechanism for the gas-phase reaction, consisting of 37 elementary reactions listed in Table 1.[19,20] Thermodynamic properties are computed from fits of JANAF tables.[21] Furthermore a multispiecies transport model based on the Curtiss-Hirschfelder approximation is used.[22]

Results

The mathematical model described above is able to simulate ignition in hydrogen–oxygen mixtures caused by hot spots including detailed chemistry and multispecies transport for one-dimensional geometries. The results presented below show ignition by hot spots for the geometry of an infinite slab. Thus, the hot spot may be either

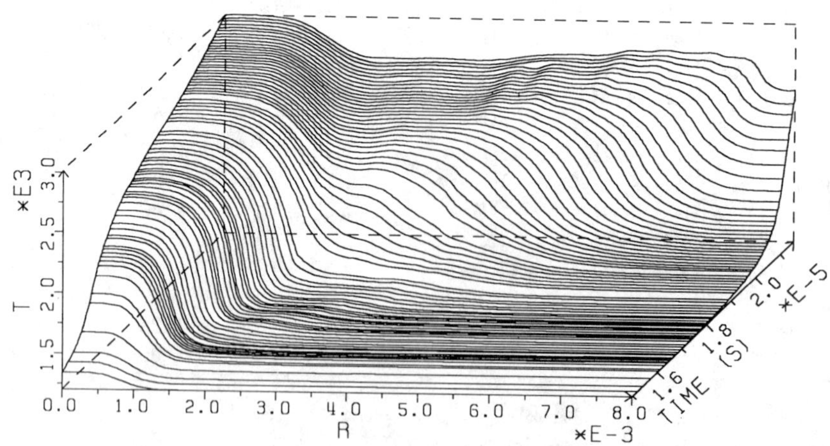

Fig. 1 Calculated temperature profiles in an igniting hydrogen-oxygen mixture (stoichiometric): infinite slab, $R_o = 8$ mm, $P_0 = 1$ bar, T_0 see text.

regarded as a layer with increased temperature in the middle of an infinite slab (leading to a symmetrical ignition) or as a layer with increased temperature at a thermally isolated wall. The shape of the hot spot is chosen quite arbitrarily, the model is not restricted to this particular shape.

Because the ignition delay is highly sensitive to the temperature, the hot spot will ignite much earlier than its surrounding. After the induction time temperature and pressure in the hot spot rise rapidly. The compression ratios associated with the ignition in the hot kernel depend strongly on the initial pressure of the mixture.[4] These compression ratios lead to the formation of pressure waves, which may perturb the local temperature and pressure in the surrounding of the hot spot [3,12,13]). Now, depending on the compression ratio in the hot spot and the initial conditions in the mixture, two extreme cases can be observed. At low compression ratios, a pressure wave begins to propagate and heats the reactive mixture. But, in fact, the heating of the mixture is not sufficient to ignite the surrounding mixture spontaneously. Instead the mixture ignites after its induction time with respect to the increased temperature. If the compression ratio in the hot spot is sufficiently high, it causes a large temperature increase in the unburnt mixture and a detonation can develop. In this case the reaction wave, which is formed, moves with a higher velocity than the initially formed pressure wave and finally merges with the latter to a detonation front.[23,24]

In order to study the effects associated with hot spot ignition, we use the numerical model described above for the simulation of the time-dependent processes at various initial conditions. Two typical results obtained with the numerical model will be shown in the following.

Figures 1-3 show the development of temperature, pressure, and oxygen atom mass fractions of a stoichiometric hydrogen-oxygen mixture at 1 bar initial pressure in a system with the geometry of an infinite slab with a width of 8 mm. The mixture is initially at rest and the temperature is given by

$$T(r, t_0) = 1200K + 50K exp\left\{-(r/1mm)^8\right\}$$

Plotted are temperature, pressure and oxygen atom mass fractions vs the spatial coordinate and the time. Only the time interval after the induction phase is shown. Excitation occurs at the hot spot first,

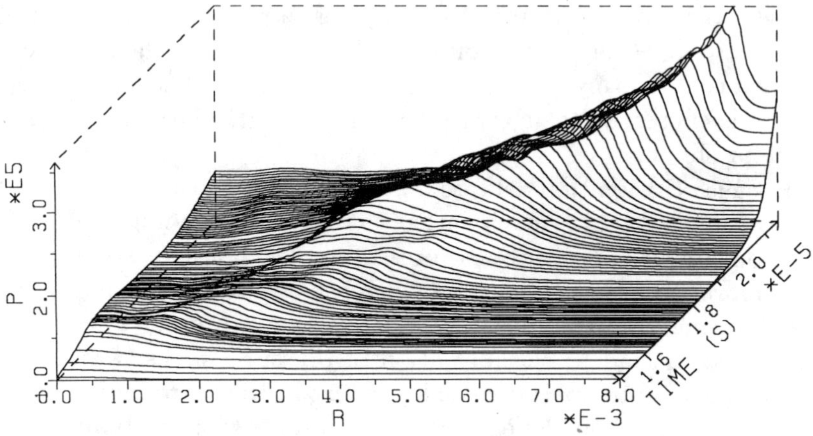

Fig. 2 Calculated pressure profiles in an igniting hydrogen-oxygen mixture (stoichiometric): infinite slab, R_o = 8 mm, P_0 = 1 bar, T_0 see text.

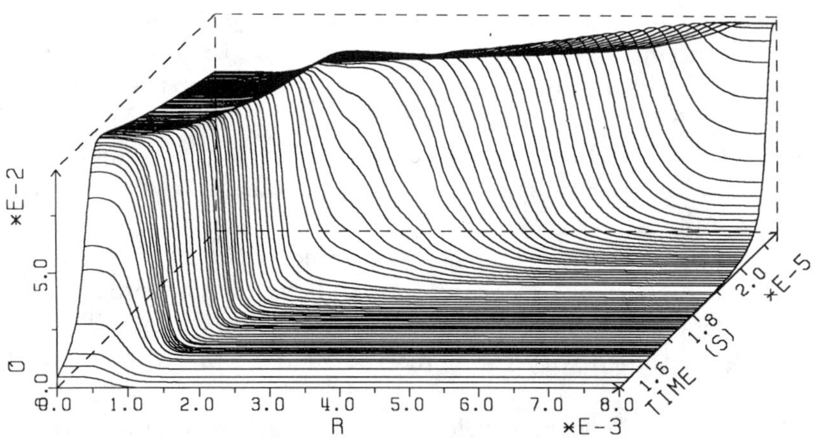

Fig. 3 Calculated profiles of oxygen-atom mass fractions in an igniting hydrogen-oxygen mixture (stoichiometric): infinite slab, R_o = 8 mm, P_0 = 1 bar, T_0 see text.

leading to a rapid temperature rise and the formation of a pressure wave, that moves to the right. After the excitation phase, a flame propagation can be seen. Some μs later, the gas heated by the pressure wave begins to ignite and forms a pressure wave, that amplifies the initial one. Due to this effect, the flame accelerates, i.e., the flame velocity is increased remarkably. The transition from normal flame

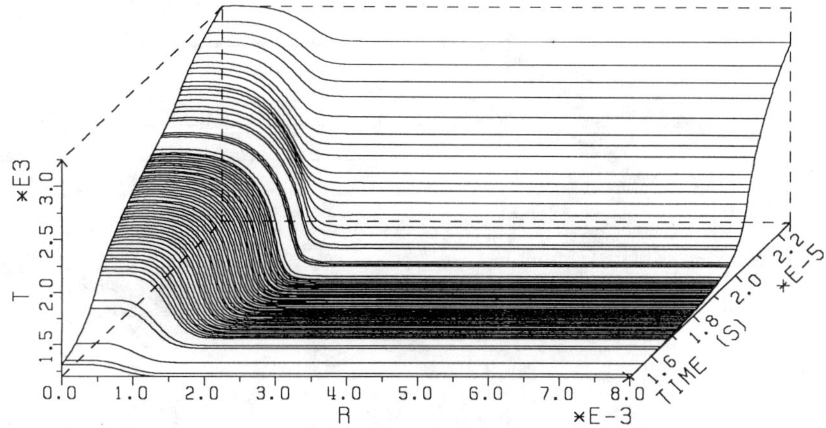

Fig. 4 Calculated profiles of temperature in an igniting hydrogen-oxygen mixture (stoichiometric): infinite slab, uniform pressure assumption, $R_o = 8$ mm, $P_0 = 1$ bar, T_0 see text.

propagation to the very fast propagation caused by preheating of the mixture by the pressure wave can be seen best in the plot of the oxygen atom mass fractions.

Figures 4 and 5 show what would have happened if the pressure increase during the excitation phase of the hot spot would have been distributed infinitely rapid over the reaction volume. These results are obtained using the "uniform pressure assumption,"which is described in detail in Maas and Warnatz.[3,13] In this case (note that a longer time interval is plotted) the flame propagation is not accelerated by local heating. Instead, after some time (i.e., the induction time of the mixture surrounding the hot spot) the mixture begins to ignite simultaneously all over the reaction volume.

Figures 6-8 show the development of temperature, pressure, and oxygen atom mass fractions of a stoichiometric hydrogen-oxygen mixture at 2 bar initial pressure in a system with the geometry of an infinite slab with a width of 1.2 cm. Again, the mixture is initially at rest. The initial temperature distribution in this example is given by

$$T\left(r,t_0\right) = 1150K + 150K exp\left\{-\left(r/1mm\right)^2\right\}$$

Plotted are temperature, velocity, and oxygen atom mole fractions after the induction phase vs the spatial coordinate and the time. Again excitation occurs at the hot spot first, leading to a rapid temperature rise and the formation of a pressure wave, that moves away from the

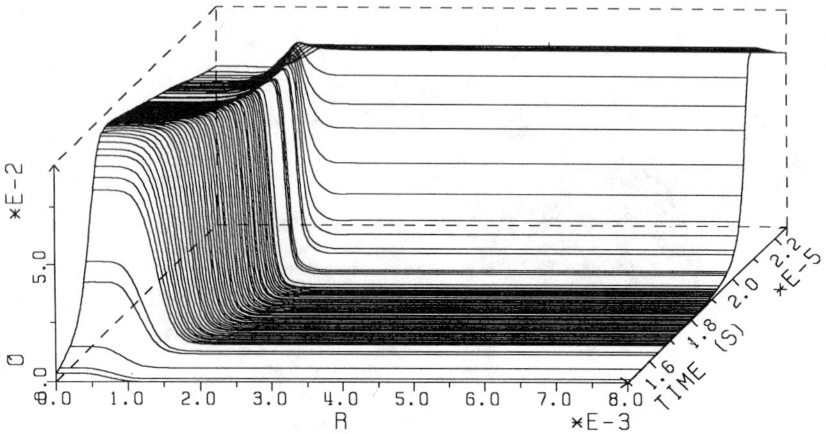

Fig. 5 Calculated profiles of oxygen-atom mass fractions in an igniting hydrogen-oxygen mixture (stoichiometric): infinite slab, uniform pressure assumption, $R_o = 8$ mm, $P_0 = 1$ bar, T_0 see text.

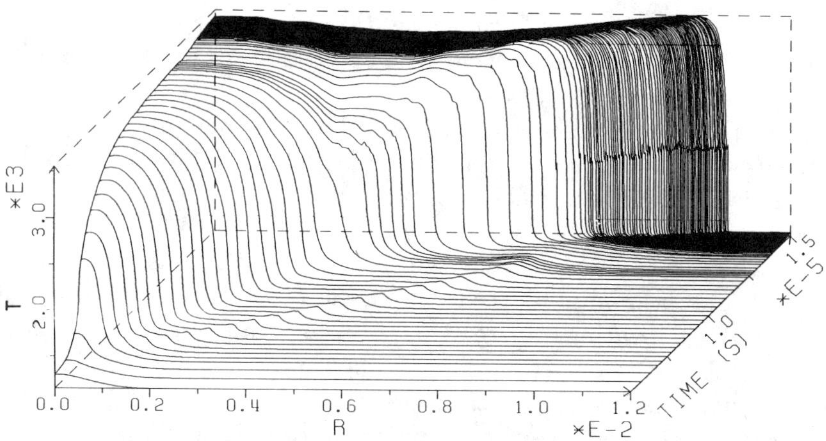

Fig. 6 Calculated temperature profiles in an igniting hydrogen-oxygen mixture (stoichiometric): infinite slab, $R_o = 12$ mm, $P_0 = 2$ bar, T_0 see text.

hot spot. Due to the higher initial pressure, the compression ratio in the hot kernel is bigger than in the first example, causing the pressure wave to be much stronger. After the excitation phase, a flame propagation can be seen. The reaction wave accelerates until it reaches a higher velocity than the pressure wave, and finally merges with the latter to form a detonation front which travels with a speed of approximately 2100 m/s. Figures 6-8 also show the structure of the

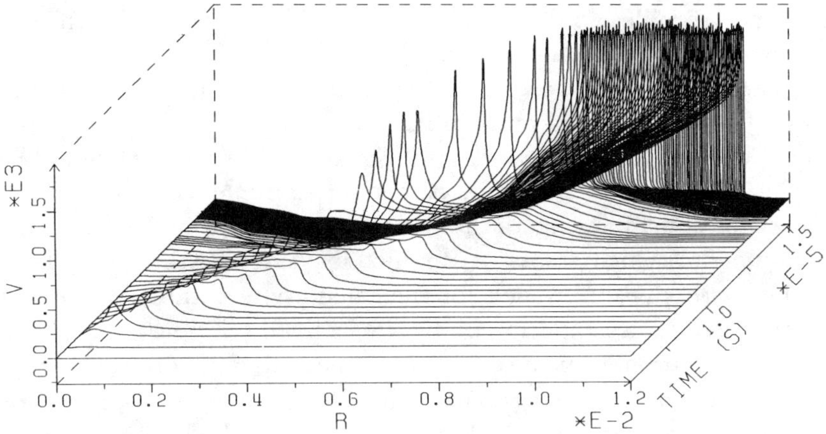

Fig. 7 Calculated velocity profiles in an igniting hydrogen-oxygen mixture (stoichiometric): infinite slab, R_o = 12 mm, P_0 = 2 bar, T_0 see text.

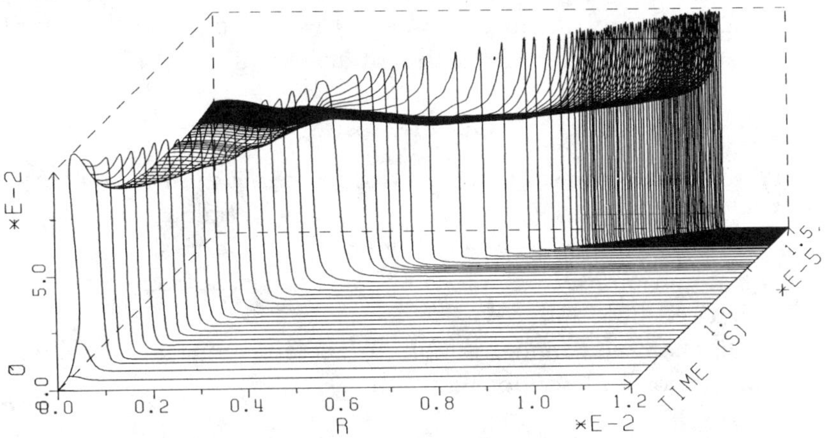

Fig. 8 Calculated profiles of oxygen-atom mass fractions in an igniting hydrogen-oxygen mixture (stoichiometric): infinite slab, R_o = 12 mm, P_0 = 2 bar, T_0 see text.

detonation wave. Characteristic are the sharp spikes of the velocity (directly related to the pressure, which shows the same behavior and, therefore, is not plotted here). The chemical processes in the detonation wave can be seen best in Fig. 8 (plot of the oxygen atom mass fractions). A spike of the O-atom profiles shows that a high amount of O is formed during the reaction. Finite reaction rates cause the O-atom mass profiles to decrease and approach their equilibrium value with respect to the conditions in the burnt gas.

These examples show that the processes associated with the ignition of hot spots have a marked influence on auto-ignition processes. Oran[23] showed that the "weak ignition regime" in shock-tube induced ignition of hydrogen-oxygen mixtures, for example, is highly sensitive to pertubations, causing the mild ignition to be a multidimensional phenomenon, which starts at distinct locations in the shock tube.[2] Furthermore, present models for the calculation of knocking behavior in Otto-engines use experimentally determined temporal temperature and pressure profiles in the end-gas.[25] The end-gas (unburnt mixture) is assumed to be uniform in space in order to allow zero-dimensional modelling. Now, because the knocking process is highly sensitive with respect to the temperature, local hot spots can induce nearly simultaneous ignition in the end-gas. Thus, even minimal peak temperatures in the end-gas of Otto-engines can determine the whole knocking process and it is evident, that a knowledge of local temperature variations (i.e., peak temperatures in the end-gas) would improve the detailed kinetic models of engine knock to a great extent.

Conclusions

1) It is possible to simulate ignition processes in hydrogen-oxygen mixtures that are caused by hot spots using detailed chemistry and multispecies transport for one-dimensional geometries.

2) Detonation phenomena as well as transition from explosion to detonation can be simulated.

3) One common model is able to explain induction, excitation, and propagation in hot spot ignition.

References

[1]Gardiner, W. C. Jr., Wakefield, C. B., and Walker, B.F., "Mathematical Methods for Modeling Chemical Reactions in Shock Waves," Shock Waves in Chemistry and Chemical Technology, edited by A. Lifshitz, Marcel Dekker, New York, 1981, pp. 319-374.

[2]Meyer, J.W., and Oppenheim, A.K., "On the Shock-induced Ignition of Explosive Gases," Proceedings of the Thirteenth Symposium (International) on Combustion, The Combustion Institute, Pittsburgh PA, 1970, p. 1153.

[3]Maas, U. and Warnatz, J., "Ignition Processes in Hydrogen-Oxygen Mixtures," Combustion and Flame, Vol. 74, 1988, pp. 53-69.

[4]Lutz, A.E., Kee, R.J., Miller, J.A., Dwyer, H.A., and Oppenheim, A.K., "Dynamic Effects of Autoignition Centers for Hydrogen and $C_{1,2}$-Hydrocarbon Fuels," Proceedings of the Twenty-second Symposium (International) on Combustion, The Combustion Institute, Pittsburgh PA, to be published

[5]Dixon-Lewis, G., and Williams, D. J., "The Oxidation of Hydrogen and Carbon Monoxide," Comprehensive Chemical Kinetics, Vol.17, Gas-Phase Combustion, edited by C.H. Bamford and C.F.H. Tipper, Elsevier, Oxford, England, 1977, pp. 1-248.

[6]Warnatz, J., "Calculation of the Structure of Laminar Flat Flames 2: Flame Velocity of Freely Propagating Hydrogen-Air and Hydrogen-Oxygen Flames," Berichte der Bunsengesellschaft für Physikalische Chemie, Vol. 82, 1978, p. 643.

[7]Warnatz, J., "Calculation of the Structure of Laminar Flat Flames 3: Structure of Burner-Stabilized Hydrogen-Oxygen and Hydrogen-Flourine Flames," Berichte der Bunsengesellschaft fr Physikalische Chemie, Vol. 82, 1978, p. 834.

[8]Warnatz, J., "Concentration-, Pressure-,and Temperature Dependence of the Flame Velocity in Hydrogen-Oxygen-Nitrogen Mixtures," Combustio Science and Technology, Vol. 26, 1981, p. 203.

[9]Peters, N., and Warnatz, J., eds., Numerical Methods in Laminar Flame Propagation, Vieweg, Braunschweig, FRG, 1982.

[10]Kee, R.J., and Miller, J.A., "Computational Modeling of Flame Structure," SAND83-8235, Sandia National Laboratories, Livermore, CA, 1983.

[11]Behrendt, F., and Warnatz, J., "The Dependence of Flame Propagation in H_2-O_2-N_2 Mixtures on Temperature, Pressure, and Initial Composition," International Journal of Hydrogen Energy, Vol. 10, 1985, p. 749.

[12]Lutz, A.E., Kee, R.J., and Dwyer, H.A., "Ignition Modeling with Grid Adaption," Progress in Astronautics and Aeronautics Dynamics of Reactive Systems, Vol. 105, Pt. 1, edited by J.R. Bowen, J.-C. Leyer, and R. I. Soloukhin, AIAA, New York, 1986, pp. 78-92.

[13]Maas, U., and Warnatz, J., "Ignition Processes in Hydrogen-Oxygen Mixtures and the Influence of the Uniform Pressure Assumption," Progress in Astronautics and Aeronautics, AIAA, (to be published).

[14]Lewis, B., and von Elbe, G., Combustion, Flames and Explosions of Gases, Academic Press, New York, San Francisco, London, 1961.

[15]Bird, R. E., Stewart, W. E., and Lightfoot, E. N., Transport Phenomena, Wiley Interscience, New York, 1960.

[16]Fritsch, F. N. and Butland, J., "A Method for Constructing Local Monotone Piecewise Cubic Interpolants," SIAM Journal on Scientific and Statistical Computing, Vol. 5, 1984, p. 300.

[17]Richtmyer, R. and Morton, K., "Difference Methods for Initial Value Problems," Tracts in Pure and Applied Mathematics, No.4, 2nd ed., edited by I. Bers, R. Courant, and J. Stoker, Wiley Interscience, New York, 1967, pp. 311-330.

[18]Deuflhard, P. and Nowak, U., "Extrapolation Integrators for Quasilinear Implicit ODEs," University of Heidelberg, Heidelberg, FRG, SFB 123 Tech. Rep. 332, 1985.

[19]Warnatz, J., "Critical Survey of Elementary Reaction Rate Coefficients in the C/H/O System," Combustion Chemistry, edited by W.C. Gardiner Jr., Chap. 5, Springer, New York, 1984.

[20]Frenklach, M., Warnatz, J., "Detailed Modeling of PAH Profiles in a Sooting Low-Pressure Acetylene Flame," Combustion Science and Technology, Vol. 51, 1987, pp. 265-283.

[21]JANAF, Thermochemical Tables, 2nd Ed., U.S. National Bureau of Standards, NSRDS-NBS 37, 1971.

[22]Hirschfelder, J.O. and Curtiss, C.F., "Theory of Propagation of Flames," Proceedings of the Third Symposium on Combustion, Flame, and Explosion Phenomena, Williams and Wilkins Cp., Baltimore, MD, 1949, pp. 121-127.

[23]Oran, E.S., Young, T.R., Boris, J.B. , Cohen, A., "Weak and Strong Ignition. I. Numerical Simulations of Shock Tube Experiments," Combustion and Flame, Vol. 48, 1982, p. 135-148.

[24]Oran, E.S. and Boris, J.B., "Weak and Strong Ignition. II. Sensitivity of the Hydrogen-Oxygen System," Combustion and Flame, Vol. 48, 1982, p. 149-161.

[25]Esser, C., Maas, U., and Warnatz, J., "Chemistry of the Combustion of Higher Hydrocarbons and Its Relation to Engine Knock," Proceedings of the International Symposium on Diagnostics and Modeling of Combustion in Reciprocating Engines, The Japanese Society of Mechanical Engineers, Tokyo, 1985, p. 335.

Ignition Phenomena in H_2-O_2 Mixtures

C. Treviño*
Universidad Nacional Autónoma de México, Mexico

Abstract

In this paper the transient process leading to ignition of a combustible mixture of hydrogen, oxygen, and argon, is studied using numerical techniques as well as asymptotic methods that allow us to deduce the most important regimes in the process as well as the characteristic times and the appropriate reduced kinetic mechanism involved. The numerical code is driven by a Computational Singular Perturbation (CSP) routine. This allows us to follow closely the temperature and species concentration evolution with time up to the ignition event. Three different regimes can be defined, depending on the initial temperature. For an initial temperature larger than the crossover temperature dictated by the competition between the chain branching Reaction $H+O_2 \rightarrow OH+O$ and the chain terminating Reaction $H+O_2+M \rightarrow HO_2+M$. The ignition event is a typical isothermal chain branching explosion, in a first approximation. For initial temperatures close to the crossover temperature, the energy release has to be considered.For lower initial temperatures a thermal explosion is obtained. For these three regimes, an analytical expression for the ignition delay time is deduced.

Nomenclature

a	nondimensional parameter defined in Eq. (14)
cp	specific heat at constant pressure of the gas mixture
cv	specific heat at constant volume of the gas mixture
C_i	concentration of species i
C_i	molar production rate of species i

---

*Professor, Department of Fluids and Thermodynamics.

h_i enthalphy of species i
K_j rate constant of Reaction (j)
M_i molecular weight of species i
N number of chemical species taking part in the process
P pressure
Q_j heat of Reaction (j) per unit mol
R universal gas constant
s nondimensional time defined as $s = \tau/\gamma_1$ and in Eq.(52)
T temperature
T_c crossover temperature
T* critical temperature defined in Eq. (45)
T_o initial temperature
t time
t_{ig} ignition time
x nondimensional concentration of hydrogen atom defined in Eq. (14)
x_{eq} nondimensional equilibrium concentration of H atom given by Eq. (22)
y normalized concentration of hydrogen atom. $y=x/x_{eq}$

Greek

α nondimensional parameter defined after Eq. (26)
α_j nondimensional parameter related to Reaction (j) defined after Eq. (54)
β_j nondimensional heat release parameter of Reaction (j) defined in Eqs. (17), (26) and (58)
ε small parameter defined by $\varepsilon = \gamma_{24}\gamma_1^2$
ϕ nondimensional temperature defined in Eqs. (17), (26), (49)
ρ_o gas density of the mixture
γ_j nondimensional parameter of Reaction (j) defined in Eq. (14), and after Eq. (48)
τ nondimensional time defined in Eq. (14)

subindex

ig denotes ignition

Introduction

The study of ignition and combustion of hydrogen/oxygen mixtures has received considerable attention in the literature. For these mixtures, the detailed reaction mechanism is very well known, representing the simplest realistic combustion process.[1-3] The understanding of the combustion of the hydrogen/oxygen mixtures also is of importance in the hydrocarbon combustion, due to the key role in the radical production zone.[4,5] Several studies have been published regarding the ignition of H_2/O_2 mixtures.[6-10] Brokaw[6] studies the ignition process using isothermal algebraic relationships. A zeroth dimensional analysis has been found to be enough to describe the ignition phenomena. Recently efforts have been conducted in order to deduce a reduced kinetic mechanism that globally contains the essential features of the studied process.[11] This procedure has been applied to the asymptotic study of premixed and diffusion flames, mainly methane-air.[4,5,12] The procedure to obtain the reduced mechanism has been improved by the CSP (Computational Singular Perturbation) technique developed by Lam and Goussis.[13] Here, the chemical reactions are reordered in reaction groups associated to the same order of magnitude of the characteristic times. These reaction groups can be active, dead, or dormant depending on the relation of their characteristic times to the actual relevant characteristic times of the process studied. The dead groups are related to fast chemical reactions from which the appropriate quasisteady and partial equilibrium assumptions can be derived. The dormant groups are related to slow chemical reactions playing no role at that moment. From the active and dormant groups the global mechanism can be obtained.

The main objective of the present study is to deduce the relevant reduced kinetic mechanism valid for the ignition of H_2/O_2 mixtures, using the CSP technique and to obtain asymptotic analytical expressions for the ignition delay time.

Governing Equations

The zeroth-dimensional governing equations of a reacting homogeneous hydrogen/oxygen mixture in a adiabatic reactor are given by

$$\rho_0 c_p\, dT/dt - dp/dt + \sum_{i=1}^{N} \dot{C}_i h_i M_i = 0 \qquad (1)$$

$$dC_i/dt = \dot{C}_i,\ i = 1,N. \qquad (2)$$

The mixture density is related to the species molar concentrations by

$$\rho_0 = \sum_{i=1}^{N} C_i M_i = \text{constant} \qquad (3)$$

Assuming a ideal gas mixture, the state equation is given as

$$P = RT \sum_{i=1}^{N} C_i \qquad (4)$$

where R is the universal gas constant, R = 8.314 KJ/(mol K). From Eqs. (1) and (4), the energy Eq. (1) transforms to

$$\rho_0 \, c_v \, dT/dt + \sum_{i=1}^{N} \dot{C}_i (h_i M_i - RT) = 0 \qquad (5)$$

The species production rate, C_i is given by

$$\dot{C}_i = \sum_{j=1}^{M} K_j(T) \, v'_{ij} \prod_{k=1}^{N} C_k^{\,v_{kj}}, \text{with } K_j(T) = A_j T^{\,n_j} \exp(-E_j/RT). \qquad (6)$$

Here, M correspond to the total number of elementary chemical reactions; v'_{ij} and v_{ij} correspond to the stoichiometric coefficients of the species i in the chemical reaction j, as product and as reactant, respectively. The constants Aj, nj and Ej are given in table 1 for all chemical reactions considered.

The solution of the governing Eqs. (2) and (5), together with the corresponding initial conditions,

$$T = T_0 \text{ and } C_i = C_{i0} \text{ at } t=0 \qquad (7)$$

is obtained using conventional codes for stiff differential equations (DGEAR). This numerical code is driven by a CSP routine, built parallel to the main program. The procedure employed in order to obtain the reduced kinetic mechanism is by dropping out those reactions with a participation index[13] lower than a critical value in the whole event leading to the ignition of the mixture. This critical value is the maximum value obtained when no appreciable change in the ignition time results after dropping these chemical reactions. The resulting set of chemical reactions can be reduced after introducing of steady-state behavior of several species. In the present study 30 elementary chemical reactions are employed as the detailed kinetic mechanism as in Mass and Warnatz.[10] The rates are presented in table 1.

Table 1. Mechanism of the H_2/O_2 mixtures

Units: A (cm, mole, s); E (KJ/mole); K - A T^n exp (-E/RT)

								A	n	E
(1,2)	H	$+O_2$		=	OH	+O		2.00 10^{14}	0.0	70.30
(3,4)	H_2	+O		=	OH	+H		5.06 10^{04}	2.7	26.30
(5,6)	H_2	+OH		=	H_2O	+H		1.00 10^{08}	1.6	13.80
(7,8)	OH	+OH		=	H_2O	+O		1.50 10^{09}	1.1	0.40
(9,10)	H	$+O_2$	+M	=	HO_2	+M		2.30 10^{18}	-0.8	0.00
(11,12)	HO_2	+H		=	OH	+OH		1.50 10^{14}	0.0	4.20
(13,14)	HO_2	+H		=	H_2	$+O_2$		2.50 10^{13}	0.0	2.90
(15,16)	HO_2	+OH		=	H_2O	$+O_2$		6.00 10^{13}	0.0	0.00
(17,18)	HO_2	+H		=	H_2O	+O		3.00 10^{13}	0.0	7.20
(19,20)	HO_2	+O		=	OH	$+O_2$		1.80 10^{13}	0.0	-1.70
(21,22)	HO_2	$+HO_2$		=	H_2O_2	$+O_2$		2.50 10^{11}	0.0	-5.20
(23,24)	OH	+OH	+M	=	H_2O_2	+M		3.26 10^{22}	-2.0	0.00
(25,26)	H_2O_2	+H		=	H_2	$+HO_2$		1.70 10^{12}	0.0	15.70
(27,28)	H_2O_2	+H		=	H_2O	+OH		1.00 10^{13}	0.0	15.00
(29,30)	H_2O_2	+O		=	OH	$+HO_2$		2.80 10^{13}	0.0	26.80
(31,32)	H_2O_2	+OH		=	H_2O	$+HO_2$		5.40 10^{12}	0.0	4.20
(33,34)	H_2	+M		=	H	+H	+M	6.76 10^{18}	-1.0	436.26
(35,36)	H	+OH	+M	=	H_2O	+M		2.20 10^{22}	-2.0	0.00
(37,38)	O	+O	+M	=	O_2	+M		2.90 10^{17}	-1.0	0.00

Asymptotic Analysis

The ignition process of hydrogen with oxygen changes dramatically depending on the initial gas temperature. A critical temperature is obtained from the competition between the chain branching Reaction (1) $H+O_2 \rightarrow OH+O$ and the chain terminating Reaction (9) $H+O_2+M \rightarrow HO_2+M$. This crossover temperature is denoted by Tc. Three different ignition regimes can be defined depending on the initial temperature, T_0. For initial temperatures T_0 larger than this critical temperature Tc, the characteristic production time for hydrogen atoms is much shorter than that of energy release. The ignition process is then a typical chain branching explosion. For initial temperatures close to the critical temperature, the characteristic time of the energy release becomes closer to that of radical production. In this case, the energy equation has to be retained in order to analyze the ignition process. However, the interaction is not strong and the ignition again is characterized as a chain branching explosion. Finally, for initial temperatures lower than another critical temperature, T* defined later, the radical concentration is maintained at a very low level and the chemical reaction is maintained alive by the very slow Reaction (14) $H_2+O_2 \rightarrow HO_2+H$. As the concentration of the intermediate species increases, a condition is reached when these recombine in such a way that an important amount of energy is released, then producing the ignition as a thermal runaway. These three regimes are analyzed in the following subsections, for the initial composition: $H_2:O_2:Ar$ = 1:0.5:1.88 in mol basis.

Regime 1: $T_0>T_c$

From the CSP data, the following elementary reactions are important in this regime. At he beginning (no radicals present), Reaction 14

(14) $H_2 + O_2 \rightarrow HO_2 + H$

is the important initiation step. Once trace amounts of radicals are present, this reaction has no more influence on the process in this high temperature limit. From this point up to the end, the three forward shuffle reactions are very important and have to be retained in the ignition process. These reactions are:

(1) $H + O_2 \rightarrow OH + O$

(3) $H_2 + O \rightarrow OH + H$

(5) $H_2 + OH \rightarrow H_2O + H$

The chain branching Reaction (9) is also important, because of its energy release

(9) $H + O_2 + M \rightarrow HO_2 + M$

The HO_2 produced by this reaction reacts very slowly at the beginning. However, as the radical concentration increases, the HO_2 is consumed by the Reactions (11) and (13)

(11) $HO_2 + H \rightarrow OH + OH$

(13) $HO_2 + H \rightarrow H_2 + O_2$

The Reaction (11) finally contributes to the production of H atoms via the rapid Reaction (5). A run of the program with only these Reactions (1,3,4,9,11,13,14) reproduces almost the same ignition time as with all the reactions included. The backward reactions (mainly radical-radical) have no influence on the ignition process. Figure 1 shows a block diagram representing the preferred paths in the ignition process of a mixture of H_2/O_2. The H_2O_2 is also represented because its importance at lower temperatures. Depending on the initial temperature, path 1 or path 9 is preferred. Path 1 is chain branching (energetically neutral), while path 9 is exothermic (chain propagation).

From the CSP data it is observed that the OH and O radicals are produced and consumed at rates much larger than the radical H. This means that the production-consumption of these radicals are contained in the dead reaction groups during the ignition time. Therefore, the reaction rate of Reactions (3) and (5) are limited by the radical pool concentration and the species O and OH can be assumed to be in steady-state. Eliminating the reaction rates (3) and (5) by the use of these steady-state assumptions, the kinetic mechanism then reduces to

$$\text{I} \quad H_2 + O_2 \xrightarrow{14} HO_2 + H$$

$$\text{II} \quad 3H_2 + O_2 \xrightarrow{1} 2H + 2H_2O$$

$$\text{III} \quad H + O_2 + M \xrightarrow{9} HO_2 + M$$

$$\text{IV} \quad 2H_2 + HO_2 \xrightarrow{11} 2H_2O + H$$

$$\text{V} \quad HO_2 + H \xrightarrow{13} H_2 + O_2$$

where the number above the arrow represents the rate of the elementary reaction associated with the global step. From the steady-state assumptions for O and OH, the concentration of these radicals are related to that of atomic hydrogen as

$$C_0 = \left[\frac{K_1 C_{0_2}}{K_3 C_{H_2}} \right] C_H \tag{8}$$

$$C_{0_H} = 2 \left[\frac{K_1 C_{0_2}}{K_5 C_{H_2}} + \frac{K_{11} C_{H_{0_2}}}{K_5 C_{H_2}} \right] C_H \tag{9}$$

For initial temperatures larger but not close to Tc, the Reaction (1) is much faster than Reaction (9). Thus, the

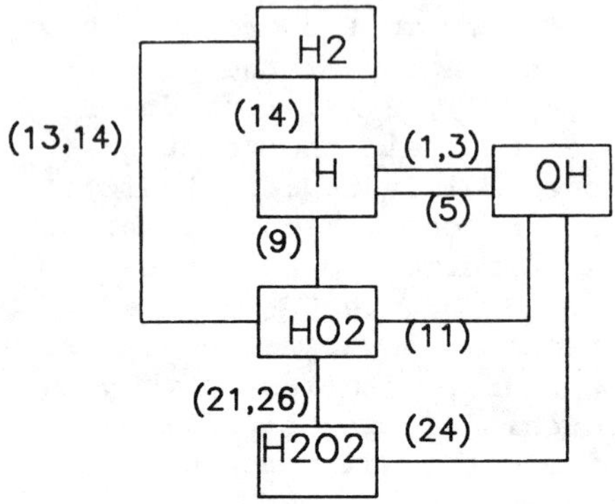

Fig. 1. Block diagram for hydrogen/oxigen ignition.

radical concentration grows without any important change in temperature, because the global step II is almost energetically neutral. In this asymptotic limit, the temperature can be assumed to be almost constant, except in evaluating the reaction rate of Reaction (14), because of its extremely high activation energy. Mass conservation Eqs. (2) for the H and HO_2 species are given by

$$dC_H/dt = K_{14}C_{H_2}C_{O_2}+2K_1C_{O_2}C_H-K_9C_{O_2}C_HC_M+(K_{11}-K_{13})C_{H_O2}C_H \quad (10)$$

$$dC_{HO_2}/dt = K_9C_HC_{O_2}C_M - (K_{11}+K_{13})C_{H_2}C_{HO_2}+ K_{14}C_{H_2}C_{O_2} \quad (11)$$

In nondimensional form, these equations take the form

$$dx/d\tau = (2\gamma_1-1+az)x + \gamma_{14} \quad (12)$$

$$dz/d\tau = \gamma_{11}[(1-z)x+\gamma_{14}] \quad (13)$$

where

$$x = C_H/(C_{H_2})_o$$

$$z = \gamma_{11}C_{HO_2}/(C_{H_2})_o$$

$$\tau = K_1(T_c)C_{O_2}t$$

$$\gamma_{14} = (K_{14})_o/K_1(T_c) \quad (14)$$

$$\gamma_{11} = \frac{(K_{11}+K_{13})C_H}{K_1(T_c)C_{O_2}}$$

$$\gamma_1 = K_1(T)/K_1(T_c)$$

$$a = (K_{11}-K_{13})/(K_{11}+K_{13})$$

Due to the very weak temperature dependence of Reactions (11) and (13), a can be assumed to be a constant close to unity. For values of $\gamma_1 >> 1$, Eq. (12) can be rewritten as

$$dx/ds = [2-1/\gamma_1(1-az)]x + \gamma_{14}/\gamma_1 \qquad (15)$$

with $s = \tau/\gamma_1$, where the second term in square brackets can be neglected to the leading order. The energy Eq. (5) can be written in nondimensional form as

$$d\phi/ds = \beta_{14}\exp(-\phi) \qquad (16)$$

where

$$\phi = E_{14}(T_o-T)/RT_o^2) \qquad (17)$$

$$\beta_{14} = [\frac{E_{14} \quad \Delta U_{14}(C_{H_2})_o(\gamma_{14})_o}{RT_o \ \rho_o \ (C_v)_o T_o \ \gamma_1}]$$

Here, the parameter β_{14} reflects the importance of the endothermicity of Reaction (14). In the limit $\gamma_1 >> 1$, Eq.(15) gives in a first approximation

$$dx/ds=2x+(\gamma_{14})_o\exp(-\phi) \qquad (18)$$

with $(\gamma_{14})_o = \gamma_{14}(T_o)$. After solving the energy Eq. (16), Eq.(18) transforms to

$$dx/ds = 2x + (\gamma_{14})_o/(1+\beta_{14}s) \qquad (19)$$

The solution to this equation, together with the initial condition $x(0) = 0$, for large nondimensional time s, is given by

$$\lim_{s\to\infty} x= \frac{(\gamma_{14})_o\exp(2s+1/\beta_{14})}{2\ \beta_{14}} E_1(1/\beta_{14}) \qquad (20)$$

where $E_1(z)$ corresponds to the exponential integral function defined by

$$E_1(z) = \int_z^{\infty} \frac{e^{-u}\,du}{u} \tag{21}$$

The concentration of atomic hydrogen grows exponentially with time as given by Eq. (20). The ignition time can be defined as the time needed to obtain partial equilibrium of Reactions (1) and its reverse (2). Using the steady-state approximations for OH and O, given by Eqs. (8) and (9), the nondimensional equilibrium concentration for H is given by

$$x_{eq} = \left(\frac{C_H}{C_{H_2}}\right)_{eq} = \left[\frac{K_3K_5C_{H_2}}{2K_1K_2C_{O_2}}\right]_o \tag{22}$$

For small values of β_{14}, which is the present case, the nondimensional time for ignition is given by

$$s_{ig} \sim (1/2)\,\mathrm{Ln}\left[\frac{2\,x_{eq}}{\gamma_{14}(1-\beta_{14}+2\beta_{14}^2)}\right], \text{ for } \beta_{14}<<1 \tag{23}$$

In dimensional units, the ignition time is

$$t_{ig} \sim \frac{1}{2(K_1)_o(C_{O_2})_o}\,\mathrm{Ln}\left[\frac{2\,x_{eq}}{\gamma_{14}(1-\beta_{14}+2\beta_{14}^2)}\right] \tag{24}$$

Regime 2: $T_o \sim T_c$

In general, the five global steps are the same as given in the previous regime. However, in this limit the characteristic time for the energy release decreases very fast, relative to the hydrogen atom production term, due to the fact that the reaction rates of (1) and (9) are the same at temperatures of T_c. Therefore, the energy equation has to be taken into account for this regime. It can be shown that its interaction with the species equation is rather weak but its inclusion gives a better approximation to the ignition delay time. As indicated in the previous section, the value of β_{14} was small. As the initial temperature is reduced, the value of β_{14} decreases strongly, thus having the importance of Reaction (14) only local character (only at the beginning) in this regime. The equations to be solved are the H and HO_2 concentration equations as well the energy equation. In nondimensional form these equations are given by Eqs. (12) and (13) and

$$d\phi/d\tau = \beta_9 x \, [1+ \alpha z] \tag{25}$$

where ϕ now is defined as

$$\phi = E_1(T-T_c)/(RT_c^2) \tag{26}$$

and

$$\beta_9 = [\frac{E_1 \, \Delta U_9 \, (C_{H_2})_o}{RT_o \, \rho_o (C_v)_o T_c}]$$

$$\alpha = (\Delta U_{11}+\Delta U_{13})/\Delta U_9$$

For small times, the species concentrations for H and HO are very small and the appropriate inner equations for this stage are given in a first approximation by

$$d\bar{x}/d\tau = (2\gamma_1-1)\bar{x} + 1 \tag{27}$$

$$d\bar{z}/d\tau = \gamma_{11}(\bar{x}+1) \tag{28}$$

where

$$\bar{x} = x/\gamma_{14}$$

$$\bar{z} = z/\gamma_{14}$$

with $\gamma_{14} \to 0$. In this first stage, the process is isothermal due to the fact that β_9 is very small. The solution to Eq. (27) together with the initial condition $x(0) = 0$, is given by

$$\bar{x} = [1/(2\gamma_1-1)] \, \{\exp \, [(2\gamma_1-1)\tau]-1\} \tag{29}$$

The solution in this stage for large nondimensional times, is the following

$$\bar{x} \to \infty \text{ as } \tau \to \infty, \text{ for } 2\gamma_1-1 > 0 \tag{30}$$

$$\bar{x} \to 1/(1-2\gamma_1) \text{ as } \tau \to \infty, \text{ for } 2\gamma_1-1 < 0 \tag{31}$$

In the regime given by Eq. (30), the hydrogen atom concentration increases monotonically reaching finally the ignition condition as a chain branching explosion. However for $2\gamma_1-1<0$, the radical concentration reaches asymptotically a constant small value because in this case the chain branching Reaction (1) is small compared with the chain terminating Reaction (9). This latter case will be analyzed in the next section. The phase-space solution of Eqs. (27) and (28) is given by

$$\bar{z} = [\gamma_{11}/(2\gamma_1-1)^2] \{(2\gamma_1-2) \text{ Ln } [1+(2\gamma_1-1)\bar{x}]+(2\gamma_1-1)\bar{x}\}$$

For large values of the nondimensional time τ, the asymptotic solution is

$$\bar{x} \sim 1/(2\gamma_1-1) \exp \{(2\gamma_1-1)\tau\}$$

$$\bar{z} \sim \gamma_{11}\bar{x}/(2\gamma_1-1), \text{ for } \tau \to \infty \qquad (32)$$

As the H and HO_2 concentration increase, the effect of the initiation Reaction (14) is small and can be neglected. In this second stage, the normalized nondimensional governing equations, are

$$dy/d\tau = (2\gamma_1-1+az)y \qquad (33)$$

$$dz/d\tau = \bar{\gamma}_{11}(1-z)y \qquad (13)$$

with $y = x/x_{eq}$ and $\bar{\gamma}_{11} = \gamma_{11}x_{eq}$. Here, the equilibrium concentration, x_{eq} is the same as defined in Eq. (22). Also in this second stage, the process can be assumed to be isothermal because β_9 is small compared with unity. The solution in phase-space is given by

$$\bar{\gamma}_{11}y = (2\gamma_1-1+a) \text{ Ln } [1/(1-z)]-az \qquad (34)$$

For values of $\bar{\gamma}_{11}$ very large compared with unity, which is the presente case, z tends very rapid to the value of unity, corresponding to the steady-state behavior of HO_2 . From Eqs. (32), it can be assumed that HO_2 reaches steady-state (z=1) at the nondimensional time given by

$$\tau \sim 1/(2\gamma_1-1) \text{ Ln } [(2\gamma_1-1)^2/(\gamma_{11}\gamma_{14})] \qquad (35)$$

The corresponding H concentration is then

$$x \sim (2\gamma_1-1)/\gamma_{11} \qquad (36)$$

In the following stage, assuming HO_2 to be in steady-state, the radical concentration and energy equations reduce to

$$dy/d\tau = [2\exp(\phi)-1+a]y \qquad (37)$$

$$d\phi/d\tau = \overline{\beta}_9 y \qquad (38)$$

with $\overline{\beta}_9 = \beta_9 x_{eq}[1+\alpha]$ and $\gamma_1 = \exp(\phi)$. The solution in phase-space is given by

$$y = 2/\overline{\beta}_9[\exp(\phi)-\exp(\phi_0)]-(1-a)(\phi-\phi_0)/\overline{\beta}_9 \qquad (39)$$

Due to the fact that ignition can be assumed to occur as y reaches the value of unity, for small β_9 the temperature increase is very small. In this case, there is a weak interaction of the energy release on the ignition event. Therefore, it is appropriate to introduce to Taylor series expansion in the exponential term of Eq. (39), obtaining

$$y = u + [e^{\phi_0}\overline{\beta}_9/(2e^{\phi_0}-1+a)]\ u^2 + 0(\overline{\beta}_9^2) \qquad (40)$$

where

$$u = [(2e^{\phi_0}-1+a)/\overline{\beta}_9]\ (\phi-\phi_0)$$

is of order unity. With the inversion of Eq. (40), Eq. (37) reduces to

$$dy/d\tau = (2e^{\phi_0}-1+a)y + [2e^{\phi_0}\overline{\beta}_9/(2e^{\phi_0}-1+a)]y^2 + 0\ (\overline{\beta}_9^{2}) \qquad (41)$$

The solution of Eq. (41) with the initial conditions dictated by Eqs. (35) and (36) is given by

$$\tau \sim \tau_0 + (1/A)\ \mathrm{Ln}\left[\frac{(A+By_0)\ y}{(A+By)\ \ y_0}\right] \qquad (42)$$

where

$$A = (2e^{\phi_0}-1+a)$$

$$B = 2e^{\phi_0}\overline{\beta}_9/2e^{\phi_0}-1+a)$$

$$\tau_0 = \frac{1}{2e^{\phi_0}-1}\ \mathrm{Ln}\left[\frac{(2e^{\phi_0}-1)}{\gamma_{11}\gamma_{14}}\right]$$

$$y = (2e^{\phi_0}-1)/\overline{\gamma}_{11}$$

The ignition is assumed to occur when y reaches the value of unity. Therefore, the nondimensional ignition time is

$$\tau_{ig} \sim \tau_0 + (1/A)\ \mathrm{Ln}\ [\frac{(A+By_0)}{(A+B)\ y_0}] \qquad (43)$$

The nondimensional ignition time given by Eq. (43) is no more valid as ϕ_0 approaches the value of

$$\phi_0 \to -\ \mathrm{Ln}(2) \qquad (44)$$

In this case the chain breaking Reaction (9) is very fast and consumes very rapid all the radicals produced mainly through the initiation step (14). The critical temperature which separates regime 3 from regime 2 is given by

$$T^* = T_c - \mathrm{Ln}(2) R T_c^{\ 2}/E_1 \qquad (45)$$

For initial temperatures below this critical temperature, regime 3 is appropriate and is discussed in the next section.

Regime 3: $T_0 < T^*$

In this regime, the chain branching Reaction (1) is slow compared with the chain termination Reaction (9), thus maintaining the active radical concentrations (O,OH,H) very low. From the CSP data, the important reactions in this regime are:

(1)	H	$+ O_2$		→	OH	+ O	
(3)	H_2	+ O		→	OH	+ H	
(5)	H_2	+ OH		→	H_2O	+ H	
(9)	H	$+ O_2$	+ M	→	HO_2	+ M	
(14)	H_2	$+ O_2$		→	HO_2	+ H	
(21)	HO_2	$+ HO_2$		→	H_2O_2	$+ O_2$	
(24)	H_2O_2	+ M		→	OH	+ OH	+ M
(26)	H_2	$+ HO_2$		→	H_2O_2	+ H	

Due to the fact that the chain terminating reaction rate is large, all active radicals (O,OH,H) can be assumed to be in steady-state in this regime, as indicated by the CSP data. Because of the low radical concentration, the chain propagation Reaction (11) has not a large influence in the ignition process in the temperature range 850 - 900 K, at one atmosphere pressure. In this regime, both HO_2 and H_2O_2 are produced at a very slow rate because of the low concentration of H atom in Reaction (9). H_2O_2 is produced at the beginning from Reaction (26) and then by Reaction (21) as the HO_2 concentration increases. As the H_2O_2 concentration increases, the decomposition Reaction (24) feeds fresh radicals to the fast Reaction (5), thus releasing heat need for the same Reaction (24). Three different stages can be defined in this ignition regime: a) initiation, b) isothermal build-up of intermediate species HO_2 and H_2O_2 and c) thermal stage, finishing with the abrupt ignition process characterized by a thermal runaway. With the steady-state assumptions for the active radicals, the reduced global mechanism valid in this regime is given by

I	$H_2 + 2O_2 \xrightarrow{14} 2HO_2$	
II	$3H_2 + 3O_2 \xrightarrow{1} 2H_2O + 2HO_2$	
III	$2HO_2 \xrightarrow{21} H_2O_2 + O_2$	
IV	$H_2O_2 + 2H_2 + 2O_2 \xrightarrow{24} 2H_2O + 2HO_2$	
V	$H_2 + O_2 \xrightarrow{26} H_2O_2$	

In this reduced mechanism, the last four global steps are exothermic. Introducing the nondimensional variables

$$x = C_H/C_{H_2}$$

$$y = C_{HO_2}/C_{H_2}$$

$$z = C_{H_2O_2}/C_{H_2}$$

the governing equations for these species are

$$dz/d\tau = \gamma_{21}y^2 + \gamma_{26}y - \gamma_{24}z \quad (46)$$

$$(1/2)\ dy/d\tau = \gamma_{24}z + \gamma_1 x - \gamma_{21}y^2 + \gamma_{14} \quad (47)$$

$$x = (\gamma_{14} + 2\gamma_{24}\ z + \gamma_{26}y)/(1 - 2\gamma_1) \quad (48)$$

where

$$\gamma_1 = [K_1(T)C_{H_2}]/[K_1(T_c)C_{O_2}]$$

$$\gamma_{21} = [K_{21}(T)C_{H_2}]/[K_1(T_c)C_{O_2}]$$

$$\gamma_{24} = [K_{24}(T)C_M]/[K_1(T_c)C_{O_2}]$$

$$\gamma_{26} = [K_{26}(T)C_{H_2}]/[K_1(T_c C_{O_2}]$$

$$\tau = K_1(T_c)C_{O_2}t$$

The last Eq. (48) comes from the steady-state assumption for the active radicals. The largest activation energy of the global reactions correspond to Reaction (24), thus retaining this reaction as the controlling step in the ignition process. This reaction is also important for the heat release. We introduce the nondimensional temperature

$$\phi = \bar{E}_{24} \ (T - T_o)/(RT_o^2) \tag{49}$$

where $\bar{E}_{24}$ is the equivalent activation energy because of the non-Arrhenius kinetics, defined by

$$\bar{E}_{24} = E_{24} - 2RT_o \tag{50}$$

Therefore, the function γ_{24} can be written as

$$\gamma_{24} = (\gamma_{24})_o \ \exp\ (\phi) \tag{51}$$

assuming the rest of the functions $\gamma_1, \gamma_{21}, \gamma_{26}$ as constants in the ignition process. Figure 2 shows the values of the parameters γ_1, γ_{21}, $(\gamma_{24})_o$, γ_{26} as a function of the initial temperature T_o. $(\gamma_{24})_o$ have values much lower than unity for all the temperatures of interest. The value of γ_1 decreases as the temperature decreases, reaching a value of order 10^{-1} at temperatures close to 800 K. In the initiation phase, radical H is generated by the initiation Reaction (14) reaching the equilibrium value of

$$x \to \gamma_{14}/(1-2\gamma_1)$$

which was given before in Eq. (31). With this very low H radical concentration, the intermediate species are slowly generated in this induction stage. At the beginning, Reaction (21) has not a big influence due to the low

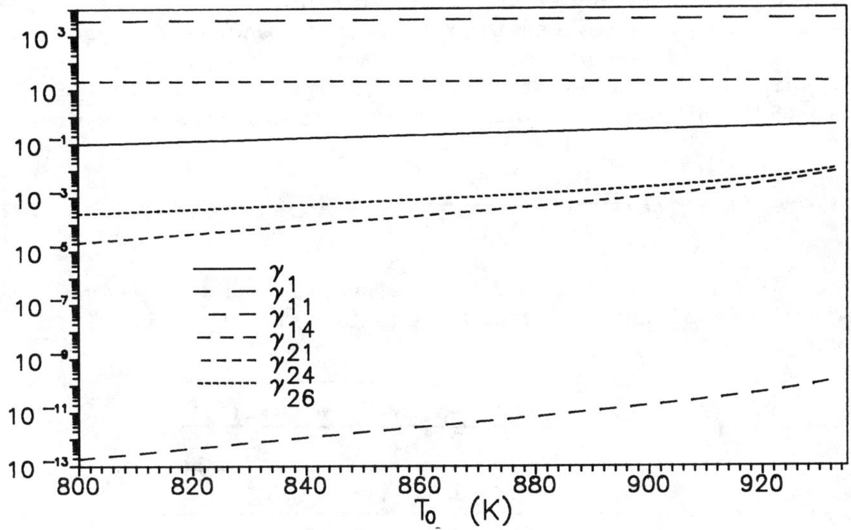

Fig. 2. Parameter variation with temperature.

concentration of HO_2. In order to analyze this starting phase it is convenient to normalize the variables as follows

$$\bar{y} = \left[\frac{\gamma_{26}\gamma_1}{\gamma_{14}(1-\gamma_1)}\right] y$$

$$\bar{z} = \left[\frac{\gamma_{24}}{\gamma_{14}(1-\gamma_1)}\right] z \qquad (52)$$

$$s = \left[\frac{2\gamma_{26}\gamma_1}{(1-\gamma_1)}\right] \tau$$

With these new variables, Eqs. (44) and (45) take the form

$$d\bar{y}/ds = 1 + \bar{y} + \bar{z} - \alpha_{21}\bar{y}^2 \qquad (53)$$

$$d\bar{z}/ds = \alpha_{24}[\alpha_{21}\,\bar{y}^2 + \bar{y} - \gamma_1\bar{z}] \qquad (54)$$

where

$$\alpha_{21} = \frac{\gamma_{21}\gamma_{14}(1-\gamma_1)}{\gamma_{26}^{\,2}\gamma_1^{\,2}} \qquad \alpha_{24} = \frac{\gamma_{24}\;(1-2\gamma_1)}{2\gamma_{26}\gamma_1^{\,2}}$$

Here, α_{21} is very small compared with unity and α_{24} is of order unity. In this starting phase, the energy release is very small and the process can be assumed to be isothermal. For very small values of α_{21}, the asymptotic solution of Eqs. (53) and (54) for large nondimensional time s is given by

$$\bar{y} \sim \bar{z}/(\lambda_2-\lambda_1) \sim \left[\frac{1+\gamma_1 \quad (1+\lambda_1)}{(\lambda_2-\lambda_1)(1+\gamma_1)}\right] \exp(\lambda_2 s), \text{ as } s \to \infty \quad (55)$$

where λ_1 and λ_2 correspond to the negative and positive eigenvalues of the homogeneous equations, respectivelly, given by

$$\lambda_1 = \frac{(1-\alpha_{24}\gamma_1) - \sqrt{(1-\alpha_{24}\gamma_1)^2 + 4\alpha_{24}(1+\gamma_1)}}{2}$$

$$\lambda_2 = \frac{(1-\alpha_{24}\gamma_1) + \sqrt{(1-\alpha_{24}\gamma_1)^2 + 4\alpha_{24}(1+\gamma_1)}}{2}$$

As the concentration of HO_2 increases further, Reaction (21) consumes this radical in an important way producing H_2O_2 needed for Reaction (24). Therefore, Reaction (21) makes HO_2 to reach rapidly the quasi-steady behavior. In order to study this second stage is convenient to renormalize the variables as follows

$$y^* = \alpha_{21}\gamma_1 \bar{y}$$

$$z^* = \alpha_{21}\gamma_1{}^2\bar{z}$$

$$\sigma = \alpha_{24}\gamma_1 s$$

The species balance equations transform to

$$dy^*/d\sigma = (1/\varepsilon)\ [e^{\phi}z^*-y^{*2}+\gamma_1 y^*] \quad (56)$$

$$dz^*/d\sigma = y^{*2}+y^*-e^{\phi}z^* \quad (57)$$

where $\varepsilon = \alpha_{24}\gamma_1{}^2 \to 0$ as $\gamma_1 \to 0$. Figure 3 show the values of the parameters α_{21}, α_{24} and ε as a function of the temperature. The nondimensional energy equation is given by

$$d\phi/\sigma = \beta_{24}\{e^{\phi}z^*[1+\frac{Q_1\gamma_1}{Q_{24}(1-\gamma_1)}]+y^*Q_{21}/Q_{24}+y^*[\frac{Q_{26}}{Q_{24}}+\frac{\gamma_1 Q_1}{Q_{24}(1-\gamma_1)}]\} \quad (58)$$

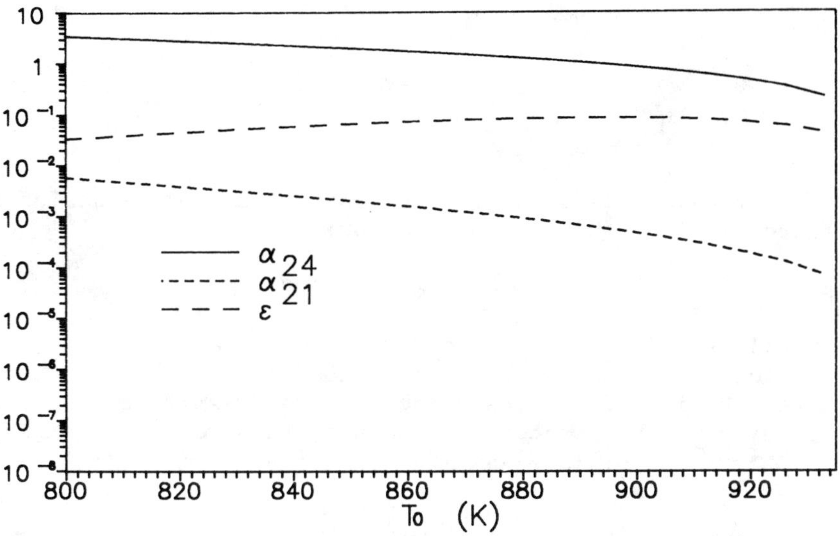

Fig. 3. Parameters α_{21}, α_{24} and ε as a function with temperature.

where

$$\beta_{24} = \frac{E_{24}}{RT_o} \; \frac{2\gamma_{26}^{\,2}}{\gamma_{21}\;(\gamma_{24})_o} \; \frac{Q_{24}C_{H_2}}{\rho c_v T_o}$$

In the asymptotic limit $\gamma_1 \to 0$ (low temperatures), with β_{24} very small compared with unity, in a first approximation, the phase-space equation, resulting from Eqs. (56) and (57), can be rewritten as

$$V\,dV/dY = 2\,Y\,(1 - V) - \varepsilon^{1/2}V \tag{59}$$

where

$$V = (z^*-y^{*2})/\varepsilon$$

$$Y = y^*/\,\varepsilon^{1/2} \tag{60}$$

The asymptotic solution of this equation is given by the transcendental equation

$$V \sim 1 - \exp\,(V-Y^2)-\varepsilon^{1/2}D(Y) \tag{61}$$

where D is the Dawson integral defined by

$$D(Y) = d^{-Y^2} \int_o^Y e^{t^2} dt.$$

Therefore V can be assumed to be unity even for values of Y of order unity. Thus, from Eqs. (60)

$$z^* \sim \varepsilon/2 + y^{*2} \qquad (62)$$

For values of z* and y* of order unity or larger, the solution can be given by $z^*=y^{*2}$, thus indicating the steady-state behavior of HO_2. Assuming the steady-state for HO_2, including the temperature dependence, this relation is given by $e^{\phi}z^*=y^{*2}$. Eqs. (57) and (58) then transform to

$$dz^*/d\sigma = e^{\phi/2} \sqrt{z^*} \qquad (63)$$

$$d\phi/d\sigma = \bar{\beta}_{24}\, e^{\phi}\, z^* \qquad (64)$$

with $\bar{\beta}_{24} = \beta_{24}\,(1+Q_{21}/Q_{24})$ in a first approximation. In phase space, these equations reduce to

$$d\phi/dz^* = \bar{\beta}_{24} e^{\phi/2} \sqrt{z^*} \qquad (65)$$

Integration gives

$$e^{-\phi/2} = 1 - (1/3)\,\bar{\beta}_{24}\,(z^{*3/2} - z_o^{*3/2}) \qquad (66)$$

where z_o^* corresponds to the reduced concentration of H_2O_2 at the end of the second stage. Thermal runaway is reached when the right hand of Eq. (66) vanishes. Therefore at ignition

$$z_{ig}^* = [z_o^{*3/2} + 3/\bar{\beta}_{24}]^{2/3} \qquad (67)$$

from Eqs. (63) and (64), the nondimensional ignition time is given as

$$\sigma_{ig} = \sigma_o + [1-(81/5184)^{1/3}]\,(24/\bar{\beta}_{24})^{1/3} \qquad (68)$$

where σ_o corresponds to the nondimensional time needed to reach the value of z_o^* at the end of the second stage. In a first approximation

$$z_o^* = 1$$

and

$$\sigma_o = \frac{\gamma_{24}}{2\gamma_{26}\gamma_1(1-\gamma_1)\lambda_2} \, \mathrm{Ln}\left[\frac{(1-2\gamma_1)(\lambda_2-\lambda_1)(1+\gamma_1)}{(\lambda_2-\lambda_1)[1+\gamma_1(1+\lambda_1)]}\right]$$

Results and Conclusions

Figures 4 and 5 show the radical and intermediate species concentrations as a function of time for two different temperatures, 800 and 1500 K and pressure of 10^5 Pa. For very low temperatures, the active radical concentrations are maintained at a low value due to the fast Reactions (5) and (9). However, the intermediate species (HO_2 and H_2O_2) increase continuously reaching the ignition condition. The intermediate species concentrations are very large at lower temperatures and negligible small at high temperatures. At the end of the isothermal induction process the steady state of HO_2 is fully justified. Figure 6 shows the phase-space trajectories (temperature-H

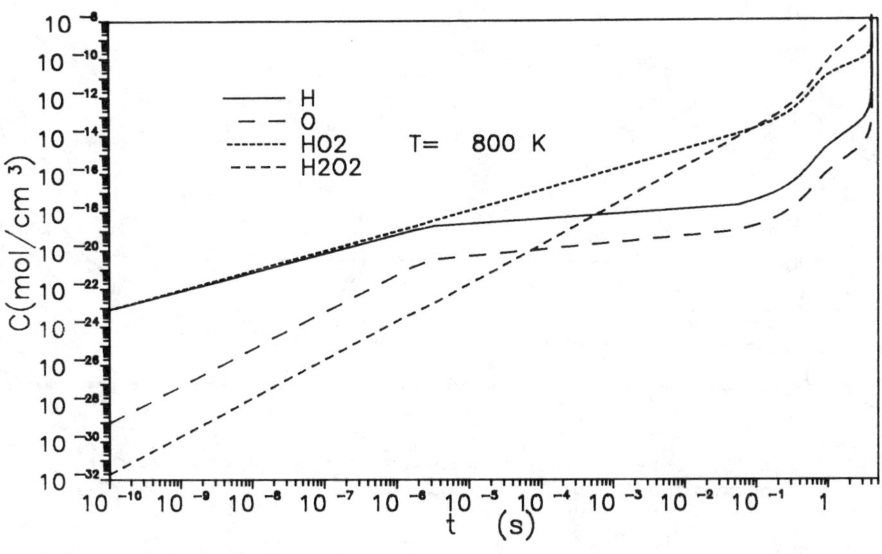

Fig. 4. Radicals and intermediate species concentration evolution for a temperature of 800 K.

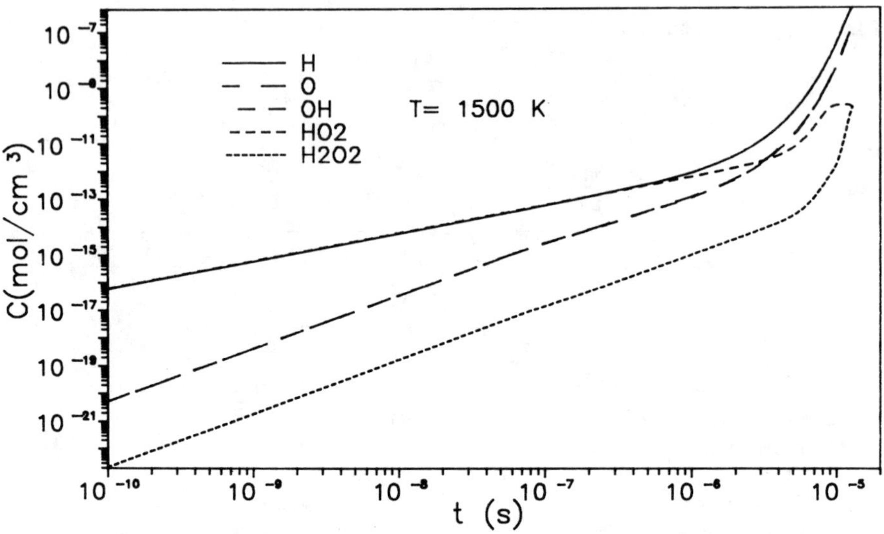

Fig. 5. Radicals and intermediate species concentration evolution for a temperature of 1500 K.

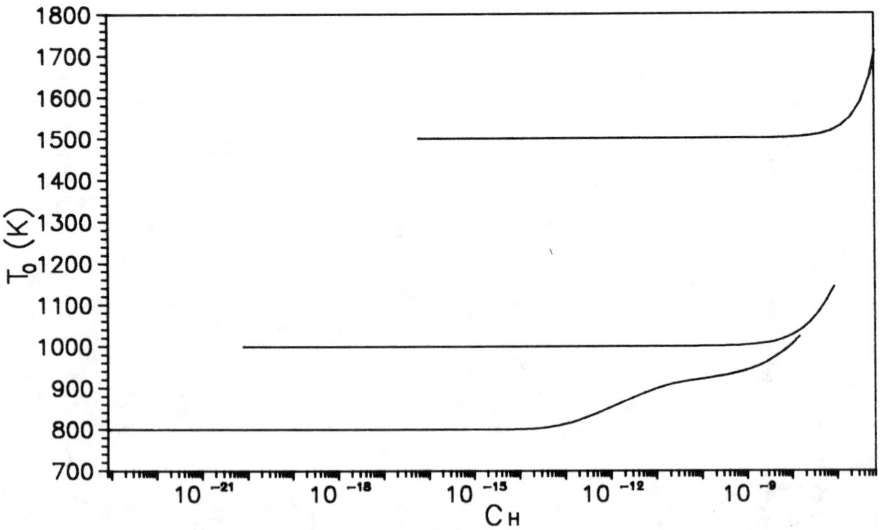

Fig. 6. Phase-space trajectories (temperature- H radical concentration), for three different initial temperatures.

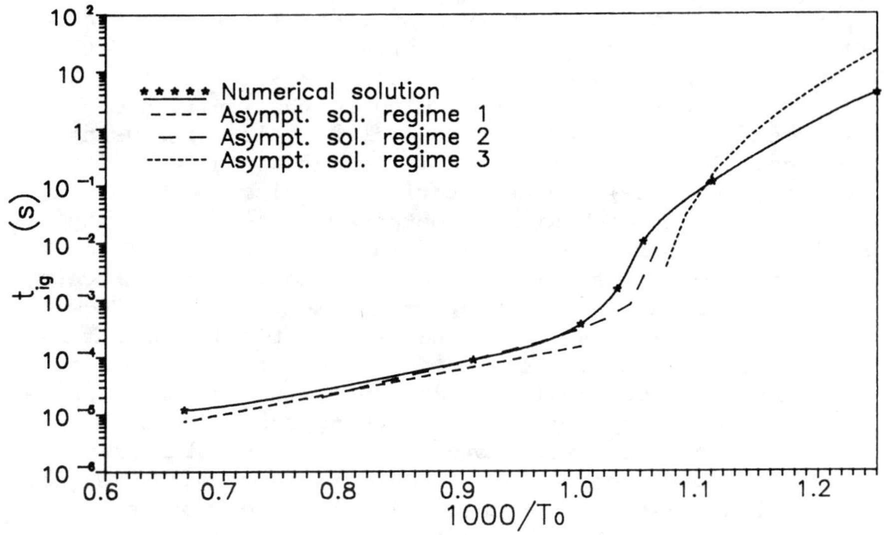

Fig. 7. Ignition time as a function of the initial temperature. Numerical and asymptotic solutions.

concentration) for three different initial temperatures. The ignition condition is achieved at lower H concentration at lower temperatures, showing the thermal ignition character. Figure 7 shows the ignition time as a function of the initial temperature. Both numerical results as well the asymptotic solutions for the three regimes are also presented. For temperatures larger than 970 K the ignition time is relatively short and the asymptotic approximation compares very well with the numerical solution. An abrupt change in the ignition time takes place at temperatures around 950 K. The asymptotic solution for regime 2 applies in this case. Finally, at lower temperatures, the ignition time increases drastically, showing a long isothermal induction period, before the thermal runaway. In this case, the ignition is obtained as mentioned before for the third regime. The asymptotic solutions for the three regimes compare very well with the numerical solution. As it is shown, Reaction (1) has not a big influence on the ignition process in the low temperature regime. Apparently, there is a contradiction with the results of Mass and Warnatz [10] obtained after applying a sensitivity analysis by increasing several times the reaction rate of the studied reaction. In this case, the increase in the reaction rate changes artificially the critical crossover temperature, thus switching artificially from regime 3 to regime 2, where the rate of Reaction (1) is very important. Therefore, the sensitivity analysis has to be done carefully in order to make no qualitative change in the process.

The transient analysis leading to the ignition of hydrogen/oxygen mixtures is studied in this paper using both numerical analysis based in Computational Singular Perturbation (CSP) and asymptotic analytical methods. The reduced kinetic mechanism is obtained after solving the evolution equations and evaluating the CSP data. Three different regimes are identified for the ignition process, depending on the initial temperature and pressure. For initial temperatures larger than the critical temperature (temperature which makes reaction rates of Reaction 1 and Reaction 9, to be equal in magnitude), the ignition can be described as a typical chain branching explosion with no heat release in a first approximation. As the temperature decreases to values close to the critical temperature, the heat release has to be retained in order to improve the ignition time. As the temperature decreases further, the chain branching reaction loses in importance and the intermediate species evolution control the process. In this last regime all three active radicals (H, OH, O) can be assumed to be in steady state. In the last stage in this regime, the decomposition Reaction of H_2O_2 (24) controls the process due to its chain-branching character. An important part of the heat release is produced by this reaction together with the fast chain propagation Reaction (5). Due to the very large activation energy of Reaction (24) a thermal runaway is produced characterizing the ignition process at low initial temperatures. In all three regimes, an analytical expression is obtained for the ignition time.

Acknowledgments

The author thanks Prof. N. Peters and G. Paczko for interesting discussions and suggestions.

References

[1]Baulch D.L., Drysdale D.D., Horne D.G., and Lloyd A.C., "Evaluated Kinetic Data for High Temperature Reactions, Vol. 1: Homogeneous Gas Phase Reactions of the H -O System", Butherworths, London, 1972.

[2]Warnatz J., "Rate Coefficients in the C/H/O System, in Gardiner W. C. (ed.), Combustion Chemistry, Springer, New York, 1984, pp. 197-360.

[3]Dixon-Lewis G. and Williams D.J. in Bamford C. H. and Tipper C.F. H. (eds.) Comprehensive Chemical Kinetics, Vol. 17: The Oxidation of Hydrogen and Carbon Monoxide, Elsevier, Oxford, 1977.

[4]Treviño C. and Williams F.A., Asymptotic Analysis of the Structure and Extinction of Methane-Air Diffusion Flames, Progress in Aeronautic and Astronautics, AIAA, Vol. 113, 1988, pp. 129-165.

[5]Seshadri K. and Peters N., "Asymptotic Structure and Extinction of Methane-Air Diffusion Flames, Combustion and Flame, Vol. 73, 1988, pp. 23-44.

[6]Brokaw R.S., Analytic solutions to the ignition kinetics of the Hydrogen-Oxygen reaction, 10th (Symposium) International on Combustion, The Combustion Institute, 1965, pp. 269-278.

[7]Gardiner W.C. Wakefield C.B. and Walker B.F. in Shock Waves in Chemistry and Chemical Technology, Lifshitz A. (ed.) Marcel Dekker, New York, 1981.

[8]Kailasaneth K., Oran E.S., Boris J. P. and Young T.R., in Peters N. and Warnatz J. (eds.): Numerical Methods in Laminar Flame Propagation, Vieweg, Braunschweig, 1982, p. 152.

[9]Wiriyawat S. and Dabora E.K., "Modeling the Chemical Effects of Plasma Ignition in one-dimensional Chamber, 20th. Symposium (International) on Combustion, The Combustion Institute, 1984, pp. 179-186.

[10]Maas U. and Warnatz J., "Ignition Processes in Hydrogen-Oxygen Mixtures", Combustion and Flame, Vol. 74, 1988, pp. 53-69.

[11]Peters N., "Numerical and Asymptotic Analysis os Systematically Reduced Reaction Schemes for Hydrocarbon Flames, in Lecture Notes in Physics (Glowinski et al. editors), Vol. 241, Springer, New York, 1985, pp. 90-109.

[12]Peters N. and Williams F.A., "The Asymptotic Structure of Stoichiometric Methane-Air Flames", Combustion and Flame, Vol. 68, 1987, pp. 185-207.

[13]Lam S.H. and Gousis D.A., "Understanding Complex Chemical Kinetics with Computational Singular Perturbation", XXII Int. Symposium on Combustion, The Combustion Institute, 1989.

Ignition of Fresh Mixtures Injected into Burned Gases

S. Kadowaki,* Y. Ohta,† M. Furutani,‡ and K. Terada§
Nagoya Institute of Technology, Nagoya, Japan
and
H. Takahashi¶
Meijo University, Nagoya, Japan

Abstract

Fresh methane-air or propane-air mixture is injected in the form of a small weak jet into high-temperature burned gases produced from flame propagation through a stoichiometric propane-air mixture. Ignition delays are measured from the beginning of mixture injection to luminous emission. Both fresh mixtures show almost identical delays, depending on the temperature of surrounding burned gases where fresh mixtures are injected. Fuel sensitivity is not shown here between methane and propane in contrast with the traditional ignition of pure fresh mixtures, where the delays of methane are usually one order of magnitude longer than those of other higher hydrocarbons. In addition, no pressure dependence is noticed. The ignition delays are shorter, and apparent activation energy is considerably smaller than in the data for ignition delays of fresh mixtures measured in shock tubes. Even when an argon-air mixture containing no fuel is injected into the burned gases, the same onset of luminosity is observed, with delays similar to those observed after fuel-air mixture injection into the burned gases. It is indicated that the ignition of fresh mixtures in which high-temperature burned gases take part would be controlled by another mechanism than the usual one.

Introduction

The autoignition of fresh mixtures or fuel sprays in the presence of burned gases seems to be considerably different from the ignitions of pure mixtures composed of fuels and oxidizers. Chemical species remaining in burned gases may have some influence on the ignition processes. This condition is carried out, as a practical example, in pulse combustors. In

---

*Associate Researcher, Department of Mechanical Engineering.
†Associate Professor, Department of Mechanical Engineering.
‡Ph.D. Candidate, Department of Mechanical Engineering.
§Professor, Department of Mechanical Engineering.
¶Professor, Department of Mechanical Engineering.

this system, combustion is induced by the mixing of fresh gases with burned gases, and the charge introduced into a combustion chamber ignites more easily than ordinary fresh mixtures. The deflagration-to-detonation transition resulting from strong turbulent mixing is also a typical example of the ignition system in which burned gases are involved. The effect of burned gases on the ignition and detailed ignition mechanism in this system are unknown.

Ignition delays of liquid fuel sprays injected into lean burned gases were reported by Siebers and Edwards.[1] The delays of injected fuels were independent of pressure and insensitive to the fuel type. These ignition characteristics are surprisingly different from those of pure mixtures. Ignition delays of gaseous methane injected into lean burned gases of methane-air were measured by Sakane et al.[2] The delays of injected methane are considerably short, and the apparent activation energy is about one-half as much as the data measured in shock tubes.[3]

In this study, we estimate the effect of burned gases on the ignition process of fresh mixtures and consider the reasons for particular characteristics of ignition in which the burned gases take part. We inject a very small amount of fresh premixed gases, methane-air or propane-air in the form of a jet into the quiescent high-temperature burned gases produced from stoichiometric propane-air flame ignited by a spark, and measure the delays from the time of injection to luminous emission. The ignition delays are compared with the published data for ignition delay times of fresh mixtures in shock tubes and those for gaseous fuel injected into hot air.

Temperature distributions in the combustion vessel are estimated numerically using the element division model for the charge in the vessel where fresh mixtures are injected.

Experiments

A concept of experimental setup is shown in Fig. 1. The combustion vessel has the shape of a rectangular prism with an inner length of $80 \times 100 \times 500$ mm and a volume of 4000 cm^3. A Pyrex glass window and a photomultiplier (Hamamatsu-Photonics 931A) are fitted on one side of the vessel to detect luminosity when light emission appears. A liquid gasoline injector (Honda 16450-PD1-023) is used to inject fresh gaseous mixtures into burned gases, for which the operation delay time is about 0.7 ms.[4] One end of the vessel is equipped with a pressure pickup (Toyoda-Koki PMS-5). A charged gaseous mixture, stoichiometric propane-air, is ignited by an electric spark fitted on the other end to prepare the burned gases. The experiment is controlled by a personal computer, and the data are acquired with it. The signals of luminosity and pressure are digitized at every 0.2-ms interval with 8-bit precision and recorded on memory chips.

After the burning is completed through a laminar flame across the vessel, the temperature and pressure of the burned gases fall gradually through heat losses to the wall. The moment the pressure attains a certain set value, the fresh mixture, methane-air or propane-air, is injected into the

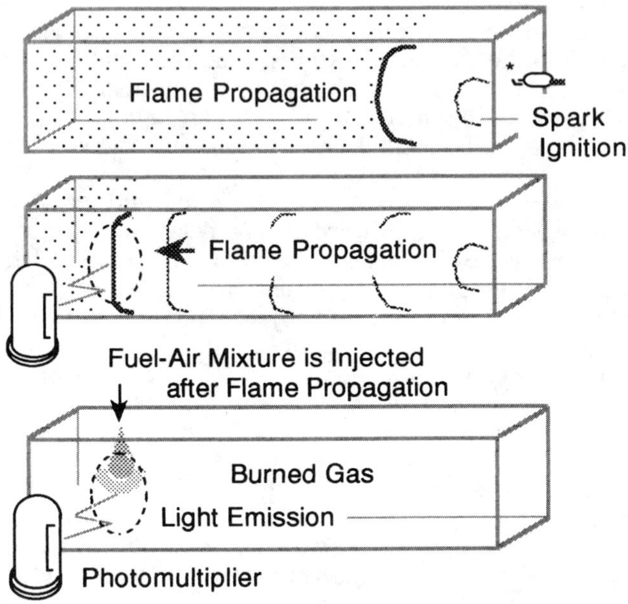

Fig. 1 Concept of the experimental arrangement.

burned gases through the injector. The injected fresh gaseous mixture contacts the burned gases, is heated up, and then ignites. Ignition is marked by luminosity. The ignition delay is defined as an interval between the beginning of injection and the onset of light emission. The temperature of burned gases in the vicinity of the injected mixture during the induction time is estimated numerically from its pressure history. The method of temperature estimation is stated in detail in the Appendix.

In the experiments, initial temperatures of the premixed charge in the vessel and of the fresh mixtures to be injected are set at 300 K. Initial pressure of the charge is varied from 0.05 to 0.08 MPa. The temperature and pressure of burned gases at the moment that fresh mixtures are injected range from 840 to 1515 K and from 0.14 to 0.40 MPa, respectively. The pressure difference between injection pressure and back pressure when fresh gases are injected into the vessel is constant at 0.05 MPa at all times to keep the same jet configuration. The volume of injected fresh gases is about 0.14 cm^3, sufficiently small compared to the volume of the burned gases. Therefore, the pressure rise in the vessel resulting from the ignition of the injected fresh mixture is not observed.

Results and Discussion

The delay time to luminous emission when stoichiometric fresh methane-air mixture is injected into burned gases is shown in Fig. 2 as an Arrhenius plot. The pressures of burned gases at the moment the fresh

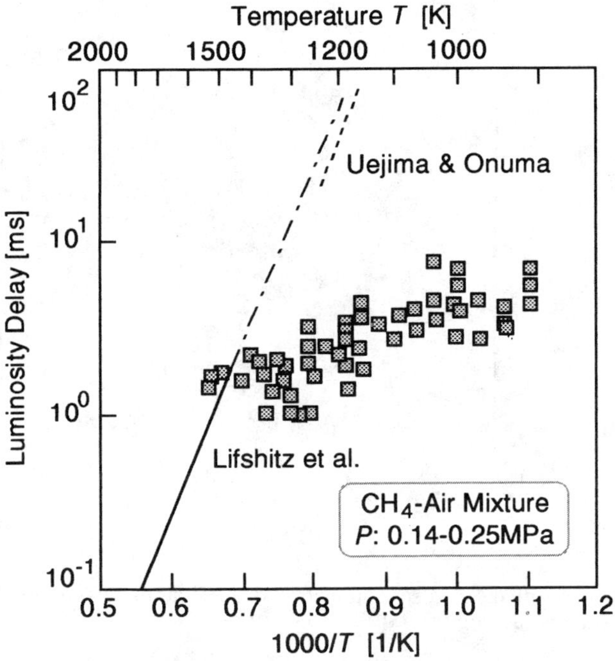

Fig. 2 Luminosity delay of stoichiometric methane-air mixture; back pressure is ranging from 0.14 to 0.25 MPa.

mixture is injected are 0.14 to 0.25 MPa. The solid line denotes the ignition delay time of methane-oxygen-argon mixture obtained from the empirical equation of shock-tube experiments by Lifshitz et al.,[3] and the dashed-and-dotted line is an extrapolation from the empirical equation. The dashed line shows the ignition delays of pure gaseous methane injected into hot air reported by Uejima et al.[5] The ignition delays in the system described here are much shorter compared to the data of Lifshitz and Uejima, and the apparent activation energy is considerably smaller. These measured data contain physical delay for heating up the fresh mixture to the surrounding temperature in addition to essential chemical delay time. The measured ignition delays reach asymptotically to a limiting value for temperatures above 1400 K. The value of about 1.5 ms is considered to be a first-order approximation of physical delay and is subtracted from the measured delays to estimate the chemical delays. Through this procedure, the activation energy for this preflame reactions up to the light emission is derived to be about 60 kJ/mol. It is about one-third of the data of fresh mixtures obtained in shock tubes (about 180 kJ/mol).

Next, we examine the effect of pressure on the ignition delays. The higher-pressure results are shown in Fig. 3, where the pressure of burned gases in the vessel is from 0.22 to 0.40 MPa. The ignition delays show no

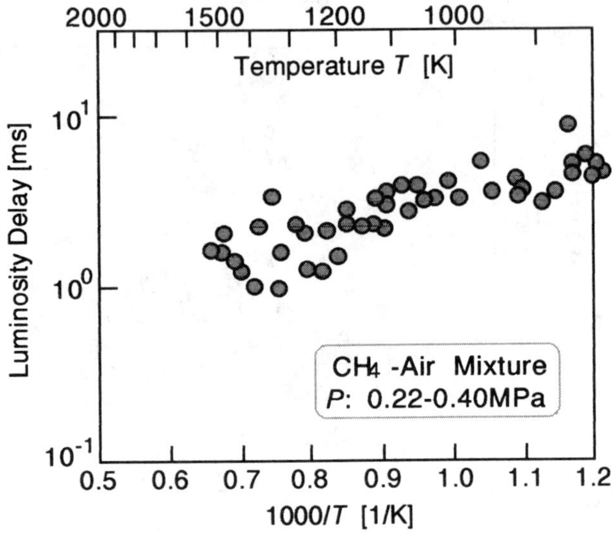

Fig. 3 Luminosity delay of stoichiometric methane-air mixture; high-pressure case; back pressure is ranging from 0.22 to 0.40 MPa.

differences between Figs. 2 and 3, and the pressure dependence on the ignition in this system is negligible.

Figure 4 is the result of stoichiometric propane-air mixture ignition. The ignition delays of propane-oxygen-argon mixture in shock tubes [6] and the ignition delays of propane injected into hot air [5,7] are also shown in the figure. The delays of methane mixture in shock tubes are usually one order of magnitude longer than those of the other higher hydrocarbon fuels, such as propane and butane.[8] However, comparing Figs. 2 and 4, we can see that the ignition delays in methane and propane mixtures are very similar to each other. No fuel sensitivity is shown here between methane and propane. This unusual result indicates that fuels in the fresh mixtures have no rate-controlling effects up to light emission and ignition.

We adopt an argon-air mixture containing no fuel as injected fresh gas in order to clarify the contribution of fuels and chemical species to the ignition mechanism in this system. The fuel in the mixture is replaced by inert argon, where the argon in fresh gas is equal to the amount of methane in the case of methane-air mixture. In general, we expect that the light emission is not obtained except for the cases in which burned gases are produced from rich mixtures. However, we observe unexpectedly the luminosity from the portion in the burned gases in which argon-air is injected. The delays for luminous emission are shown in Fig. 5. No strong difference in delays can be found between this no-fuel case and the previous fuel-contained cases. In comparison to the results of fuel-air mixture cases, the intensity of luminosity is slightly stronger and the ignition delay is somewhat shorter in relatively higher-temperature regions.

High-temperature burned gases produced from stoichiometric hydrocarbon-air include CO and H_2 and many other radicals, and the total heat of formation of them is larger than that of the burned gases in simple stoichiometry combustion. The burned gases that are established when the flame propagation has finished attain roughly equilibrium state depending on the temperature. The addition of oxygen through the injection will break down this state and make it shift to another state having a lower equivalence ratio. Equilibrium calculation shows that this procedure will lead CO and H_2 to decrease rapidly. It means that CO_2 and H_2O are produced and that some heat is released at the same time; then the reaction in this transition is always exothermic. This heat release is considered one of the sources of the luminous emission observed in our experiments.

To clarify the effect of this heat release on the fresh mixture ignition, we calculate the reaction process using the detailed chemical kinetic mechanism that involves 23 chemical species and 78 elementary reactions. The data of reaction rates are taken from Westbrook et al.[9] It is assumed that the burned gases maintain equilibrium state up to injection and that the fresh gas is uniformly mixed with the burned gases. The initial temperature of mixtures of fresh gas with burned gases is the same as that of burned gases for calculations. The initial concentration of each species is determined through the uniform mixing of fresh mixture with burned gases in several ratios. We employ the explicit Adams-Bashforth method which is second-order accurate in time. We do not reproduce numerically the reactions up to ignition though we simulate this problem in detail and in many cases. Because the concentration of CO, H_2, and other radicals remaining in burned gases are very low, the heat release due to the addition of oxygen is too little for the temperature rise. We do not explain these unusual experimental results by the concept of the uniform mixing of fresh mixture with burned gases.

In our experiments, the fresh mixture injected in the form of a weak jet contacts high-temperature burned gases, is heated up, and starts to react. The reaction region may be placed in the vicinity of the contact surface of the fresh gas jet and burned gases. The contact surface moves and widens rapidly in the burned gases since the pressure of the fresh mixture is slightly higher than the back pressure. This means that the elements of burned gases that are directly concerned with the chemical reaction in the surface are replaced, one after another. The concentration of chemical species taking part in the heat release is very low in the burned gases; however "fresh" burned gases attack successively the fresh mixture in the reactions. In this way, radicals involved in the preflame reaction for the contact surface are not always a small amount. Carbon monoxide, hydrogen, and other radicals rather than hydrocarbon are active in the reaction. These active species are supplied from burned gases to the reacting surface and support the process of unusually fast ignition reaction, which makes the ignition delays short and the apparent activation energy small. The experimental results shown in Figs. 2-4 are explained through these considerations.

The luminosity emitted after the injection of the fresh mixture may have causes other than the addition of oxygen. To remove all doubt, we

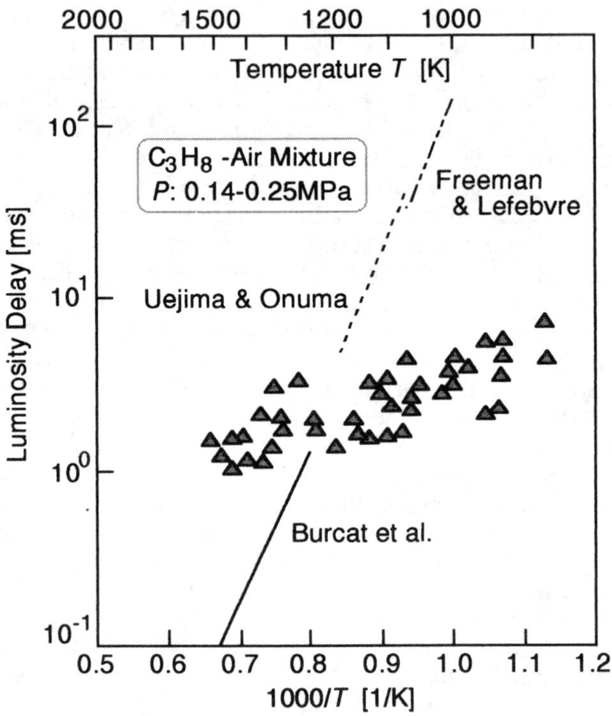

Fig. 4 Luminosity delay of stoichiometric propane-air mixture; back pressure is ranging from 0.14 to 0.25 MPa.

carried out the experiment using pure argon as the injected gas. We observed no luminosity in the case of the inert gas only. Temperature decrease due to the gas injection is not the reason for light emission. We used pure methane as injected gas to confirm the influence of the fuel itself on the ignition. Much as in the case of argon, we did not observe luminosity. This result indicates that the luminous ignition is not initiated by fuels.

It is considered that the luminosity delays of the argon-air mixture, which are shorter than those of the hydrocarbon-air mixture in relatively high-temperature regions, result from the shorter heat-up time that stems from lower heat capacity.

The ignition process of fresh mixtures in this experiment is also identical to the autoignition in diesel engines. In diesel engines, fuel ignites through injection into hot compressed air containing several residual gases. It is well known that liquid fuels injected into pure hot air have considerably longer ignition delays than those in practical engine cylinders. The effect of burned gases on the ignition of fuels is not negligible. The ignition mechanism proposed in this study would give a more appealing

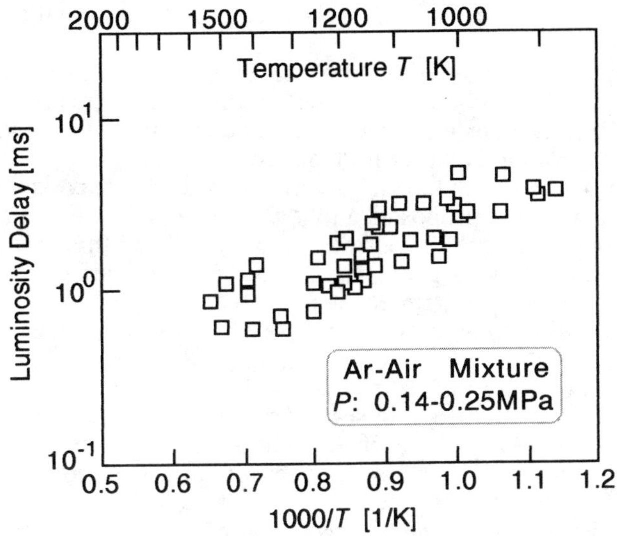

Fig. 5 Luminosity delay of argon-air mixture; back pressure is ranging from 0.14 to 0.25 MPa.

explanation of the ignition characteristics in the reactive systems in which burned gases take part.

Conclusion

The ignition of fresh mixtures injected into burned gases has been investigated. The burned gases are produced from the spark flame of a stoichiometric propane-air mixture. The ignition delays of a methane-air mixture injected into the burned gases show no difference from those of a propane-air mixture. In addition to the absence of fuel sensitivity, the absence of pressure dependence is confirmed in this ignition system. The ignition delays are short, and the apparent activation energy is surprisingly small compared to those of pure mixtures obtained in shock tubes. The light emission is observed even when an argon-air mixture is injected into the burned gases, and delays in this case are similar to those in the fuel-air mixture case. Ignition systems concerned with burned gases should have another ignition mechanism than the normal fuel oxidations. We propose a new mechanism for this ignition system that the fresh-mixture jet successively contacts renewed burned gases and that the ignition reactions occur in the vicinity of the contact surface, supported by a continuous supply of CO, H_2, and other active radicals from burned gases.

Acknowledgment

Y. Ohta gratefully acknowledges the financial support provided by the Nitto Foundation.

Appendix: Estimation of Temperature Distributions in the Vessel

We use the element division model to calculate the temperature distributions in the combustion vessel. In this model, the total mass of gas is divided into minute mass elements, and the quantity of state in each element is determined by thermodynamic relations. The improved model, including heat transfer in the vicinity of a wall and heat conduction between mutual elements, was proposed, and it was reported that this model obtains good agreement with the experimental results.[10] In this study, we adopt this model, improve it slightly, and calculate the temperature distributions of contained gases in each time

For calculating the temperature distributions, we assume the following:

1) The burning rate is known.
2) Pressure in the combustion vessel is homogeneous.
3) The state of gas in each element is instantaneously homogeneous.
4) There is no matter transfer between mutual elements.
5) Equation of state for ideal gas is realized.
6) Combustion is isoenthalpy burning.
7) The reaction-zone thickness is infinitesimal.
8) The flame front propagates from the surface containing the spark plug to the end of the vessel in maintaining a plane structure.
9) Temperature dependence of heat conductivity and specific heat are taken into account.

From the energy conservation and the equation of state, we obtain the relations satisfied in each element of burned and unburned regions as follows:

$$C_v dT_i = -PdV_i + dQ_i \tag{A1}$$

$$dQ_i = -hF_i(T_i - T_w)dt - kS_i(2T_i - T_{i+1} - T_{i-1})/x_i \cdot dt \tag{A2}$$

$$PV_i = G_i R T_i \tag{A3}$$

where

C_v = specific heat at a constant volume
F = contact area of an element with a wall
G = mole number
h = coefficient of heat transfer
k = heat conductivity
P = pressure
Q = heat quantity
R = universal gas constant
S = cross section of an element
T = temperature
T_w = temperature of a wall
t = time
V = volume
x = width of an element

i = element number

We solve numerically the preceding equations. From Eqs. (A1)-(A3), we obtain the temperature and the volume of each element in burned and unburned regions corresponding to the assumed coefficient of heat transfer h. On the other hand, for elements in the burning region, we find the adiabatic flame temperature from the state at forward time using the chemical equilibrium calculation[11] and obtain the volume of each element in the burning region from the equation of state. Now the unknown h is determined as a convergent value from the total volume of elements being constant. After the calculation, we obtain the temperature distributions of contained gas in each time and the gas temperature history at the place where the fresh mixture is injected.

Moreover, we have calculated the gas temperature using the modified model, which includes the trapezoidal temperature distribution in the direction of a wall in each element. We confirmed that the temperature distributions for the centered region obtained by this model are equivalent to those obtained by the method mentioned earlier, since the inner size of the vessel is large enough. Therefore, for efficiency, we adopt the method mentioned first and evaluate the temperature of burned gases surrounding the injected fresh mixtures.

References

[1]Siebers, D. L., and Edwards, C. F., "Autoignition of Methanol and Ethanol Sprays under Diesel Engine Conditions," SAE Paper 870588, 1987.

[2]Sakane, A., Murakami, S., Toshioka, S., and Obara, K., "Ignition Delays of Methane Jets," Journal of the Marine Engineering Society in Japan, Vol. 23, No. 6, 1988, pp. 341-346 (in Japanese).

[3]Lifshitz, A., Scheller, K., and Skinner, G. B., "Shock-Tube Investigation of Ignition in Methane-Oxygen-Argon Mixtures," Combustion and Flame, Vol. 16, 1971, pp. 311-321.

[4]Nomura, T., Takaoka, Y., Tanino, T., and Irino, H., "Flow Characteristics of Electronic Fuel Injector," Journal of the Society of Automotive Engineers of Japan, Vol. 40, No. 9, 1986, pp. 1128-1133 (in Japanese).

[5]Uejima, M., and Onuma, Y., "Ignition Delay of Unsteady Jet in Hot Air," 21th Combustion Symposium (Japanese Section), Combustion Society of Japan, Tokyo, 1986, pp. 97-99 (in Japanese).

[6]Burcat, A., Lifshitz, A., Scheller, K., and Skinner, G., "Shock-Tube Investigation of Ignition in Propane-Oxygen-Argon Mixtures," Thirteenth Symposium (International) on Combustion, The Combustion Institute, Pittsburgh, 1971, pp. 745-755.

[7]Freeman, G., and Lefebvre, A. H., "Spontaneous Ignition Characteristics of Gaseous Hydrocarbon-Air Mixtures," Combustion and Flame, Vol. 58, 1984, pp. 153-162.

[8]Burcat, A., Scheller, K., and Lifshitz, A., "Shock-Tube Investigation of Comparative Ignition Delay Times for C_1-C_5 Alkanes," Combustion and Flame, Vol. 16, 1971, pp. 29-33.

[9]Westbrook, C. K., Dryer, F. L., and Schug, K. P., "Numerical Modeling of Ethylene Oxidation in Laminar Flames," Combustion and Flame, Vol. 52, 1983, pp. 299-313.

[10]Tasaka, H., Matsuoka, S., and Takano, M., "Thermodynamic Model of Premixed Combustion in Chamber," Transactions of the Japan Society of Mechanical Engineers, B, Vol. 40, No. 9, 1982, pp. 2355-2363 (in Japanese).

[11]Harker, J. H., "The Calculation of Equilibrium Flame Gas Compositions," Journal of the Institute of Fuel, Vol. 40, No. 316, 1967, pp. 206-213.

Analysis of Ignition Mechanism of Combustible Mixtures by Short-Duration Sparks

M. Kono,* K. Ishii,† K. Niu,‡ T. Tsukamoto,§ and Y. Ujiie¶
University of Tokyo, Tokyo, Japan

Abstract

The formation process of flame kernels produced by short duration sparks (less than 0.5 µs) in a combustible gas is modeled by using a set of partial differential equations with unsteady and two-dimensional cylindrical coordinates and an overall reaction equation. These equations are solved numerically. The correctness of the present simulation is verified by the experimental results. The simulation shows that the gas flow pattern governs the kernel configuration and it is influenced by the spark gap width and the tip end configuration of the spark electrode. The minimum ignition energy obtained agrees quantitatively with the experimental results.

Nomenclature

A	= pre-exponential factor, $m^3/(mol \cdot s)$
a	= order of reaction
b	= order of reaction
C_p	= specific heat at const. pressure, $J/(kg \cdot K)$
D	= diffusion coefficient, m^2/s
E	= activation energy, J/mol
e	= specific internal energy, J/kg
h	= specific enthalpy, J/kg
k	= reaction rate constant, $m^3/(mol \cdot s)$
M	= molecular mass, kg/mol
p	= pressure, Pa

*Professor, Department of Aeronautics.
†Doctoral Course Student, Department of Aeronautics.
‡Engineer, Mitsubishi Heavy Industry Ltd.
§Research Associate, The Tohoku University.
¶Associate Professor, The Nihon University.

Q = heat release, J/mol of fuel
q = energy flux, J/(mol·s)
$\dot{q}$ = heat release rate, J/(m^3·s)
R = gas constant, J/(kg·K)
$\bar{R}$ = universal gas constant, J/(mol·K)
r = radial coordinate, m
T = temperature, K
t = time, s
v = velocity, m/s
Y = mass fraction
z = axial coordinate, m
κ = thermal conductivity, J/(m^2·s·K)
μ = viscosity, kg/(m·s)
ν = kinematic viscosity, m^2/s
ρ = density, kg/m^3
$\dot{\rho}$ = production rate, kg/(m^3·s)
τ = stress tensor, Pa
ω = reaction rate, mol/m^3·s)

<u>Subscripts</u>

c = chemical
C_3H_8 = propane
i = species
O_2 = oxygen
r = radial component
z = axial component
θ = angle in spherical coordinate

Introduction

Two important contributions have been given in recent fundamental studies of the ignition of a combustible mixture by short-duration sparks; one concerns capacitance sparks[1,2] and the other breakdown sparks.[3] The ignition mechanism by the former has been investigated because such sparks play an important role as a trigger of the following component of composite sparks,[4,5] which are conventional as ignition sparks for automobile engines. The latter is recommended as an ignition spark with a good ability for automobile engines.[6] It is reported in these works that the flame kernel produced by such short-duration sparks has a toroidal shape[1-3,7] in its incipient stage, namely, the ignition process, which is estimated to be less than 200 μs after spark discharge.[2] Therefore, the toroidal development of flame kernels has a close relation with the ignition process, but the mechanism of formation of such toroidal kernels is unknown except for some estimates based on gas movement[2,3] and ionic wind effect.[7]

In the present work, the formation process of the flame kernel is modeled and numerically simulated, with emphasis on its mechanism. The most important assumption in the present model is that the elementary reaction of radical species is neglected. Although the radical species has significant effects in some cases,[8-10] in most cases of spark ignition, physical aspects have more dominant effects on the ignition process. Therefore, in the present work, the physical effects of flame kernel shapes[2], spark electrode configuration[1], gap width[1], mixture movement, and a heat loss to the spark electrodes on the flame kernel formation and the minimum ignition energy are examined by using the overall reaction.

As is well known, the spark electrode affects the minimum ignition energy (see, for example, Fig. 1). The result has been obtained by using the ignition unit described in Ref. 1. The spark duration is less than 0.5 µs. For a gap width larger than the quenching distance (3.0 mm), the minimum ignition energy is constant for various diameters of the spark electrode. For smaller values (1.5 and 1.0 mm), the minimum ignition energy increases in a complicated manner as the diameter of the spark electrode increases. The gap width less than the quenching distance

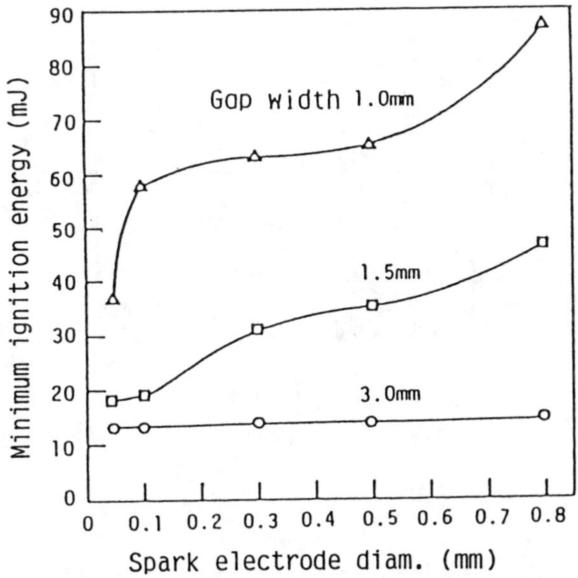

Fig. 1 Effect of spark electrode size and gap width on minimum ignition energy (experiment) for propane 3% (vol.)-air mixture.

is employed widely in practical ignition systems, so clarification of the mechanism of spark electrode effect seems to be very important to realize an ignition system of higher igniting ability.

Model Description

After the instantaneous spark discharge, the growth of the flame kernel is modeled by a set of partial differential equations in an unsteady, two-dimensional state. The coordinate system, the computing region, and the location of the spark electrode are shown in Fig. 2. Since the flame kernel is assumed to be symmetric with respect to the z and r axes, the calculation is made on a region as shown in Fig. 2. The region is 2 and 4 mm in the z and r directions, respectively. These values are determined based on the experimental fact that the final size of the practical flame kernels in the ignition process is 7 mm in diameter, at most.[4]

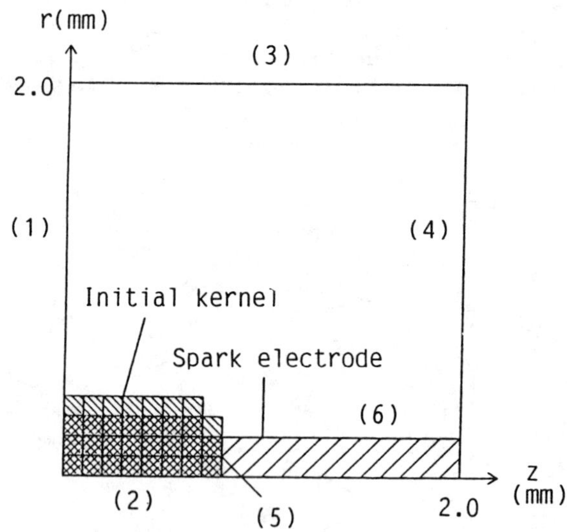

	T(K)	P(MPa)	ρ(kg/m³)
(cross-hatched)	2301	0.7906	1.199
(single-hatched)	1420	0.4615	1.199

Fig. 2 Region of simulation, initial kernel structure, and notation of boundary conditions.

To simplify the treatment of the model, the following assumptions are made.

(1) Natural convection is neglected.

(2) The gas consists of propane, oxygen, nitrogen, and the product, and the overall irreversible reaction is used.

(3) Heat conduction to the spark electrode is assumed for constant temperature of the electrode surface.

(4) Prandtl number is unity.

(5) C_p and ν of the gas are given as a function of temperature.

$$C_p/R = 3.33 - 3.61 \times 10^{-4}T + 5.06 \times 10^{-7}T^2 - 1.04 \times 10^{-11}T^3$$

$$\nu = -9.69 \times 10^{-6} + 6.32 \times 10^{-8}T + 6.99 \times 10^{-11}T^2 \quad (m^2/s)$$

With these assumptions, the following equations, written in cylindrical coordinates with axial symmetry and zero azimuthal velocity, are obtained.

Mass conservation:

$$\frac{\partial \rho}{\partial t} + \frac{1}{r}\frac{\partial \rho v_r r}{\partial r} + \frac{\partial \rho v_z}{\partial z} = 0$$

Momentum conservation:

$$\frac{\partial \rho v_r}{\partial t} + \frac{1}{r}\frac{\partial \rho v_r^2 r}{\partial r} + \frac{\partial \rho v_z v_r}{\partial z} = -\frac{\partial p}{\partial r} - \frac{1}{r}\frac{\partial \tau_{rr} r}{\partial r} + \frac{\tau_{\theta\theta}}{r} - \frac{\partial \tau_{rz}}{\partial z}$$

$$\frac{\partial \rho v_z}{\partial t} + \frac{1}{r}\frac{\partial \rho v_r v_z r}{\partial r} + \frac{\partial \rho v_z^2}{\partial z} = -\frac{\partial p}{\partial z} - \frac{1}{r}\frac{\partial \tau_{rz} r}{\partial r} - \frac{\partial \tau_{zz}}{\partial z}$$

Energy conservation:

$$\frac{\partial \rho e_t}{\partial t} + \frac{1}{r}\frac{\partial \rho v_r e_t r}{\partial r} + \frac{\partial \rho v_z e_t}{\partial z} = -\frac{1}{r}\frac{\partial q_r r}{\partial r} - \frac{\partial q_z}{\partial z} - \frac{1}{r}\frac{\partial p v_r r}{\partial r}$$

$$-\frac{\partial p v_z}{\partial z} - \frac{1}{r}\frac{\partial(\tau_{rr} v_r r + \tau_{rz} v_z r)}{\partial r} - \frac{\partial(\tau_{rz} v_r + \tau_{zz} v_z)}{\partial z} + \dot{q}_c$$

Species conservation:

$$\frac{\partial \rho_i}{\partial t} + \frac{1}{r}\frac{\partial \rho_i v_r r}{\partial r} + \frac{\partial \rho_i v_z}{\partial z}$$

$$= \frac{1}{r}\frac{\partial}{\partial r}(\rho D \frac{\partial Y_i}{\partial r} r) + \frac{\partial}{\partial z}(\rho D \frac{\partial Y_i}{\partial z}) + (\dot{\rho}_i)_c$$

where

$$\tau_{rr} = -\mu(2\frac{\partial v_r}{\partial r} - \frac{2}{3}(\frac{1}{r}\frac{\partial v_r r}{\partial r} + \frac{\partial v_z}{\partial z}))$$

$$\tau_{\theta\theta} = -\mu(2\frac{v_r}{r} - \frac{2}{3}(\frac{1}{r}\frac{\partial v_r r}{\partial r} + \frac{\partial v_z}{\partial z}))$$

$$\tau_{zz} = -\mu(2\frac{\partial v_z}{\partial z} - \frac{2}{3}(\frac{1}{r}\frac{\partial v_r r}{\partial r} + \frac{\partial v_z}{\partial z}))$$

$$\tau_{rz} = -\mu(\frac{\partial v_r}{\partial r} + \frac{\partial v_z}{\partial z})$$

$$q_r = -\kappa\frac{\partial T}{\partial r} - \sum_{i=1}^{n} \rho_i h D \frac{\partial Y_i}{\partial r}$$

$$q_z = -\kappa\frac{\partial T}{\partial z} - \sum_{i=1}^{n} \rho_i h D \frac{\partial Y_i}{\partial z}$$

$$e_t = e + \frac{1}{2}(v_r^2 + v_z^2)$$

$$p = \rho R T$$

Overall reaction:

$$C_3H_8 + O_2 + 18.8N_2 \rightarrow 7\text{Product} + 18.8N_2$$

where the product consists of $3CO_2 + 4H_2O$. The reaction rate, k, for 4.0% (vol.) propane-air mixture is assumed as

$$k = 0.4 \times 10^8 \exp(-E/RT)[C_3H_8]^{\frac{1}{2}}[O_2]^{\frac{1}{2}}$$

The boundary conditions are as follows: In Fig. 2, for "1" and "2", the gradients of all variables are zero except that $v_z = 0$ and $v_r = 0$, respectively. For "3" and "4", the gradients of all variables are zero. On the surface of the spark electrodes, "5" and "6", the no-slip condition is assumed.

As for the initial condition, the temperature and pressure of the quiescent gas are 300 K and 0.1 MPa, respectively, except for within the initial kernel which is assumed to be formed by an instantaneous spark. The temperature and thermodynamic property distributions as well as the size of the initial kernel are shown in Fig. 2. The initial kernel occupies several cells, and its length is the same as the spark gap width. The value of the pressure was determined from the velocity of the shock wave, which was measured experimentally beforehand, based on the simple shock wave theory, where the specific heat ratio is assumed to be 1.4. The temperature distribution is roughly determined so that the calculation may be carried out.

To determine the minimum ignition energy, initial energy (spark energy) must be varied. From the observation of practical spark discharge, with increasing spark energy, it is found that the luminous part of the spark path becomes thick. Therefore, the number of initial kernel cells as shown in Fig. 2 is changed accordingly to a given spark energy.

The preceding partial differential equations are solved numerically by using a portion of the RICE code.[11] Briefly, at a given time step, the conservation equations for mass and momentum are solved in an iterative manner using a first-order implicit scheme. Then the conservation equation for energy is solved, and the values of internal energy and density are updated. From the internal energy, temperature is obtained from the following equation, and the reaction rate is calculated accordingly.

$$T = 64.8 + 1.28 \times 10^{-7}e - 1.83 \times 10^{-18}e^2 + 9.71 \times 10^{-30}e^3 \text{(K)}$$

Using the equation of state, the final pressure is determined. The numerical stability is satisfied by considering the discussion mentioned in Ref. 11. Based on this, the mesh size and the time step are chosen to be 0.05 mm and 0.02 μs.

Flame Kernel Structure

Figures 3 and 4 show the calculated results of temperature profiles for the same spark electrode size and spark gap width. The former was calculated under a no-heat-release condition, and this corresponds to a phenomenon of a hot kernel formation in an inert gas. From a comparison of the results, there exists many common features. Namely, initially, the hot kernel is an ellipsoid and the maximum temperature region is located in the center of the spark gap. Afterward, at t = 0.02-0.04 ms, the hot kernel has developed into a torus and, accordingly, the highest

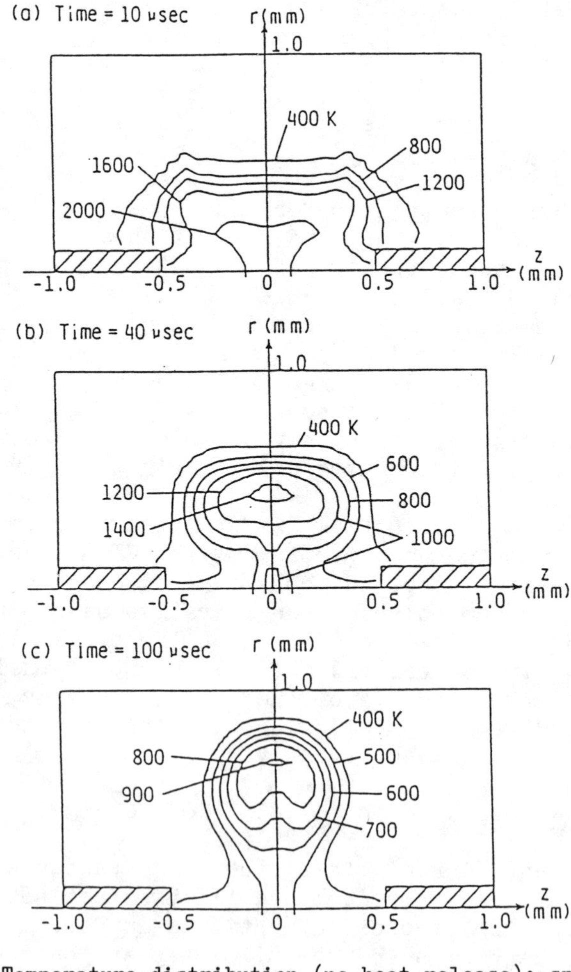

Fig. 3 Temperature distribution (no heat release): spark gap width, 1.0 mm; spark electrode, 0.2 mm in diameter.

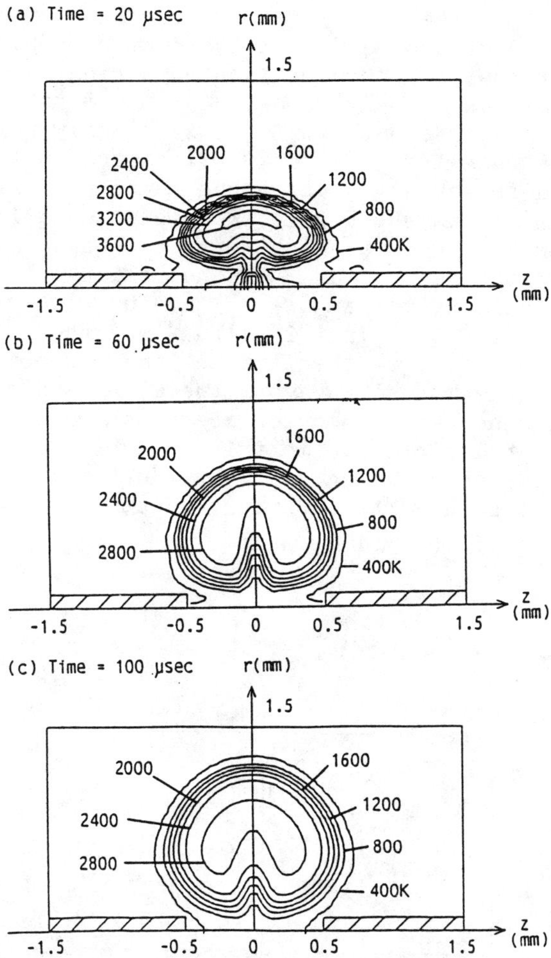

Fig. 4 Temperature distribution: spark gap width, 1.0 mm; spark electrode, 0.2 mm in diameter.

temperature region has moved into the ring of the torus. Then the torus ring grows, and its center extends outward. The torus structure has been verified by using an interferometry for the same condition as in Fig. 3.[12] Therefore, it is estimated that such a structure is also formed under the condition that heat release exists, as in Fig. 4.

Figure 5 shows a calculated velocity distribution using the the same conditions as for Fig. 3, but this distribution is essentially the same as in Fig. 4. It is indicated that at t = 0.04 ms there is a pair of vortices with rotations in opposite directions, and that the center of each vortex is located at the periphery of the torus ring (see Fig. 3b). This vortex has already appeared at t = 8 μs. The gaseous motion that causes such vortices is closely related to shock-wave behavior. The spark discharge initially produces a spherical shock wave, which can be seen in Fig. 5a, and on the inside of the shock wave the gas moves quickly outward. This movement causes a slight over expansion after the shock wave separates rather far from the residual hot kernel; the over expansion raises an inward gaseous flow near the spark gap, as in Fig. 5a. In this situation, the momentum of the z component is larger than that of the r component. Therefore, the mutual inward flows along the z axis collide with each other at the center of the spark gap and deviate toward the r direction. In this manner, a pair of vortices are produced as shown in Fig. 5b. Furthermore, the flow that causes such vortices brings about the toroidal development of the hot kernels.

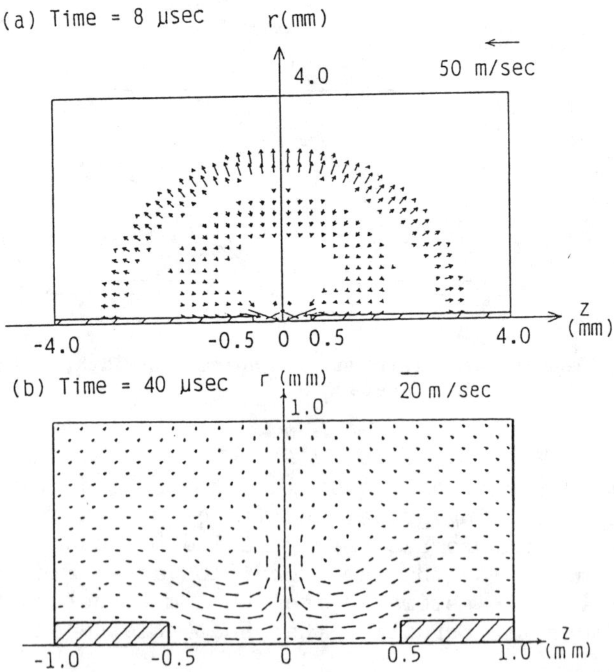

Fig. 5 Velocity distribution: spark gap width, 1.0 mm; spark electrode, 0.2 mm in diameter.

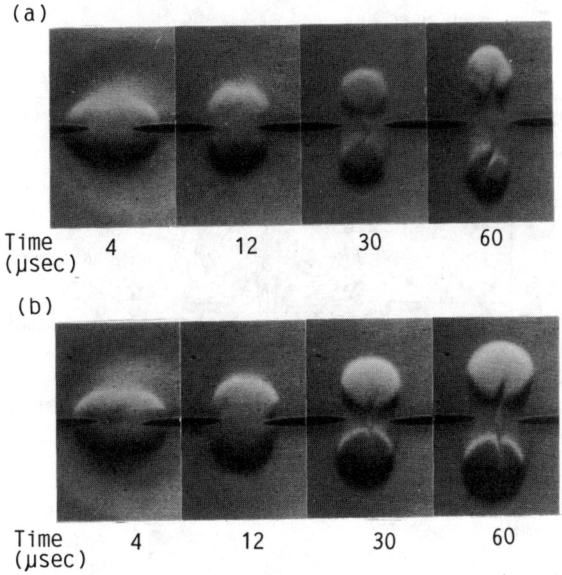

Fig. 6 Schlieren photographs of hot air kernel and flame kernel: a) air; b) propane-air 4(vol.)%.

Figures 3 and 4 illustrate the difference between kernel sizes with and without heat release; namely, the kernel size with heat release is larger than that in an inert gas. The correctness of such calculated results is verified from the schlieren photographs shown in Fig. 6.

Effect of Spark Electrode

From the present simulation, it is found that the pattern and strength of the gas flow are affected by several factors, such as spark energy per unit length of the spark path, configuration of the tip end of the electrode, and spark gap width. The flame kernel changes in its shape by these factors. Figures 7 and 8 show the effect of gap width and spark electrode diameter on the temperature distribution of the flame kernel. Comparison of Fig. 4b and especially Fig. 8 shows a distinctive difference on the flame kernel. For the flame kernel in Fig. 8, a high-temperature region is formed not only in the ring of the torus, but also in the center of the spark gap. As the spark electrode diameter increases further, this feature becomes much more prominent. The reason why two high-temperature regions appear is explained by the existence of stagnation flow in the

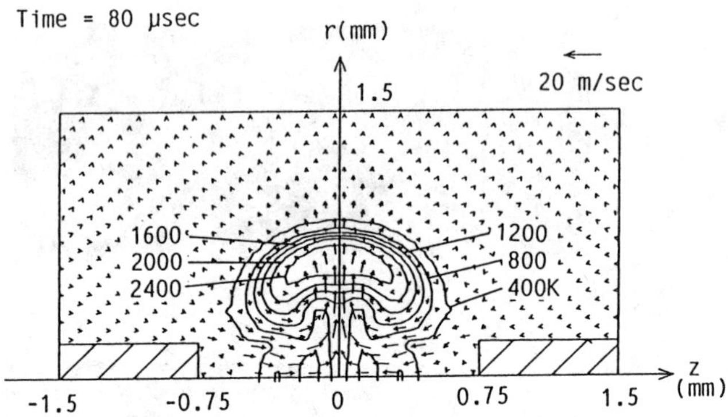

Fig. 7 Temperature and velocity distributions: spark gap width, 1.5 mm; spark electrode, 0.4 mm in diameter.

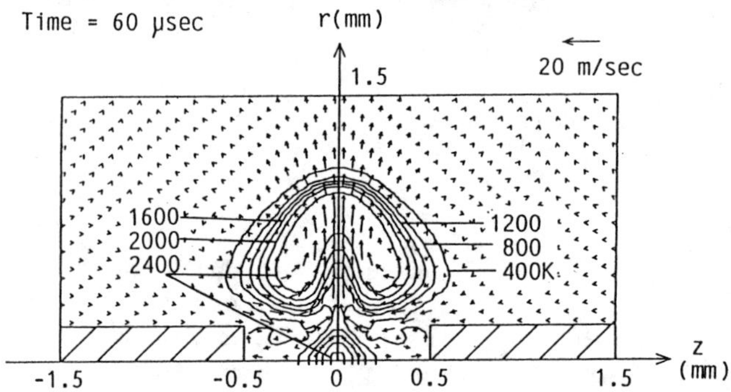

Fig. 8 Temperature and velocity distributions: spark gap width, 1.0 mm; spark electrode, 0.4 mm in diameter.

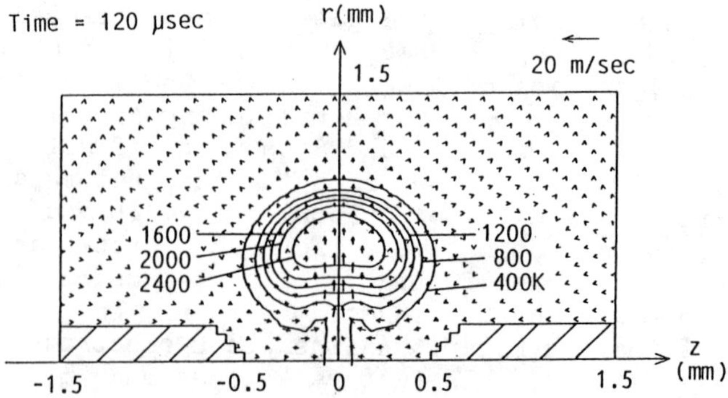

Fig. 9 Temperature and velocity distributions: spark gap width, 1.5 mm; spark electrode, 0.4 mm in diameter; half cone angle, 45 deg.

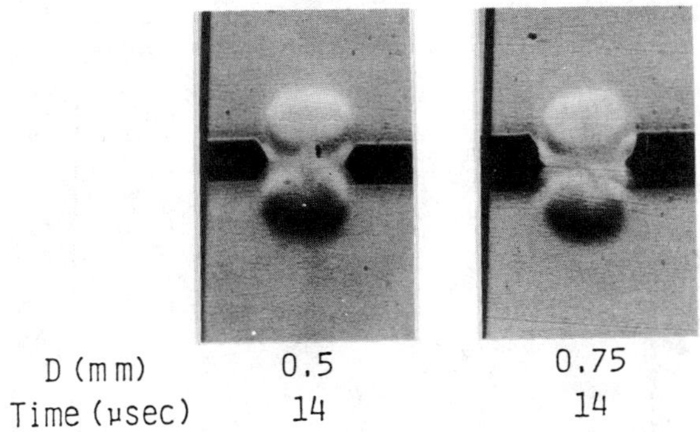

Fig. 10 Schlieren photographs of hot gas kernel. Spark gap width, 1.0 mm; D, diameter of spark electrode.

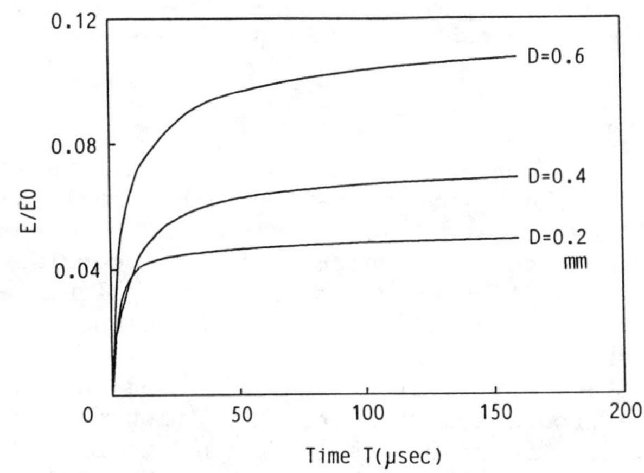

Fig. 11 Ratio of heat loss (E) to spark electrode to spark energy (EO). D is diameter of spark electrode.

vicinity of the tip end of the spark electrodes, where inward flow of the surrounding gas separates in opposite directions. The flow directed toward the center forms a pair of recirculation zones between the two tip ends of the spark electrodes, as shown in Fig. 8. These recirculation zones capture the hot gas that originally existed in the spark gap. With an increased electrode diameter, the amount of the hot gas captured in the recirculation zone increases, so that the highest temperature region has a tendency to move into the spark gap. As easily estimated, if spark electrodes with a conical shape at the tip end are employed, formation of such a recirculation zone seems to be

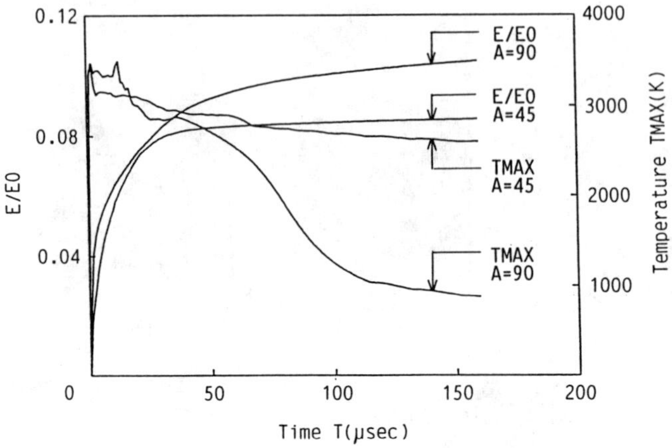

Fig. 12 Ratio of heat loss (E) to spark electrode to spark energy (E0) and maximum temperature of flame kernel. Spark gap width, 1.0 mm; spark electrode, 0.4 mm in diameter; A, half cone angle (deg).

suppressed. Figure 9 illustrates that this estimation is correct. As shown, the high-temperature region at the center of the spark gap is not discernible because the inward flow sweeps the center region and carries hot gas into the torus. The correctness of the calculated results shown in Figs. 8 and 9 is verified by the schlieren photographs shown in Fig. 10.

The flame kernel structure mentioned earlier is of much importance from a practical point of view on ignition. For instance, for the spark electrode used in Fig. 8, larger amounts of the spark energy remain in the recirculation zones where the quenching effect is rather strong on the flame kernel. Furthermore, the recirculation flow promotes the heat transfer onto the tip end surface of the spark electrode by forced convection. In Fig. 11, which shows the calculated heat loss to the spark electrode, it is indicated that, as the diameter of the spark electrode is increased, the amount of heat loss increases rapidly. This corresponds to the formation of the recirculation zone near the tip end of the spark electrode, as mentioned earlier. Figure 12 shows that, even for the same diameters of the spark electrode, the heat loss is affected substantially by its tip end configuration; accordingly, the maximum kernel temperature changes. As shown, it is estimated that the flame kernel for the spark electrode tipped with a 45-deg half-cone angle leads to successful ignition, and with 90 deg it leads to misfire or extinction. Figure 13 shows the

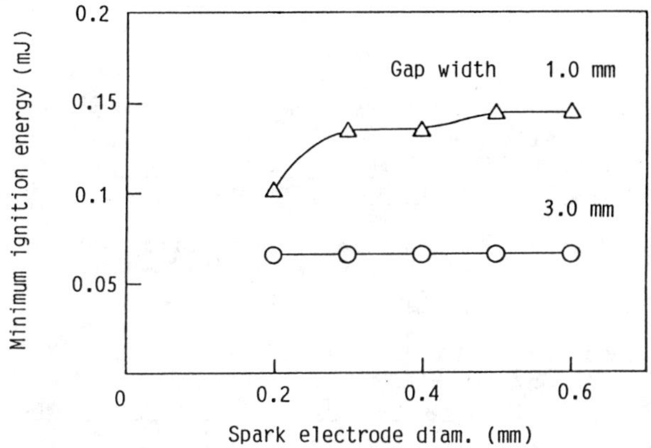

Fig. 13 Effect of spark electrode size and gap width on minimum ignition energy (calculated) for propane 4 (vol.)%-air mixture.

minimum ignition energy obtained from the present simulation. This figure should be compared with the experimental results shown in Fig. 1. In comparison, the whole behavior agrees well and, therefore, the present simulation seems to be less impractical. Also of importance here is that such simulation is very useful for understanding the ignition process and for development of ignition systems of higher igniting ability.

Conclusion

With reference to the development of flame kernels in spark ignition of combustible mixtures, their formation process by a short-duration spark in propane-air mixtures was modeled by using an overall reaction scheme and was solved numerically. The conclusions follow.

(1) The gas flow produced by the secondary effect of the shock wave is the main cause for specific kernel configurations.

(2) The simulated structure of the flame kernel agrees well with experimental results.

(3) In the present simulation, the spark electrode and gap width were found to be related closely to the flame kernel configuration and, thus, the minimum ignition energy.

(4) The present modeling and simulation are very useful for understanding the ignition process. They also provide useful suggestions for enhancing the igniting ability of electric sparks.

References

[1]Kono, M. and Hatori, K., "IGNITION OF LOW FLOW VELOCITY MIXTURES BY SHORT DURATION SPARKS," Eighth International Conference on Gas Discharges and Their Applications, 1985, p. 500.

[2]Kono, M., Kumagai, S. and Sakai, T., "Ignition of Gases by Two Successive Sparks with Reference to Frequency Effect of Capacitance Sparks," Combustion and Flame, Vol. 27, 1976, p. 850.

[3]Maly, R. and Vogel, M., "INITIATION AND PROPAGATION OF FLAME FRONTS IN LEAN CH_4-AIR MIXTURES BY THE THREE MODES OF THE IGNITION SPARK," Seventeenth Symposium (International) on Combustion, The Combustion Inst., Pittsburgh, PA, 1979, p. 821.

[4]Kono, M., Kumagai, S. and Sakai, T., "THE OPTIMUM CONDITION FOR IGNITION OF GASES BY COMPOSITE SPARKS," Sixteenth Symposium (International) on Combustion, The Combustion Inst., Pittsburgh, PA, 1977, p. 757.

[5]Kono, M., Hatori, K. and Iinuma, K., "INVESTIGATION ON IGNITION ABILITY OF COMPOSITE SPARKS IN FLOWING MIXTURES," Twentieth Symposium (International) on Combustion, The Combustion Inst., Pittsburgh, PA, 1985, p. 133.

[6]Ziegler, G. F. W., Wagner, E. P., Saggau, B., Maly, R. and Herden, W., "Influence of a Breakdown Ignition System on Performance and Emission Characteristics," SAE Paper 840992, 1984.

[7]Olsen, H. L., Edmonson, R. B. and Gayhart, E. L., "Microchronometric Schlieren Study of Gaseous Expansion from an Electric Spark," Journal of Physics, Vol. 23, 1952, p. 1157.

[8]Dixon-Lewis, G. and Shepherd, I. G., "SOME ASPECTS OF IGNITION BY LOCALIZED SOURCES, AND OF CYLINDRICAL AND SPHERICAL FLAMES," Fifteenth Symposium (International) on Combustion, The Combustion Inst., Pittsburgh, PA, 1981, p. 1736.

[9]Sloane, T., "A Computational Study of Ignition by Oxygen Dissociation," Combustion Science and Technology, Vol. 34, 1983, p. 317.

[10]Refael, S. and Sher, E., "A Theoretical Study of the Ignition of a Reactive Medium by Means of an Electrical Discharge," Combustion and Flame, Vol. 59, 1985, p. 17.

[11]Rivard, W. C., Farmer, O. A. and Butler, T. D., "RICE: A Computer Program for Multicomponent Chemically Reactive Flows at All Speeds," National Technical Information Service, Rept. LA-5812, 1975.

[12]Kono, M., Niu, K., Tsukamoto, T. and Ujiie, Y., "MECHANISM OF FLAME KERNEL FORMATION PRODUCED BY SHORT DURATION SPARKS," Twenty-Second Symposium (International) on Combustion, The Combustion Inst., Pittsburgh, PA, 1989, p. 1643.

Flat Plate Boundary Layer Ignition with Fuel Thermal Diffusion

Pedro L. García-Ybarra* and Jose L. Castillo*
Universidad Nacional de Educación a Distancia, Madrid, Spain

Abstract

The influence of fuel thermal diffusion in gas flow ignition by a hot plate is studied. Using boundary layer approximation and high activation energy asymptotics, the fuel concentration profile is obtained by matched asymptotic expansions in the inverse of the fuel Schmidt number (low fuel mobility). Noticeable increases in the ignition distance are predicted with respect to the results of previous analyses where thermal diffusion was neglected.

Introduction

Cross mass transport by thermal diffusion (Soret effect) is proportional to the local temperature gradient. When concentration gradients are present, thermal diffusion is often a negligible mechanism masked by the direct mass transport (i.e., Fick diffusion). For instance, it has been shown that thermal diffusion does not significantly affect the propagation velocity of a plane premixed flame,[1] although it may produce measurable distortions in the usual profiles of the flame internal structure.[2] On the other hand, under some circumstances thermal diffusion can play a major role.[3] Thus in problems related with the stability properties of the plane flame propagation,[1,4] Fick mass flow effect is counterbalanced by Fourier heat flow effect and a change in this balance, even by the small amounts induced by thermal diffusion, may drastically alter the flame stability. The aim of this paper is to address a different problem where, again, the influence of thermal diffusion becomes crucial in determining the behavior of the system, namely, the problem of a reactive gas mixture ignition by a hot body. Specifically, the case studied here will be the ignition of a premixed combustible gas flowing on a hot flat plate. Of course, the occurrence of ignition depends on temperature and fuel-oxidizer relative concentration near the plate but, due to thermal diffusion, the gas concentration is not homogeneous in the chemically frozen zone prior to ignition. Far from the plate no concentration gradients are present in the gas mixture and Fick diffusion can be important only in the thermal boundary layer around the

---

*Associate Professor, Departamento de Física Fundamental.

plate, by trying to smooth out the concentration profiles induced by thermal diffusion. Thus, the cross mass transport mechanism acts *at least* on the same scale than the direct transport. Even more, in many cases (i.e., heavy fuels diluted in air) the gas mixture is close to the so-called quasi-Lorentzian gas limit and, whereas, the fuel diffusion coefficient D takes on *small* values, the product of this coefficient times the fuel thermal diffusion factor α_T (a measure of the ratio of Soret flux to Fick flux) approaches a positive limiting value.[5] In that particular case, a *blowing* force acts on the fuel molecules pushing them away from the hot plate. Therefore, a fuel lean gas layer around the plate is produced. Consequently, an increase in the minimum distance from the leading edge of the plate required for the onset of ignition is predicted (for light fuels, like hydrogen, the phenomenon should be reversed). Using the well-known boundary layer approximations to describe the gas flow and the classical first order activation energy asymptotic methods to locate ignition,[6-8] the fuel concentration near the plate is obtained in the asymptotic limit of high Schmidt number ($Sc=\nu/D \gg 1$, ratio of momentum to fuel mass diffusivities very large). To leading order in the Schmidt number expansion, diffusion is negligible and the balance between thermal diffusion and convection gives birth to a fuel stagnation line (to every algebraic order, there is no fuel mass transport across this line and a region of exponentially small fuel mass fraction appears over the plate). However, diffusion becomes important around this line on a scale of the order of $Sc^{-1/2}$. Taking this inner scale, the fuel mass fraction equation is solved in terms of parabolic cylinder functions[9] and the inner fuel mass fraction determined through matching with the outer solutions.

In the next Sec. the problem is formulated and the method of solution is presented in the third Sec. In Analysis of the Mass Transport Equation the fuel concentration in the reaction zone is determined. Finally, in Discussion of Results the corresponding ignition distance is compared with the predictions of previous analyses where thermal diffusion was neglected.

Formulation

Let us consider a reactive gas mixture with mainstream temperature T_∞ and velocity u_∞ flowing over a hot flat plate kept at a constant temperature T_w ($T_w > T_\infty$). The fuel and the oxidizer are very diluted in an inert gas and can react according to the following one-step overall reaction

$$\nu_F[F] + \nu_O[O] \rightarrow [P] + Q \tag{1}$$

whose kinetic is controlled by an Arrhenius law such that the reaction rate per unit volume is

$$w = Y_O^p Y_F^q T^{(r-p-q)} B \exp(-E_a/kT) \tag{2}$$

In the above equations Q is the heat of reaction, ν_F and ν_O are the stoichiometric coefficients, Y_i (i=O,F) are the reactant mass fractions, p and q the orders of the reaction, T the gas temperature, the product BT^r accounts for the frequency factor of the reaction among other constant parameters, k is the Boltzmann constant, and E_a the overall activation energy.

Near the plate the gas velocity, temperature, and species mass fractions obey the classical set of equations of a fully developed reacting boundary layer.[10,11] Let x and y be, respectively, the longitudinal and transverse coordinates along the plate; then, in terms of the nondimensional independent variables η, ζ and the stream function $\Phi(x,y)$, defined by

$$\eta = \left(\frac{\rho_\infty u_\infty}{2\,\mu_\infty x}\right)^{1/2} \int_0^y (\rho/\rho_\infty)\, d\tilde{y} \tag{3a}$$

$$\zeta = x \frac{2\,B\,\nu_F M_F}{\rho_\infty u_\infty T_\infty} \left(\frac{Q\,Y_{F\infty}}{c_p \nu_F M_F}\right)^{(r-p-q+1)} \tag{3b}$$

$$\Phi(x,y) \equiv (2\,x\,\rho_\infty \mu_\infty u_\infty)^{1/2} f(\eta) \tag{3c}$$

with $f(\eta)$ being the Blasius function, the governing equations for the temperature and the fuel mass fraction can be written as

$$\frac{\partial^2\Theta}{\partial\eta^2} + Pr\, f \frac{\partial\Theta}{\partial\eta} - 2\,Pr\,\zeta\, f' \frac{\partial\Theta}{\partial\zeta} = - \zeta\, Pr\, Y_O^p\, Y_{F\infty}^{q-1}\, \Psi^q\, \Theta^{(r-p-q+1)} \exp(-\Theta_a/\Theta) \tag{4}$$

$$\frac{\partial^2\Psi}{\partial\eta^2} + Sc\, f \frac{\partial\Psi}{\partial\eta} + \alpha_T \frac{\partial}{\partial\eta}\left(\frac{\Psi}{\Theta}\frac{\partial\Theta}{\partial\eta}\right) - 2\,Sc\,\zeta\, f' \frac{\partial\Psi}{\partial\zeta} = \zeta\, Sc\, Y_O^p\, Y_{F\infty}^{q-1}\, \Psi^q\, \Theta^{(r-p-q+1)} \exp(-\Theta_a/\Theta) \tag{5}$$

Here, $\Theta \equiv c_p \nu_F M_F T/(Q\, Y_{F\infty})$, $\Psi \equiv Y_F/Y_{F\infty}$, $\Theta_a \equiv c_p \nu_F M_F E_a / (k\, Q\, Y_{F\infty})$, M_F is the fuel molar mass and ρ, c_p, and μ are, respectively, the density, specific heat, and dynamic viscosity of the gas mixture. Subscript ∞ refers to values in the oncoming mainstream and the prime denotes the total derivative with respect to η. The Prandtl number ($Pr \equiv$ ratio of momentum diffusivity to temperature diffusivity), the Schmidt number (Sc

≡ ratio of momentum diffusivity to fuel mass diffusivity), the fuel thermal diffusion factor (α_T) and the product $\rho\mu$ have been considered constant throughout the boundary layer and equal to their respective mainstream values.

The following boundary conditions are imposed:

at $\eta \to \infty$

$$\Theta \to \Theta_\infty \tag{6a}$$

$$\Psi \to 1 \tag{6b}$$

at $\eta=0$

$$\Theta = \Theta_w \tag{7a}$$

$$\frac{\partial \Psi}{\partial \eta} + \alpha_T \Psi \frac{\partial \ln \Theta}{\partial \eta} = 0 \tag{7b}$$

that is, at the plate, the temperature is kept constant and the fuel mass flux vanishes.

The effect due to the thermal (Soret) diffusion is proportional to α_T and appears in the boundary condition (7b) and in Eq. (5). These extra terms couple the concentration and the temperature distributions, even in the chemically frozen zone where the reaction rate is negligible. An equation similar to Eq. (5) can be written for the oxidant mass fraction. However, it is inessential for the subsequent analysis, because we will disregard oxidant thermal diffusion effects (the thermal diffusion factor of oxygen in nitrogen, the most practical case, is very small compared to those of the usual fuels) and we will neglect oxygen consumption by assuming a far from stoichiometry fuel lean concentration for the fresh oncoming gas mixture. Then $Y_O=Y_{O\infty}$ everywhere inside the boundary layer.

In the next section we proceed to develop the method of solution of Eqs. (4) and (5) with boundary conditions (6) and (7), in the double asymptotic limit of high activation energy of the chemical reaction and low mobility of the fuel molecules (small fuel molecular diffusivity).

Method of Solution

When the activation energy E_a takes on *large* values, ignition problems can be treated by using $\varepsilon=\Theta_w^2/\Theta_a$, ($\varepsilon \ll 1$), as an expanding parameter.[10] In space-dependent problems, there exists a solution with an induction region, prior to ignition, where the chemical reaction term is non-negligible only in a *thin* zone close to the region of highest temperature i.e., the heated wall in the case of interest here. In this thin zone (whose thickness is of order ε with respect to the thermal boundary layer) a balance between heat chemical generation and heat diffusion is established. Close to the plate, the gas temperature rises continuously in the stream-wise direction until a point is reached, the ignition point, beyond which the ignition-like solution no longer exists and the premixed flame solution takes over. Out from this *inner* zone chemical reaction can be neglected and the problem is equivalent to that of an inert boundary layer with a varying wall temperature.

As thermal diffusion does not enter in the temperature equation, Eq. (4), the classical analysis remains valid up to this point.[12,7,8] We will sketch this analysis below although extended to general values of the Prandtl number.

In the outer nonreactive zone the dimensionless temperature is expanded as

$$\Theta_{out} = \Theta_f(\eta) + \varepsilon\, \theta_{out}(\zeta,\eta) + o(\varepsilon) \tag{8}$$

where $\Theta_f(\eta)$ is the chemically frozen solution

$$\Theta_f(\eta) = \Theta_w - (\Theta_w - \Theta_\infty)\left(1 - \frac{\int_\eta^\infty (f'')^{Pr} d\eta}{\int_0^\infty (f'')^{Pr} d\eta}\right) \tag{9}$$

given by the classical Pohlhausen profile (see, for instance, the book by Schlichting[13]). The term $\theta_{out}(\zeta,\eta)$ accounts for the varying temperature at the boundary with the inner region, where, due to the chemical heat release, the temperature increases with ζ.

To study the inner region an adequate stretched variable is

$$\chi = \left[\frac{(\Theta_w - \Theta_\infty)\, [f''(0)]^{Pr}}{\int_0^\infty (f'')^{Pr} d\eta}\right] \frac{\eta}{\varepsilon} \tag{10}$$

and the temperature is expanded as

$$\Theta_{inn} = \Theta_w + \varepsilon\, [\, \theta_{inn}(\zeta,\chi) - \chi\,] + o(\varepsilon) \tag{11}$$

The term θ_{inn} is obtained by a direct integration of the heat equation, Eq. (4), written in terms of the inner variable χ and by using the matching condition with the outer solution, Eq. (8), as the boundary condition when $\chi \rightarrow \infty$. On the wall the condition (7a) is used, i.e., $\theta_{inn}(\zeta,0)=0$. The result is a two-valued function of the relevant Damköhler number

$$\Delta = 2\,\mathrm{Pr}\,\varepsilon\,\zeta\, Y_O^p\, Y_{F\infty}^{q-1}[\Psi(0)]^q\, \Theta_w^{(r-p-q+1)} \left\{ \frac{\int_0^\infty (f'')^{\mathrm{Pr}} d\eta}{(\Theta_w - \Theta_\infty)[f''(0)]^{\mathrm{Pr}}} \right\}^2 \exp(-\Theta_a/\Theta_w) \tag{12}$$

as far as $\Delta \leq 1$. This solution corresponds to the lower half of the characteristic ignition-extinction S-curve (see, for instance, the review paper by Merzhanov and Averson[14]). No solution exists for $\Delta > 1$ and the condition $\Delta = 1$ locates the ignition point.

The main purpose of this paper is to point out the relevance of thermal diffusion in the ignition of some (heavy) fuels. A first estimation of this effect can be obtained by neglecting fuel consumption, that is, without consideration of the inner reactive zone for the fuel mass fraction. Then, the value on the wall, $\Psi(0)$, appearing in Eq. (12) is just a parameter determined by the nonreactive conservation equation. If thermal diffusion were not taken into account, the fuel mass fraction would be constant and then $\Psi(0)=1$. But when thermal diffusion is considered, $\Psi(\eta)$ must be determined by solving Eq. (5) (without the reaction term in the r.h.s.) together with conditions (6b) and (7b).

The next section is devoted to the calculation of $\Psi(0)$. The Eq. (5) will be solved in the limit of high fuel Schmidt number, $\mathrm{Sc} \gg 1$, but keeping $\alpha_T/\mathrm{Sc} = O(1)$ in such a way that, to leading order, fuel thermal diffusion and convection remain as the dominant transport mechanisms in the outer zones, far from the plate and close to the plate, respectively, where diffusion is negligible. At some intermediate value of η large variations in fuel concentration appear and diffusion becomes of a comparable order of magnitude. Using a suitable stretched scale, the inner profile is determined by matching with the outer solutions.

Analysis of the Mass Transport Equation

Assuming that the reduced fuel mass fraction $\Psi(\eta,\zeta)$ is everywhere of order unity in the ε-expansion, to leading order, we can neglect the chemical consumption term in Eq. (5) and look for a self-similar solution corresponding to the chemically frozen flow equation

$$\mathrm{Sc}^{-1}\,\Psi'' + [f + \alpha\,(\ln \Theta_f)']\,\Psi' + \alpha\,(\ln \Theta_f)''\,\Psi = 0 \tag{13}$$

with boundary conditions (6) and (7). In the above equation the parameter $\alpha \equiv \alpha_T/Sc$ has been introduced. When Sc»1, but $\alpha = O(1)$, the second derivative term, that takes into account Fick diffusion, becomes negligible and we are faced to a singular perturbation problem. This equation has been studied by Goren[15] in a different context. He pointed out that in the case of interest here, $T_w > T_\infty$ and $\alpha_T > 0$, the coefficient of Ψ' vanishes at a value of the similarity variable. For very large η ($\Theta_f' \to 0$, $f \to \infty$) the coefficient is positive, whereas, for small η ($\Theta_f' < 0$, $f \to 0$) it is negative, thus, it vanishes at some intermediate value $\eta = \eta_c$. As a consequence for η close to η_c, the diffusion term cannot be neglected any more and an inner sublayer develops around this critical value η_c where diffusion becomes comparable to both convection and thermal diffusion. The location of this critical line is then implicitly given by the relation

$$[f + \alpha\,(\ln \Theta_f)']_{\eta=\eta_c} = 0 \tag{14}$$

which, for small values of α, gives

$$\eta_c \approx \left\{ 2\,\alpha \,\frac{\Theta_w - \Theta_\infty}{\Theta_w}\, \frac{[f''(0)]^{Pr-1}}{\int_0^\infty (f'')^{Pr} d\eta} \right\}^{1/2} \quad ; \quad \alpha \ll 1 \tag{15}$$

In general, Eq. (14) is easily solved for α and some typical values, obtained numerically for Pr=1 and $\Theta_w = 3\Theta_\infty$, are shown in Table 1. The figures clearly point out that for characteristic values of α (0.1 to 0.5), the fuel stagnation line is located away from the plate, well inside the thermal boundary layer.

Out from this inner diffusive sublayer ($\eta \neq \eta_c$), asymptotic expansions in the limit $Sc^{-1} \to 0$ for the solutions of equations like Eq. (13) were given by Horn[16] in the form of two series, corresponding to algebraic and exponential orders, respectively. To leading order, the solution can be written as (see, for instance, the book by Nayfeh[17])

$$\Psi_{out} = C_1 \exp\left(-\int_{\eta_0}^{\eta} \frac{H(\tilde\eta)}{G(\tilde\eta)}\, d\tilde\eta \right)$$

$$+ C_2\, G(\tilde\eta)^{-1} \exp\left(\int_{\eta_0}^{\eta} \frac{H(\tilde\eta)}{G(\tilde\eta)}\, d\tilde\eta \right) \exp\left(-Sc \int_{\eta_0}^{\eta} G(\tilde\eta)\, d\tilde\eta \right) \tag{16}$$

Table 1

η_c	0.14	0.28	0.42	0.57	0.71	1.00	1.41
α	0.01	0.05	0.12	0.20	0.30	0.55	1.06
δ	0.02	0.05	0.07	0.09	0.11	0.10	−.02

where we have introduced the functions

$$G(\eta) \equiv f + \alpha\,(\ln \Theta_f)' \tag{17a}$$

$$H(\eta) \equiv \alpha\,(\ln \Theta_f)'' \tag{17b}$$

We will use superscript (+) to denote the solution in the upper region ($\eta > \eta_c$) and superscript (−) for the lower region ($0 \leq \eta < \eta_c$).

In the upper region, we take $\eta_0 = \infty$ in Eq. (16). Then, C_2^+ must be taken equal to zero to eliminate the Sc-exponentially divergent term and $C_1^+ = 1$ to fulfill the boundary condition (6b), resulting[15]

$$\Psi_{out}^+ = \exp\left[\int_\eta^\infty \frac{H(\tilde{\eta})}{G(\tilde{\eta})}\,d\tilde{\eta}\right] + o(Sc^0) \tag{18}$$

It is interesting to note the identity

$$H(\eta) \equiv (1 - \alpha\,Pr)\,f\,(\ln \Theta_f)' - [f + \alpha\,(\ln \Theta_f)']\,(\ln \Theta_f)' \tag{19}$$

When this relation is used in Eq. (18) the contribution of the second term can be integrated giving the factor Θ_f/Θ_∞, whereas, the first term gives a contribution in the exponential whose sign depends on the product αPr. If this product were greater than the unity, a rather unusual situation, the integral would take on large and positive values when $\eta \rightarrow \eta_c$ and Ψ_{out} would show a divergence close to the critical layer, i.e., there would be accumulation of fuel at this position. On the contrary, when αPr is less than the unity, the most interesting case, the integral of Eq. (18) takes on large and negative values and the fuel concentration vanishes (algebraically in the Sc-expansion) in the proximity of the locus $\eta = \eta_c$. In the following we will consider only this last possibility.

The integrand of solution (18) exhibits a singularity at the critical value $\eta \rightarrow \eta_c$ where the denominator, $G(\eta)$, goes to zero. However, this outer solution can be rewritten as

$$\Psi^+_{out} = \left(\frac{\eta-\eta_c}{\eta^*-\eta_c}\right)^{\delta} \exp\left[\int_{\eta^*}^{\infty} \frac{H(\eta)}{G(\eta)}\, d\eta + \int_{\eta}^{\eta^*} F(\tilde{\eta})\, d\tilde{\eta}\right] \tag{20}$$

where η^* is any value greater than η_c and we have introduced the function

$$F(\eta) = \frac{\alpha\,(\ln \Theta_f)''}{f + \alpha(\ln \Theta_f)'} + \frac{\delta}{\eta-\eta_c} \equiv \frac{H(\eta)}{G(\eta)} + \frac{\delta}{\eta-\eta_c} \tag{21}$$

with

$$\delta = \left[\frac{-\alpha\,(\ln \Theta_f)''}{f' + \alpha\,(\ln \Theta_f)''}\right]_{\eta=\eta_c} \equiv \frac{-H_c}{G'_c} \tag{22}$$

the subscript (c) denoting the value at $\eta=\eta_c$. Characteristic values of this parameter are given in Table 1.

In this way the exponential term appearing in Eq. (20) does not contain any singularity because the function defined by Eq. (21) has a regular behavior at $\eta=\eta_c$. In fact, its expansion around this point is

$$F(\eta \approx \eta_c) = \frac{H'_c + \delta\, G''_c/2}{G'_c} + O(\eta-\eta_c) \tag{23}$$

The expression (20) can be used to compute numerically Ψ^+_{out} at any value of η. Therefore, the inner limit form of this outer solution is

$$\Psi^+_{out}(\eta \to \eta_c) \approx C\,(\eta-\eta_c)^{\delta} \tag{24}$$

where

$$C = (\eta^*-\eta_c)^{-\delta} \exp\left[\int_{\eta^*}^{\infty} \frac{H(\eta)}{G(\eta)}\, d\eta + \int_{\eta_c}^{\eta^*} F(\eta)\, d\eta\right] \tag{25}$$

depends on the thermophysical properties of the gas mixture and on the imposed temperature ratio, T_∞/T_W, but does not really depend on the chosen value for η^*.

Within the lower region ($0 \leq \eta < \eta_c$) we take η_0=0 in Eq. (16). The vanishing flux boundary condition at the wall, Eq. (7b), implies that $C_1^- = 0$ and using again the function F(η) to take the divergency out from the integral, it results

$$\Psi_{out}^- = C_2^- \left(\frac{\eta_c}{\eta_c-\eta}\right)^{\delta} G(\eta)^{-1} \exp\left\{\int_0^{\eta} [F(\tilde{\eta}) - Sc\, G(\tilde{\eta})]\, d\tilde{\eta}\right\} \tag{26}$$

By considering the Taylor expansion of G(η) around η_c

$$G(\eta \rightarrow \eta_c) = (\eta-\eta_c)\, G'_c + O(\eta-\eta_c)^2 \tag{27}$$

we get the inner limit form of solution (26)

$$\Psi_{out}^-(\eta \rightarrow \eta_c) \approx D \frac{\exp\left[\dfrac{-Sc(\eta_c-\eta)^2 G'_c}{2}\right]}{(\eta_c-\eta)^{1+\delta}} \tag{28}$$

where

$$D = -\, C_2^- \, \eta_c^{\delta}\, (G'_c)^{-1} \exp\left\{\int_0^{\eta_c} [F(\eta) - Sc\, G(\eta)]\, d\eta\right\} \tag{29}$$

To investigate the structure of the inner zone we must look for an appropriate stretching variable in the form $\Lambda = Sc^{\gamma}(\eta-\eta_c)$. By taking into account the Taylor expansion of the singular coefficient around η_c, Eq. (27), it appears clearly that to have a non-negligible contribution, in Eq. (13), from the diffusion term to lowest order, we must take γ=1/2. Thus, the relative thickness of the inner sublayer is $O(Sc^{-1/2})$.

So, it appears that the inner form of Eq. (13) to leading order is

$$\frac{d^2\Psi_{inn}}{d\Lambda^2} + G'_c\, \Lambda \frac{d\Psi_{inn}}{d\Lambda} + H_c\, \Psi_{inn} = 0 \tag{30}$$

Solutions of this equation are the parabolic cylinder functions[18] and the general solution can be written as the following linear combination involving the Whittaker function, U

$$\Psi_{inn} = e^{-G'_c \Lambda^2/4}\left[E_1\, U\left(\frac{1}{2}+\delta, \sqrt{G'_c}\,\Lambda\right) + E_2\, U\left(\frac{1}{2}+\delta, -\sqrt{G'_c}\,\Lambda\right)\right] \tag{31}$$

where the constants E_1 and E_2 must be determined by matching with the outer solutions.

To perform the matching we first investigate the behavior of solution (31) in the limit $\Lambda \rightarrow \pm\infty$, by using the asymptotic form of the Whittaker functions

$$\Psi_{inn}(\Lambda \rightarrow +\infty) \rightarrow \begin{cases} E_1 \left[\dfrac{(2\pi)^{1/2}}{\Gamma(1+\delta)} \dfrac{(1/2+\delta)^{1/2+\delta}}{(G'_c)^{1+\delta}} \right]^{1/2} \dfrac{e^{-G'_c \Lambda^2/2}}{\Lambda^{1+\delta}} \\ + E_2 \left[\dfrac{(2\pi)^{1/2}}{\Gamma(1+\delta)} \dfrac{(G'_c)^{\delta}}{(1/2+\delta)^{1/2+\delta}} \right]^{1/2} \Lambda^{\delta} \end{cases} \tag{32}$$

$$\Psi_{inn}(\Lambda \rightarrow -\infty) \rightarrow \begin{cases} E_1 \left[\dfrac{(2\pi)^{1/2}}{\Gamma(1+\delta)} \dfrac{(G'_c)^{\delta}}{(1/2+\delta)^{1/2+\delta}} \right]^{1/2} (-\Lambda)^{\delta} \\ + E_2 \left[\dfrac{(2\pi)^{1/2}}{\Gamma(1+\delta)} \dfrac{(1/2+\delta)^{1/2+\delta}}{(G'_c)^{1+\delta}} \right]^{1/2} \dfrac{e^{-G'_c \Lambda^2/2}}{(-\Lambda)^{1+\delta}} \end{cases} \tag{33}$$

where Γ is the usual gamma function.

Then, the matching with the outer solution in the upper region, Eq. (24), determines the value of E_2

$$E_2 = \left[\frac{\Gamma(1+\delta)}{(2\pi)^{1/2}} \frac{(1/2+\delta)^{1/2+\delta}}{(G'_c Sc)^{\delta}} \right]^{1/2} C \tag{34}$$

whereas, the matching with the lower region solution, Eq. (28), imposes

$$E_1 = 0 \tag{35}$$

and

$$E_2 = \left[\frac{\Gamma(1+\delta)}{(2\pi)^{1/2}} \frac{(G'_c Sc)^{1+\delta}}{(1/2+\delta)^{1/2+\delta}} \right]^{1/2} D \tag{36}$$

By equating Eq. (36) with Eq. (34) the remaining unknown is determined

$$C_2^- = - \frac{(G'_c)^{1/2-\delta}}{[(\eta^*-\eta_c)\,\eta_c]^{\delta}} \left(\frac{1/2+\delta}{Sc}\right)^{1/2+\delta}$$

$$\exp\left\{\int_0^{\eta_c} [Sc\, G(\eta) - F(\eta)]\, d\eta + \int_{\eta_c}^{\eta^*} F(\eta)\, d\eta + \int_{\eta^*}^{\infty} \frac{H(\eta)}{G(\eta)}\, d\eta\right\} \tag{37}$$

With this result, we can finally write the expression for the concentration profile in the region close to the wall

$$\Psi^-_{out}(\eta) = \left\{ \frac{(G'_c)^{1/2-\delta}}{(\eta^*-\eta_c)^{\delta}} \left(\frac{1/2+\delta}{Sc}\right)^{1/2+\delta} \exp\left[\int_{\eta_c}^{\eta^*} F(\eta)\, d\eta + \int_{\eta^*}^{\infty} \frac{H(\eta)}{G(\eta)}\, d\eta\right]\right\}$$

$$\frac{\exp\left\{\int_{\eta}^{\eta_c} [Sc\, G(\tilde{\eta}) - F(\tilde{\eta})]\, d\tilde{\eta}\right\}}{[-G(\eta)\,(\eta_c-\eta)^{\delta}]} \tag{38}$$

Eq. (38) evaluated at $\eta=0$ provides the value of $\Psi(0)$ determining the ignition distance through Eq. (12). As the function $G(\eta)$ is negative in this region, our result shows that the relative fuel concentration on the wall is exponentially small in the Schmidt number. Implications of this result are discussed in the next section.

Discussion of Results

From Eq. (12) with $\Delta=1$ and using Eq. (3), the ignition distance, x_{ig} can be written as

$$\frac{x_{ig}}{[x_{ig}]_{\alpha_T=0}} = [\Psi^-_{out}(0)]^{-q} \tag{39}$$

where $[x_{ig}]_{\alpha_T=0}$ stands for the corresponding value when thermal diffusion is neglected and then $\Psi(0)=1$. This result and Eq. (38) show how this distance is expected to increase notably in cases where Sc»1. Nevertheless, to quantify the effect reasonably physical values must be assigned to the parameters α_T, Sc, δ, and η_c (Gökoglu and Rosner[19] have reported numerical studies and correlations for the mass fraction profile in this problem).

Diffusion coefficients of heavy fuel vapors in air are not well-known but we can estimate them as being of order 10^{-2} cm^2/s at room temperature. Thermal diffusion factors for these species are even less known although values above the unity ($\alpha_T \approx 3$) can be expected (for estimates of both diffusion coefficients and thermal diffusion factors of long-chain paraffin hydrocarbons see García-Ybarra and Rosner[20]). As the kinematic viscosity of air is close to 10^{-1} cm^2/s under the same conditions, we get Sc$\approx$10 and α_T/Sc$\approx$0.3. These results agree with our assumptions of low fuel mobility, Sc»1, and quasi-Lorentzian gas ($[\alpha_T/\text{Sc}]_{QL}=3/4$ for hard spheres[5]).

For illustration, let us fix α=0.3 and Pr=1. Then, η_c=0.7071 and δ=0.1074. From Eq. (38), using the tabulated values of the Blasius function (note that our definitions for η and f(η) are $2^{-1/2}$ times the tabulated ones in Schlichting's[13] book) we get

$$\Psi(0) = 8.870\left(\frac{1/2+\delta}{\text{Sc}}\right)^{1/2+\delta} \exp\left\{\int_0^{\eta_c} [\text{Sc}\, G(\eta) - F(\eta)]\, d\eta\right\} \tag{40}$$

Due to the fact that G(η) is negative (for $0 \le \eta \le \eta_c$), relation (40) shows that the fuel concentration on the wall decays exponentially with increasing values of the Schmidt number. For moderate values of Sc, the large prefactor overlooks the effect of fuel concentration reduction. However, as Sc increases the exponential decay becomes overwhelming. For instance, when the Schmidt number takes on the values Sc=10, 20, 30; Eq. (40) provides $\Psi(0)$=0.914, 0.376, and 0.184, respectively. Then, by taking q=1, the ignition distance x_{ig}, given by Eq. (39), is increased by about 10, 170, and 400 %, respectively, compared to the ignition distance when thermal diffusion is neglected.

The validity of the theory developed here is limited in both extremes of small and large thermal diffusion effect. In the former case, α_T/Sc«1 or $(T_w-T_\infty)/T_w$«1, Fick diffusion and Soret effect act on the same scale and η_c becomes of the same order of magnitude as the inner layer thickness. Therefore, there is no lower outer zone because the inner zone extends down to the wall. The diffusion scale is still small with respect to the momentum boundary layer thickness because Sc»1 and it is legitimate to replace the Blasius function as well as the temperature profile by its expansions around the origin in the mass equation, Eq. (13). This type of analysis has been performed by Friedlander et al.[21] in problems related to dust deposition on hot walls.

To deal with the opposite extreme of very large thermal diffusion effect, it is physically more reasonable to assume $Sc^{-1}=O(\varepsilon)$. In that case, chemical consumption and fuel mass fraction are of a comparable exponential order in the whole region below η_c, then fuel is completely consumed around η_c and the ignition is expected to occur at this value of η instead of on the wall. The extension of the present analysis to account for this case will be the object of follow-on studies.

Acknowledgments

This work has been supported by the Dirección General de Investigación Científica y Técnica (DGICYT) Spain, under contracts PB87-0588 and PB88-0159. We are very much indebted to Professor Daniel Rosner for his valuable comments during our research stays in the HTCRE Laboratory (Yale University, Deptartment of Chemical Engineering) where part of this work was completed.

References

[1]García-Ybarra, P. L., and Clavin, P., "Cross-Transport Effects in Nonadiabatic Premixed Flames," Progress in Astronautics and Aeronautics, Vol. 76, 1981, p. 463-481.

[2]Fristrom, R. M., and Monchick, L., "Two Simple Approximations to the Thermal Diffusion Factor and Their Applications to Flame Studies," Combustion and Flame, Vol. 71, 1988, p. 89-99.

[3]Rosner, D. E., "Thermal (Soret) Diffusion Effects on Interfacial Mass Transport Rates," PhysicoChemical Hydrodynamics, Vol. 1, 1980, p. 159-185.

[4]García-Ybarra, P. L., Nicoli, C., and Clavin, P., "Soret and Dilution Effects on Premixed Flames," Combustion Science and Technology, Vol. 42, 1984, p. 87-109.

[5]Mason, E. A., "Higher Approximation for the Transport Properties of Binary Gas Mixtures: II. Applications," Journal of Chemical Physics, Vol. 27, 1957, p. 782-790.

[6]Liñán, A., and Williams F. A., "Ignition of a Reactive Solid Exposed to a Step in Surface Temperature," SIAM Journal of Applied Mathematics, Vol. 36, 1979, p. 587-603.

[7]Law, C. K., and Law H. K., "Thermal-Ignition Analysis in Boundary-Layers Flows," Journal of Fluid Mechanics, Vol. 92, 1979, p. 97-108.

[8]Méndez, F., Treviño, C., and Liñán, A., "Premixed Combustion in Boundary Layers for Moderate Values of the Zeldovich Numbers," Combustion Science and Technology, Vol. 48, 1986, p. 129-149.

[9]Abramowitz, M., and Stegun, I., Handbook of Mathematical Functions, Dover, New York, 1972, p. 504.

[10]Williams, F. A., Combustion Theory, 2nd ed., Benjamin/Cummings Publishers, Menlo Park, CA, 1985, Chap. 12 and p. 284.

[11]Rosner, D. E., Transport Processes in Chemically Reacting Flow Systems, Butterworth Publishers, Boston, MA, 1986, Chap. 6.

[12]Zeldovich, Ya. B., Barenblatt, G. I., Librovich, V. B., and Makhviladze, G. M., The Mathematical Theory of Combustion and Explosions, Plenum Publishers, New York, 1985, Chap. 3.

[13]Schlichting, H., Boundary Layer Theory, McGraw-Hill, New York, 1968, p. 280.

[14]Merzhanov, A. G., and Averson, A. E., "The Present State of the Thermal Ignition Theory: An Invited Review," Combustion and Flame, Vol. 16, 1971, p. 89-124.

[15]Goren, S. L., "Thermophoresis of Aerosol Particles in the Laminar Boundary Layer on a Flat Plate," Journal of Colloid and Interface Science, Vol. 61, 1977, p. 77-85.

[16]Horn, J., "Ueber Eine Lineare Differentialgleichung Zweiter Ordnung mit Einem Willkurlichen Parameter," Mathematische Annalen, Vol. 52, 1899, p. 271-292.

[17]Nayfeh, A., Perturbation Methods, John Wiley and Sons, New York, 1973, p. 317.

[18]Abramowitz, M., and Stegun, I., Handbook of Mathematical Functions, Dover, New York, 1972, p. 687.

[19]Gökoglu, S. A., and Rosner, D. E., "Prediction and Rational Correlation of Thermophoretically Reduced Particle Mass Transfer to Hot Surfaces Across Laminar or Turbulent Forced-Convection Gas Boundary Layers," Chemical Engineering Communications, Vol. 44, 1986, p. 107-119.

[20]García-Ybarra, P. L., and Rosner D. E., "Thermophoretic Properties of Nonspherical Particles and Large Molecules," AIChE. Journal, Vol. 35, 1989, p. 139-147.

[21]Friedlander, S. K., Fernández de la Mora, J., and Gökoglu, S.A., "Diffusive Leakage of Small Particles Across the Dust-Free Layer Near a Hot Wall," Journal of Colloid and Interface Science, Vol. 125, 1988, p. 351-355.

Chapter II. Diffusion Flames and Shear Effects

Analysis of the Structure of Counterflow Hydrogen-Air Diffusion Flames

V. Tangirala,* and K. Seshadri†
University of California, San Diego, La Jolla, California
and
C. Treviño,‡
Universidad Naciónal Autonoma de México, Mexico
and
M. D. Smooke§
Yale University, New Haven, Connecticut

Nomenclature

A_j	=	frequency factor characterizing the rate constant of reaction j (see Table 1)
a	=	rate of strain
C_i	=	molar concentrations of species i per unit volume
c_p	=	specific heat at constant pressure
D	=	Damkohler number
E_j	=	activation energy in the rate constant of reaction j (see Table1)
$(-h_i)$	=	heat of formation per mole of species i nondimensionalized by division by the overall heat release rate Q_T
$(K_e)_j$	=	A function of concentrations of species in reaction j.
$(K_T)_j$	=	Equilibrium constant for reaction j.
k_j	=	rate constant of reaction j (see Table 1)
L_i	=	Lewis number of species i

* Research Engineer, Department of AMES.

† Associate Professor,Department of AMES.

‡ Professor, Depto. Fluidos Y Termica.

§ Associate Professor, Dept. of Mechanical Engineering.

N	=	number of species
n_j	=	temperature exponent in the rate constant of reaction j (see Table 1)
P	=	pressure
Q_j	=	heat release associated with reaction j
Q_T	=	overall heat release for the one step reaction $H_2 + \frac{1}{2}O_2 \rightarrow H_2O$
q_j	=	Q_j / Q_T
R^o	=	universal gas constant
T	=	temperature
T_{max}	=	maximum value of temperature
W_i	=	molecular weight of species i
$\bar{W}$	=	average molecular weight of the mixture
w_j	=	reaction rate of reaction j
X_i	=	mole fraction of species i
Y_i	=	mass fraction of species i
y_i	=	normalized mass fraction i defined in Eq. (9)
Z	=	a conserved scalar as defined in Eq. (1)
Z_m	=	value of Z_x at $T = T_{max}$
Z_x	=	Conserved scalar based on mole fraction of nitrogen.

Greek Symbols

θ	=	normalized temperature defined in Eq. (10)
ν_{ij}	=	stoichiometric coefficients denoting the number of molecules of species i participating in reaction j

Subscripts

1	=	denotes conditions at the ambient fuel stream
2	=	denotes conditions at the ambient oxidizer stream
c	=	denotes conditions at $Z = Z_c$
f	=	denotes forward rate of reaction
b	=	backward rate of reaction
i	=	index denoting various species
j	=	index denoting reactions 1-11 in Table 1

Abstract

Numerical calculations were performed to determine the structure of diffusion flames stabilized between counterflowing, laminar streams of hydrogen and air. The calculations were performed at different values of the rate of strain and at a value of the absolute pressure equal to one atmosphere. The detailed kinetics of oxidation of hydrogen was represented by 11 elementary reactions involving seven species. Temperature profiles, concentration profiles of various stable species and radicals, and rates of various elementary chemical reactions were calculated for a number of values of the rate of strain. Results are plotted at a low value of the rate of strain ($= 30\ \text{sec}^{-1}$) and at a high value of the rate of strain ($= 1440\ \text{sec}^{-1}$). At the low value of the rate of strain, a number of reactions were found to be in partial equilibrium over a major part of the flow field. Using approximations suggested by the results of the numerical calculations, the equilibrium structure of the flame was determined and the results were compared with the results of numerical calculations performed at a value of the rate of strain equal to 30 sec^{-1}, and good agreement was obtained for the concentration profiles of H_2, O_2, H_2O, and H. However, the shapes of the temperature profiles were different and they are attributed to nonequilibrium of the major chain branching reaction $H + O_2 = OH + O$, and neglect of the three body termination reaction $H + H + M = H_2 + M$ in the analysis.

Introduction

Detailed studies of the oxidation of hydrogen is of practical interest because hydrogen is considered as a fuel for propulsion of supersonic aircraft. Since elementary reactions associated with the oxidation of hydrogen in air are a significant part of the main chain branching processes for ignition, extinction, and high temperature combustion of hydrocarbon fuels,[1–7] a detailed knowledge of the structure of hydrogen-air flames is useful for improving models describing the combustion of hydrocarbon fuels. Because of the relative simplicity of the chemical system, it is possible to give a complete picture of the structure of hydrogen-air flames. Previous studies of hydrogen-air reactions systems have been primarily concerned with premixed flames.[8–10] Dixon-Lewis et al.,[3] have performed numerical calculations to determine the structure of a counterflow hydrogen-air diffusion flame. Although extinction of the flame was found to occur at a strain rate of 1.3×10^4 sec,$^{-1}$ Dixon-Lewis et al.[3] note that this result may not be valid because it may not be possible to obtain such high rates of strain in low speed flows. Here an attempt is made to provide a more detailed numerical and analytical study of the structure of a steady laminar, stretched, counterflow, hydrogen-air diffusion flame, at an absolute value of pressure equal to one atmosphere. However, no experimental results to date are available for comparison with the results of numerical and analytical study presented here.

The results obtained in this paper are useful for future asymptotic analyses of the structure of hydrogen-air, and hydrocarbon-air diffusion flames using multistep overall chemical kinetic models.[4–7] In these models the hydrocarbon fuel is presumed to react with radicals to form H_2 and CO to H_2O and CO_2. In modeling the oxidation of H_2, different reactions were assumed to be in partial equilibrium, and the concentrations of different species were assumed to be in steady state.[4–7] These different assumptions yielded fundamentally different results. The purpose of this investigation is to examine these approximations in detail, and to guide future asymptotic analyses of combustion of hydrocarbon fuels.

In next Sec. the results of numerical calculations are summarized. Using approximations deduced from the results of the numerical study, the equilibrium structure of the flame is obtained in the third Sec., and results are compared with those in the second Sec. The major conclusions are presented in Summary and Conclusions.

Numerical Calculations

Formulation

The system considered is the diffusion flame stabilized near the stagnation region formed between two steady, laminar, infinitely wide, axisymmetric counterflowing jets of hydrogen and air. The governing equations for mass, momentum, chemical species, and energy, and the method for numerical integration of these equations is described in detail elsewhere.[11,12] The chemical kinetic mechanism used in the calculations is shown in Table 1. The rate constants for the reactions are assumed to be represented in the Arrhenius form $k_j = A_j T^{n_j} exp(-E_j/(RT))$, where the notation is described in the list of symbols. The values for A_j, n_j, and E_j shown in Table 1 for all reactions except 5b and 5c were obtained from Ref. 1. For reactions 5b and 5c the values for A_j, n_j, and E_j were obtained from Ref. 12. The data for k_j, shown in Table 1, represents the forward rate of reaction j. In the numerical calculations, the reactions were allowed to proceed in the forward as well as in the backward direction. The backward rate of the reaction is calculated from the equilibrium constant for the reaction, obtained by using the thermochemical data shown in JANAF tables.[13] Elementary reactions involving the formation and consumption of hydrogen peroxide (H_2O_2) were not considered here because the numerical calculations were performed at a value of the total pressure equal to one atmosphere. The reactions involving H_2O_2 become important only at higher values of the absolute pressure.[1] Soret and Dufour effects have been neglected in the present analysis. Numerical calculations were performed for a number of values of the strain rate a, between 30 sec^{-1} and 1440 sec^{-1} (using the Cray-XMP Supercomputer at San Diego). The strain rate a, is defined as the radial gradient of the radial component of the flow velocity. Since the flow is axisymmetric, the absolute value of the axial gradient of the axial component of the flow velocity is 2a. At each value of a, converged solutions were obtained in approximately 5-9 min. CPU units. For very high values of a, the corresponding flow velocities would be high. Since the governing equations used in the numerical calculations are valid only for values

Table 1: The elementary reaction mechanism and associated rate coefficients

NO.	Reaction	A_j	n_j	E_j
1	$H + O_2 = O + OH$	1.20 x 10^{17}	-0.91	16507
2	$H_2 + O = H + OH$	1.50 x 10^{7}	2.00	7549
3	$H_2O + O = OH + OH$	1.50 x 10^{10}	1.14	17248
4	$H_2 + OH = H_2O + H$	1.00 x 10^{8}	1.60	3296
5a	$H + O_2 + M^a = HO_2 + M^a$	7.00 x 10^{17}	-0.80	0
5b	$H + O_2 + O_2 = HO_2 + O_2$	6.70 x 10^{19}	-1.42	0
5c	$H + O_2 + N_2 = HO_2 + N_2$	6.70 x 10^{19}	-1.42	0
6	$HO_2 + OH = H_2O + O_2$	2.00 x 10^{13}	0.00	0
7	$HO_2 + O = OH + O_2$	2.00 x 10^{13}	0.00	0
8	$HO_2 + H = OH + OH$	1.50 x 10^{14}	0.00	1003
9	$HO_2 + H = H_2 + O_2$	2.50 x 10^{13}	0.00	693
10	$H + H + M^b = H_2 + M^b$	6.40 x 10^{17}	-1.00	0
11	$H + OH + M^c = H_2O + M^c$	7.00 x 10^{21}	-2.00	0

The rate constant $k_j = A_j T^{n_j} exp(\frac{-E_j}{RT})$, and the units are cc, mol, cal and K.

[a] Third body efficiencies[12]: k_5 (H_2O) = 21 k_5 (Ar), k_5 (H_2) = 3.3 k_5 (Ar), k_5 (N_2) = k_5 (O_2) = 0.

[b] Third body efficiencies[12]: k_{10} (H_2O) = 6 k_{10} (Ar), k_{10} (H) = 2 k_{10} (Ar), k_{10} (H_2) = 3 k_{10} (Ar).

[c] Third body efficiencies[12]: k_{11} (H_2O) = 20 k_{11} (Ar).

a)

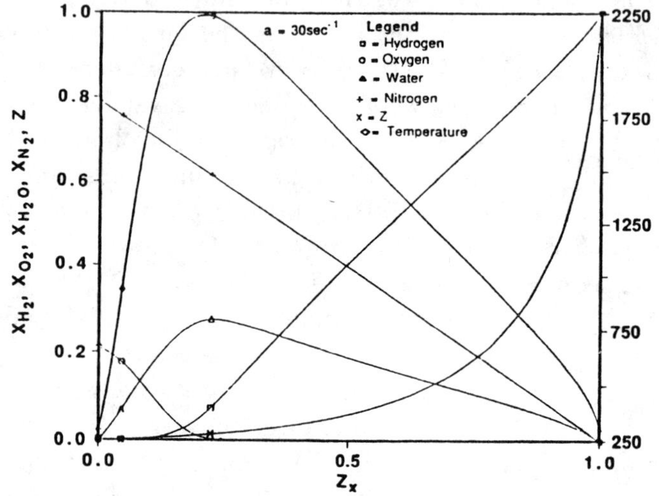

b)

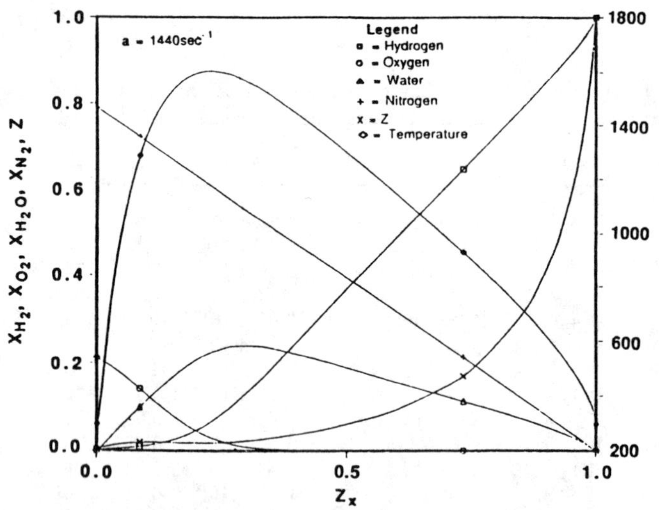

Fig. 1 a) major species for a=30 sec^{-1}, b) major species for a=1440 sec^{-1}, c) minor species for a=30 sec^{-1} and D) minor species for a=1440 $sec.^{-1}$ Computed structure of a counterflow, hydrogen-air diffusion flame.

c)

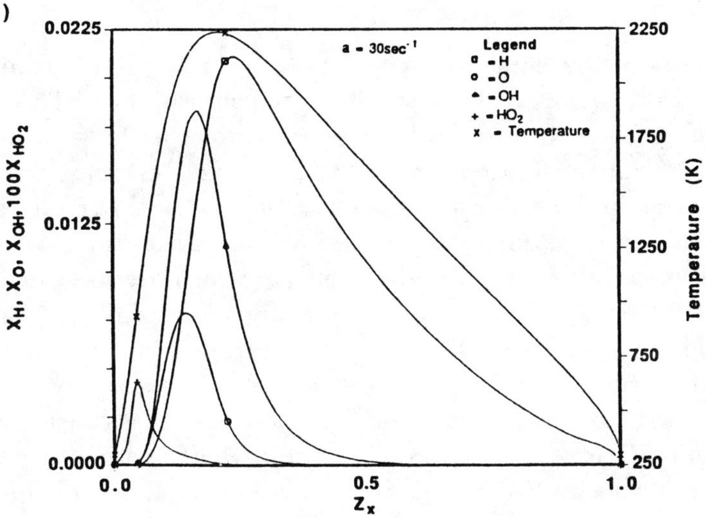
0.0225
0.0125
0.0000
2250
1750
1250
750
250
0.0
0.5
1.0
Z_x
X_H, X_O, X_{OH}, $100X_{HO_2}$
Temperature (K)
a = 30sec⁻¹
Legend
= H
= O
= OH
= HO_2
x = Temperature

d)

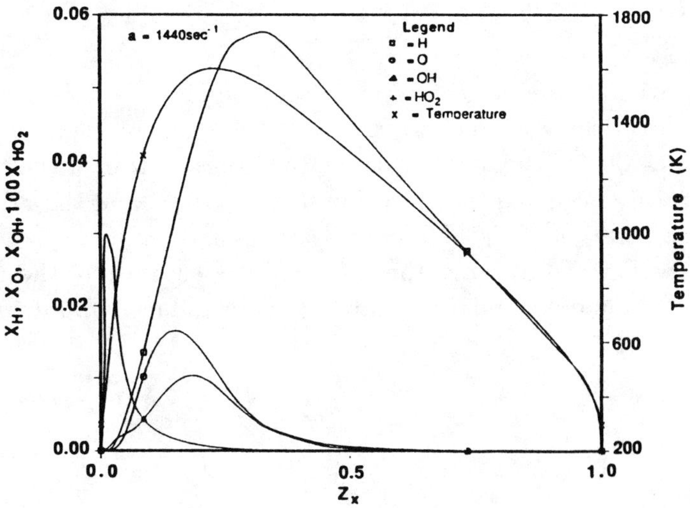
0.06
0.04
0.02
0.00
1800
1400
1000
600
200
0.0
0.5
1.0
Z_x
X_H, X_O, X_{OH}, $100X_{HO_2}$
Temperature (K)
a = 1440sec⁻¹
Legend
= H
= O
= OH
= HO_2
x = Temperature

of the Mach number much less than unity, the structure of the flame at high values of a were not caluculated.

Results

Results of the numerical calculations are interpreted using a conserved scalar quantity Z as the independent variable. This permits analysis to be performed without reference to specific flow configurations. Previous definitions of Z[3,14,15] were based on mass fractions of the elements present in the fuel and/or the oxidizer. For Z to satisfy a source free conservation equation, it is necessary that the Lewis number of all the species containing the elements which are used in the definition of Z be equal to unity. For a hydrogen-air diffusion flame, the definition of Z based on the element hydrogen is not appropriate because the structure of the flame contains relatively large concentrations of H_2 and H, and the Lewis number of these species is considerably less than unity. Since the Lewis number of nitrogen is very nearly equal to unity, and because this species is inert, we define Z as

$$Z = \frac{Y_{N_2,2} - Y_{N_2}}{Y_{N_2,2} - Y_{N_2,1}} \tag{1}$$

Equation(1) implies that Z = 0 in the ambient oxidizer stream, and Z = 1 in the ambient fuel stream. If the numerical results are plotted using Z as the independent variable, then the position of the maximum temperature occurred in the region $0 < Z < 0.03$ and the gradients of the concentration of all the chemical species and temperature were steep in this region. To improve the clarity of presentation of the numerical results, a new independent variable Z_x is introduced and defined as

$$Z_x = \frac{X_{N_2,2} - X_{N_2}}{X_{N_2,2} - X_{N_2,1}} \tag{2}$$

In Figs. 1a-1d the calculated structure of the flame is shown for a = 30 sec^{-1} and for a = 1440 sec^{-1}, respectively. At a = 30 sec^{-1} the structure of the flame is nearly in chemical equilibrium, whereas, at a = 1440 sec^{-1} the structure of the flame is far from chemical equilibrium. Figures 1a and 1b show that the value of Z increases monotonically with increasing values of Z_x, therefore, there is no fundamental inconsistancy in plotting the results of numerical calculations in terms of Z_x for the chemical system considered here. Dixon-Lewis et al.[3] presented the results of their calculations of the structure of hydrogen-air diffusion flame using a mixture fraction defined using the elements present in the chemical system. This definition of the mixture fraction resulted in negative values for this quantity between its boundary values of 0 and 1. The definition of Z shown in Eq. (1) was found to increase monotonically from 0 to 1, thus negative values of Z were not obtained. Figures 1a-1d show that the maximum value of the flame temperature T_{max} with a value around 2250 K is attained at a value of Z_m around 0.25 for a = 30 sec^{-1}, and a maximum value of the flame temperature T_{max} with a value around 1600 K is attained at a value of Z_m around 0.28 for a = 1440 sec^{-1}. For convenience, the region $Z_x < Z_m$ will be called the oxidizer side of the flame and the region $Z_x > Z_m$ will be called the fuel side of the flame. The value of Z_m and the concentration of H_2 and O_2 at $Z_x = Z_m$ are higher for a = 1440 sec^{-1} when compared with the values of these quantities at a = 30 sec^{-1}. This implies that as the value of a is increased, the flame moves towards the fuel stream and the leakage of the reactants increases. Figures 1c and 1d show that the profiles of the radicals H, O, and OH are steeper on the oxidizer side when compared to the fuel side, because the predominant mechanism for destruction of the radicals is via the three body recombination reactions 5 shown in Table 1. Since the concentration of O_2 is small on the fuel side of the flame, the rates of the reactions 5 are much smaller on the fuel side when compared to the rates of these reactions on the oxidizer side. Consequently, the profiles of the radicals H, OH, and O are steeper on the oxidizer side of the flame. Dixon-Lewis et al.[3] obtained similar results for a = 100 sec^{-1}.

To examine if certain reactions are in partial equilibrium in Fig. 2a, the quantity $(K_e)_j/(K_T)_j$ is plotted as a function of Z_x for reactions 1-4 for a = 30 sec^{-1}. At a given value of Z_x, $(K_e)_j$ is the ratio of the product of the concentration of the species appearing on the r.h.s. of reaction j, to the product of the concentration of the species appearing on the l.h.s. of reaction j, and $(K_T)_j$ is the equilibrium constant of reaction j. From an empirical fit of the thermochemical data,[13] for values of T between 298 K and 2400 K, the following expressions for $(K_T)_j$ were obtained

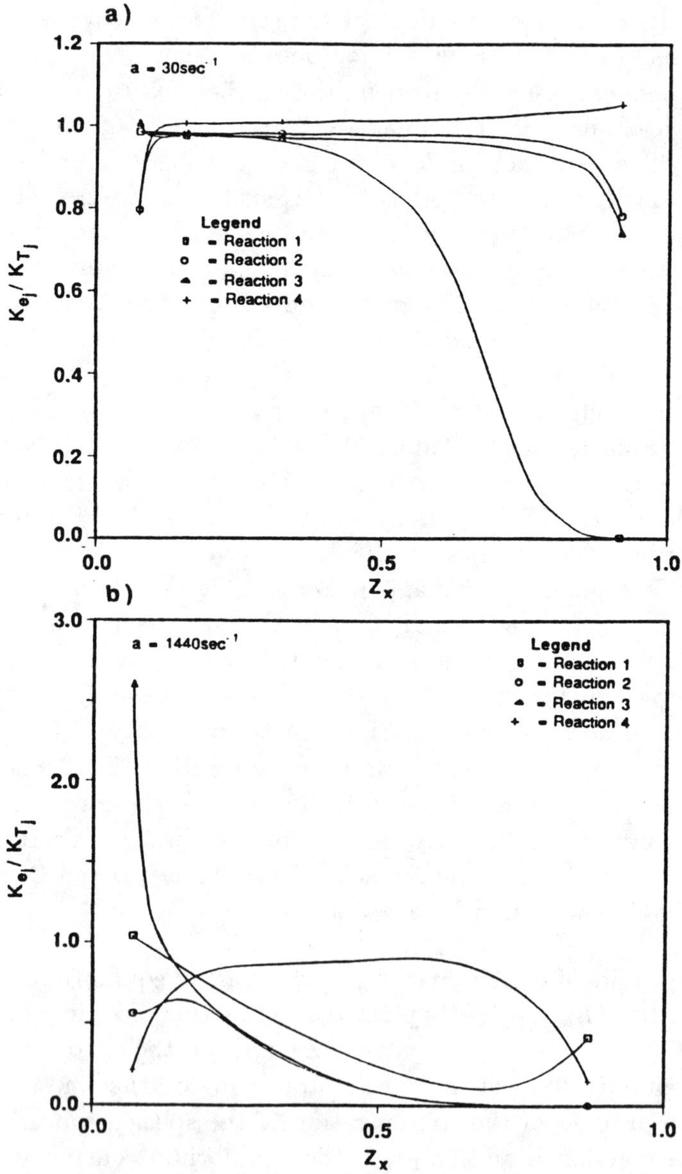

Fig. 2 a) a=30 sec^{-1} and (b) a=1440 $sec.^{-1}$ $(K_c)_j$ / $(K_T)_j$ versus Z_x.

$$(K_T)_1 = 16.0exp(\frac{-8397}{T}) \tag{3}$$

$$(K_T)_2 = 2.23exp(\frac{-995}{T}) \tag{4}$$

$$(K_T)_3 = 10.46exp(\frac{-8687}{T}) \tag{5}$$

$$(K_T)_4 = 0.2133exp(\frac{7693}{T}) \tag{6}$$

The results plotted in Fig. 2a show that the reactions 2-4 are in partial equilibrium everywhere because the value of the ratio $(K_e)_j/(K_T)_j$ is nearly equal to unity. Reaction 1 also appears to be in partial equilibrium near Z = Z_m, however, on the fuel side of the flame the ratio $(K_e)_j/(K_T)_j$ decreases rapidly, and is related to the rapid decrease in the product of the concentrations of the radicals O and OH in comparison to the product of the concentration of the radicals H and O_2. In Fig. 2b similar results are plotted for a = 1440 sec^{-1} and show that reactions 1-4 are not in partial equilibrium. In Figs. 3a and 3b the ratio of the forward rates of the three body reactions 5,10, and 11 to the forward rates of reaction 1 at T = T_{max} is plotted as a function of Z_x for a = 30 sec^{-1} and a = 1440 sec^{-1}, respectively. The data for reaction 5 shown in Figs. 3a and 3b represents the sum of the forward rates of reactions 5a, 5b, and 5c shown in Table 1. Figures 3a and 3b show that on the oxidizer side of the reaction zone chain termination occurs predominantly via reacton 5. On the fuel side of the flame chain termination occurs predominantly via reaction 10 because the concentration of molecular oxygen is small in this region. Finally, in Fig. 4 the quantities T_{max}, Z_m, and the value of the mole fractions of H_2 and O_2 at Z = Z_m are plotted as a function of a. Figure 4 shows that with increasing value of a, T_{max} decreases, the leakage of the reactants through the reaction zone increases, and the flame moves toward the fuel side.

Equilibrium Structure

In this Sec. the equilibrium structure of the flame is determined in the limit a $\rightarrow$ 0, using the approximations suggested by the results of the previous section. It is useful to compare the equilibrium structure

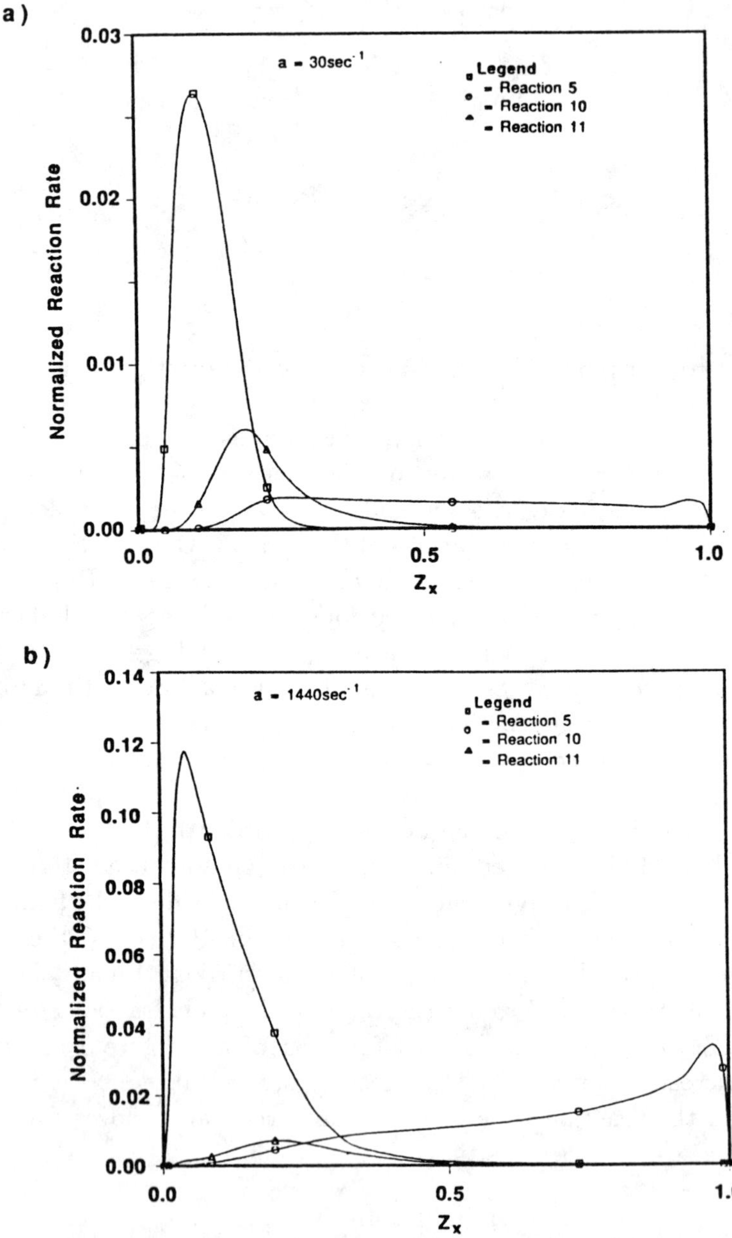

Fig 3 Normalized rates of rection 5,10 and 11, nondimensionalized by division with the forward rates of reaction 1 at T = T_{max} versus Z_x.

of the hydrogen-air diffusion flame described here with the flame sheet approximation used in the literature.[16,17] Both of these structures are fundamentally the same because they are obtained mathematically in the singular limit of $D \to \infty$, or $a \to 0$, where D is the characteristic Damkohler number of the reaction. In the flame sheet approximation used in the literature,[16,17] the reactants directly react with each other to form products by an overall one-step, irreversible chemical reaction. In the limit $D \to \infty$ the rate of the chemical reaction is infinitely fast, therefore, the reactants cannot coexist. Hence, the chemical reaction is confined to a thin layer, and outside this layer the concentration of one of the reactants is zero. The maximum flame temperature in this limit is equal to the adiabatic flame temperature, and the products exist everywhere in the flowfield. For the hydrogen-air diffusion flame considered here, multistep and reversible chemical reactions are considered. The equilibrium structure is obtained mathematically in the singular but distinguished limit of $D_{f,j} \to \infty$ and $D_{b,j} \to \infty$ with $D_{f,j}/D_{b,j} \to O(1)$ for reversible chemical reactions. Here, subscripts f,j and b,j denote the forward and backward directions of reaction j, respectively. The equilibrium structure obtained in this distinguished limit can have nonzero values for the concentrations of the reactants, in-

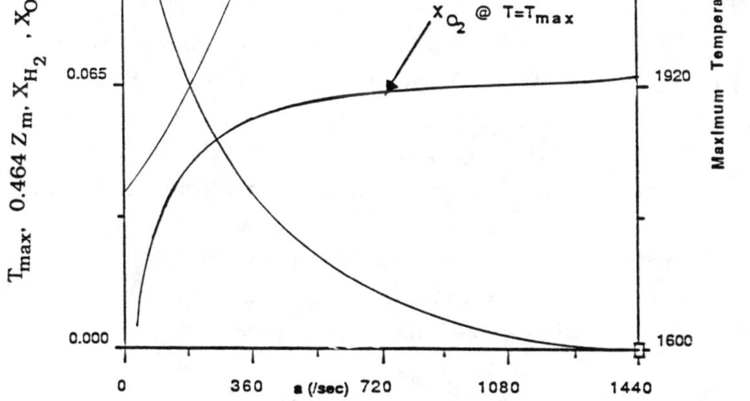

Fig 4 T_{max}, Z_m and values of X_{H_2} and X_O at $T = T_{max}$ as a function of a.

termediate species, and products everywhere in the flowfield. The value of the maximum flame temperature in this limit would be less than the value of the adiabatic flame temperature. In the present study the equilibrium structure of the hydrogen-air diffusion flame is obtained in this limit.

Based on the results shown in Fig. 2a, it is reasonable to assume that reactions 2-4 are in partial equilibrium everywhere in the flowfield, consequently, only two of these three reactions are independent. Without loss of generality reaction 3 will not be considered in the limit a $\rightarrow 0$. Results of numerical calculations show that reactions 5-11 proceed primarily in the forward direction, hence the backward rates of these reactions will be neglected. Numerical calculations show that the destruction of HO_2 occurs primarily via reaction 6, consequently reactions 7-9 are not considered in the analysis, and reaction 6 is assumed to proceed irreversibly only in the forward direction. Under a well-known set of approximations[7] the species balance equation and the energy equation can be written as

$$\frac{d^2 y_i}{dZ^2} = -\sum_{j=i}^{6} v_{ij} w_j \quad i = 1, \ldots N,\ j \neq 3 \tag{7}$$

$$\frac{d^2 \theta}{dZ^2} = -\sum_{j=1}^{6} q_j w_j \quad , j \neq 3 \tag{8}$$

$$y_i = \frac{C_i W_{H_2} R^o T L_{H_2}}{L_i Y_{H_{2,1}} p \bar{W}} \tag{9}$$

$$\theta = \frac{(T - T_2) c_p W_{H_2} L_{H_2}}{Q_T Y_{H_{2,1}}} \tag{10}$$

where the quantity Q_T is the heat release associated with the one-step reaction $H_2 + \frac{1}{2}O_2 \rightarrow H_2O$ and $q_j = Q_j/Q_T$ with Q_j representing the heat release associated with the reaction step j. In writing Eq. (8) it has been assumed that the factor Le_i-1 is small. Although this assumption is somewhat questionable for hydrogen-air flames, nevertheless, it is used here for simplicity. If the concentration of the species HO_2 is assumed to be in a dynamic steady state[17] then Eq. (7) implies that $w_5 = w_6$. From the reaction scheme 1,2,4,5,6,10, and 11 and Eqs. (7) and (8) the following equations are obtained

$$\frac{d^2 y_{O_2}}{dZ^2} = w_1 \tag{11}$$

$$\frac{d^2(y_{H_2} + y_H - y_{O_2})}{dZ^2} = w_5 + w_{10} + w_{11} \qquad (12)$$

$$\frac{d^2(2y_{H_2O} + 2y_{H_2} + y_H + y_{OH}}{dZ^2} = 0 \qquad (13)$$

$$\frac{d^2(y_{H_2O} + 2y_{O_2} + y_O + y_{OH})}{dZ^2} = 0 \qquad (14)$$

$$\frac{d^2(\theta - \sum_{i=1}^{N} h_i y_i)}{dZ^2} = 0 \qquad (15)$$

where (-h$_i$) is the effective heat of formation of species i nondimensionalized by division by Q_T. It follows that $q_j = \sum_{i=j}^{N} \nu_{ij} h_i$. Boundary conditions for Eqs. (11-15) are

$$\theta = 0,\ y_{O_2,2},\ y_i = 0\ \ for\ i \neq O_2,\ at\ Z = 0 \qquad (16)$$

$$\theta = \theta_1,\ y_{H_2} = 1,\ y_i = 0\ \ for\ i \neq H_2,\ at\ Z = 1 \qquad (17)$$

Integration of Eqs. (13), (14), and (15) using boundary conditions (16) and (17) yields the result

$$2y_{H_20} + 2y_{H_2} + y_H + y_{OH} = 2Z \qquad (18)$$

$$y_{H_20} + 2y_{0_2} + y_O + y_{OH} = 2y_{O_2,2}(1 - Z) \qquad (19)$$

$$\theta - \sum_{i=1}^{N} h_i y_i = \theta_1 Z \qquad (20)$$

Since reactions 2-4 are in partial equilibrium, it follows that

$$y_H y_{H_20} = K_H y_{H_2} y_{OH} \qquad (21)$$

$$y_O y_{H_2O} = K_O y_{OH}^2 \qquad (22)$$

where $K_H = (K_T)_4 L_{H_2} L_{OH}/(L_H L_{H_2O})$, and $K_O = ((K_T)_4/(K_T)_2)$

$(L_{OH}^2/(L_O L_{H_2O}))$. It can be shown that the nondimensional reaction rate appearing in Eq. (11) is given by the expression $w_1 = B_1(y_{OH}/y_{H_2O})(y_{OH}^2 - K_I y_{H_2} y_{O_2})$ where B_1 is a function of temperature and is inversely proportional to the strain rate,[7] and the quantity K_I is given by the expression

$$K_I = \frac{(K_T)_1 (K_T)_2 L_{O_2} L_{H_2}}{L_{OH}^2} \tag{23}$$

In the limit $a \rightarrow 0$, it follows that $B_1 \rightarrow \infty$. Hence, for w_1 to be finite $y_{OH}^2 - K_I y_{H_2} y_{O_2} = 0$, or

$$y_{OH} = \sqrt{K_I y_{H_2} y_{O_2}} \tag{24}$$

Table 1 shows that the rate of reaction 1 decreases exponentially with decreasing values of temperature. On the lean side of the flame where the concentration of oxygen is not small, following previous analyses of Trevino and Williams,[7] a critical temperature T_c attained at $Z = Z_c$ can be identified where the reaction rate of the major chain branching reaction 1 (forward rate only) is equal to the reaction rate of major chain termination reaction 5. This equality implies that $k_{1f} = k_{5f} p/(R^O T)$. For $p = 1$ atm, using the data shown in Table 1, the result T_c is obtained. The corresponding value of θ_c can be calculated from Eq. (10). Since the quantity $k_{5f} p/(R^O T)$ decreases with increasing values of T, it can be assumed that $w_5 \approx 0$ in the region $Z > Z_c$. Similar to previous analysis[7] it is postulated that in the limit $a \rightarrow 0$ reaction 5 occurs in a thin "radical destruction layer" around $Z=Z_c$. Since the rate of reaction 5 is proportional to the product $y_H y_{O_2}$, in the region $Z < Z_c$, $y_H = 0$ in the limit $a \rightarrow 0$ because $y_{O_2} \neq 0$. Hence, Eqs. (21-22), and (24) yield the result[7]

$$y_{H_2} = y_H = y_O = y_{OH} = 0, \; 0 \leq Z \leq Z_c \tag{25}$$

The values of $y_{O_2,c}$, $y_{H_2O,c}$, and Z_c can be obtained using Eqs. (18-20) which yield the relations $\theta = (\theta_1 + h_{H_2O})Z_c$, $y_{H_2O,c} = Z_c$ and $y_{O_2,c} = y_{O_2,2}(1 - Z_c) - Z_c/2$. Also, Eqs. (18-20) can be used to determine the profiles of y_{H_2O}, y_{O_2}, and θ in the region $0 \leq Z \leq Z_c$. In the region $Z_c \leq Z \leq 1$ integration of Eq. (12) with $w_5 + w_{10} + w_{11} \approx$

a)

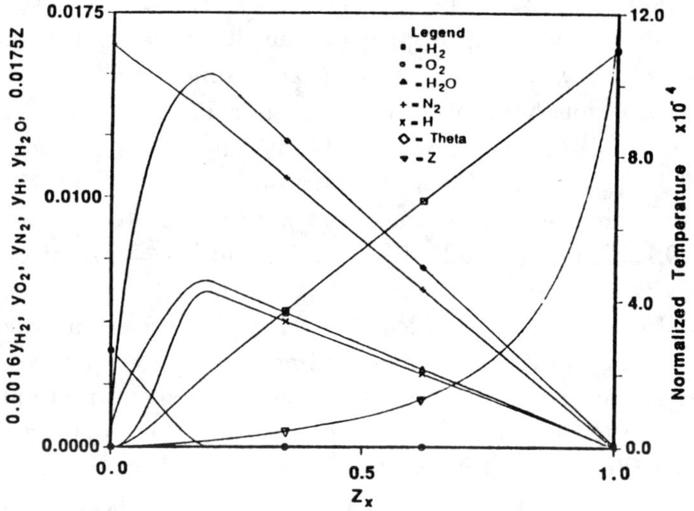

b)

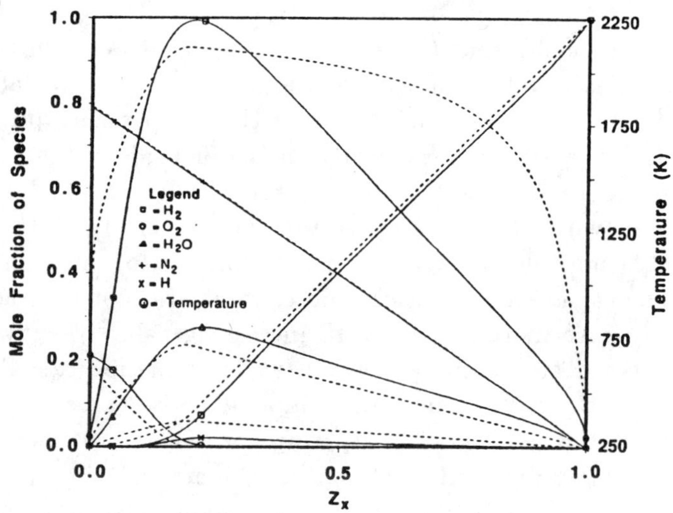

Fig 5 a) Equilibrium structure, (b) comparison between the equilibrium structure (dotted lines) and the computed structure (solid lines).

0, together with the boundary conditions Eqs. (17) and (25) yield the result

$$y_{H_2} + y_H - y_{O_2} = \frac{(1 + y_{O_2,c})Z}{1 - Z_c} - \frac{Z_c + Y_{O_2,c}}{1 - Z_c} \tag{26}$$

Equations (18-22), (24), and (26) represent seven equations to determine the profiles of the seven unknown quantities y_{H_2}, y_{O_2}, y_{H_2O}, y_{OH}, y_O, y_H, and θ as a function of Z. The set of algebraic equations was solved numerically using the Newton-Raphson technique. In the calculations the Lewis numbers were chosen to be $L_{H_2} = 0.3$, L_{O_2}, $= 1.05$, $L_{H_2O} = 0.74$, $L_{OH} = 0.73$, $L_O = 0.73$, and $L_H = 0.2$, the thermochemical constants were chosen to be $h_{H_2O} = 1.0$, $h_{OH} = -0.1632$, $h_H = -0.827$, $h_O = -1.03$, and $h_{H_2} = h_{O_2} = 0$, and the boundary values were chosen to be $Y_{H_2,1} = 1$, $Y_{O_2} = 0.23$, and $\theta_1 = 0$. Results of the equilibrium calculations for y_i and θ are plotted in Fig. 5a as a function of Z. The corresponding values of the temperature T, calculated from Eq. (10) and the values of the mole fractions X_i calculated from the relation $X_i = (\bar{W} L_i Y_{H_2,1} y_i)/(W_{H_2} L_{H_2})$ are shown as dotted lines in Fig. 5b. For purposes of comparison, the solid lines in Fig. 5b represent the results of the numerical calculations for a = 30 $\sec^{-1}$.

Figure 5b shows that there is good agreement between the numerically calculated values of X_i and the values of X_i calculated in the limit a $\rightarrow$ 0. However, the shapes of the temperature profiles show substantial disagreement especially on the fuel side of the flame. This is attributed to nonequilibrium of reaction 1 on the fuel side of the flame as shown in Fig. 2a, where the value of $(K_e)_j/(K_T)_j$ decreases with increasing value of Z_x. This conclusion is also consistent with the results shown in Fig. 1c where the concentration of the radicals OH and O decrease more rapidly with increasing value of Z_x when compared to the decrease in the concentration of the radicals OH and O, especially in the region where the temperature is decreasing, can cause the reaction $OH + O \rightarrow H + O_2$ to freeze. Therefore, there will exist a transition layer around $Z = Z_{TF}$ with reaction 1 in equilibrium in the region $Z < Z_{TF}$ and frozen in the region $Z > Z_{TF}$. Analysis of the structure of this transition layer can be performed using previously established procedures.[18] Trevino and Williams[7] have also postulated in their asymptotic analysis of the structure of methane-air diffusion flames that the exothermic reaction $OH + O \rightarrow H + O_2$ can freeze because of a rapid decrease in the concentration of the radicals OH and O. Dixon-Lewis et al.,[3] from their analysis, have postulated that the extinction of the flame may be caused by a breakdown of reaction 1. In addition the chain termination reaction 10 which is neglected in the analysis becomes important on the rich side of the flame. Recently Chung and Williams[19] have identified a critical temperature T_h attained at $Z = Z_h$ on the rich side of the flame by equating the backward rate of reaction 2 to the forward rate of reaction 10. In the region $Z > Z_h$ Chung and Williams[19] presumed that all chemical reactions are frozen. If these factors are taken into account better agreement can be attained between the results of numerical calculations and asymptotic analyses. Figure 5b also shows small differences in the peak values of T, and X_i between the results of the numerical calculations, and the results of the equilibrium analysis and is attributed to inaccuracies associated with the values of the Lewis numbers used in the analysis for the various species.

Summary and Conclusions

Results of a numerical and analytical study of the structure of a counterflow hydrogen-air diffusion flame is outlined here. The calculations were performed at various values of the rate of strain, a, and at a value of the absolute pressure equal to 1 atm. The detailed kinetics of oxidation of hydrogen shown in Table 1 was used in the numerical calculations. The structure of the flame is shown in Figs. (1a-1d) for a $= 30\ sec^{-1}$ and for $a = 1440\ sec^{-1}$.

Results plotted in Fig. 2a show that for $a = 30\ sec^{-1}$ reactions 2-4 are in partial equilibrium over a major part of the reaction zone and the major chain branching reaction 1 is in partial equilibrium near the position of the maximum temperature, but not in partial equilibrium on the fuel side of the flame. Results plotted in Fig. 2b show that reactions 1-4 are not in partial equilibrium for $a = 1440\ sec^{-1}$. Results plotted in Figs. 3a and 3b for $a = 30\ sec^{-1}$ and $a = 1440\ sec^{-1}$, respectively, show that three body chain termination reaction occurs predominantly via reaction 5 on the oxidizer side of the flame and via reaction 10 on the fuel side of the flame. Figure 4 shows that as the value of a is increased the value of T_{max} decreases, and leakage of the reactants H_2 and O_2 increases, and the flame moves towards the fuel stream.

The equilibrium structure of the flame was analyzed in the limit of $a \to 0$. Only reactions 1-6, 10 and 11were considered in this analysis. Figure 5b shows comparison between the results of numerical computations and the equilibrium structure. There is good agreement for the profiles of species H_2, O_2, H_2O, and H. However, the shapes of the temperature profiles do not agree and this is mainly attributed to the nonequilibrium of reaction 1 on the fuel side of the flame as shown in Fig. 2a and also due to neglect of the chain termination reaction 10. The approximations suggested by the results of the numerical calculations can be used to reduce the detailed chemical kinetic mechanism for oxidation of hydrogen to a few global steps similar to that performed previously for hydrocarbon flames.[4–7] The reduced mechanism and the results of the equilibrium can be used to perform an asymptotic analysis to characterize the structure and to predict critical conditions of extinction of hydrogen-air diffusion flames similar to those performed recently for methane-air diffusion flames.[6,7] Results of the asymptotic analysis can then be used to predict the multidimensional and time-dependent combustion of hydrogen in air. Studies along these lines are currently in progress by the present authors.

Finally, it is useful to compare the structure of hydrogen-air flame with the structure of the "H_2-CO oxidation region" of methane-air diffusion flames[6,7]. It has been suggested[6,7] that in methane-air diffusion flames there exists a "fuel consumption region" where methane reacts with the radicals H, OH, and O to form H_2 and CO. On the oxidizer side of the fuel consumption region there exists a "H_2-CO oxidation region" where H_2 and CO react with radicals to form H_2O and CO_2. The oxidizer side of the structure of the hydrogen-air diffusion flame is similar to the oxidizer side of the structure of the methane-air diffusion flame. However, for methane-air diffusion flames the reaction of methane with the radicals acts as a strong chain breaking reaction for the radicals in addition to the three body recombination reactions. For hydrogen-air diffusion flames on the fuel side of the reaction zone chain breaking occurs predominantly via the three body recombination reaction 10.

Acknowledgments

This work was supported by the U.S. Army Research Office, Contract No. DAAL03-86-K-0001. Dr. David Mann is the technical monitor of this project. Authors would like to thank F. A. Williams and E.Gutheil for their helpful technical discussions.

References

[1]Warnatz, J., Combustion Chemistry, edited by W.C. Gardiner, Springer, New York, 1984, p. 197.

[2]Westbrook, C.K., and Dryer, F.L., Prog. Energy and Combust. Sci., vol. 10, 1984, p. 1.

[3]Dixon-Lewis, G., David, T., and Gaskell, P.H., "Structure and Properties of Hydrogen-Air Counterflow Diffusion Flames," Archivum Combustionis, vol. 6(1), 1986, p.3

[4]Bilger, R.W., and Kee, R.J., "Simplified Kinetics for Diffusion Flames of Methane in Air," Paper No. 6A-087, Joint Conference, Western States and Japanese Sections, The Combustion Institute, 1986, p. 277.

[5]Paczko, G., Lefdal, P.M., and Peters, N., "Reduced Reaction Schemes for Methane, Methanol and Propane Flames," Prceedings of the Twenty-First Symposium (International) on Combustion, The Combustion Institute, Pittsburgh, PA, 1986, p. 739.

[6]Seshadri, K., and Peters, N., "Asymptotic Structure and Extinction of Methane-Air Diffusion Flames," Combust. Flame vol. 73, 1988, p. 23.

[7]Trevino, C., and Williams, F.A., "An Asymptotic Analysis of the Structure and Extinction of Methane-Air Diffusion Flames," Dynamics of Reactive Systems, edited by A.L. Kuhl, J.R. Bowen, J.L. Leyer, and A. Borisov), Progress in Astronautics and Aeronautics, vol. 113, AIAA, Washington, D.C., 1988, p. 321.

[8]Kailasnath, K., Oran, E.S., and Boris, J., "A Theoretical Study of Ignition of Premixed Gases," Combust. Flame, vol. 47(2), 1982, p. 173.

[9]Kailasnath, K., Oran, E.S., and Boris, J., "A One Dimensional Time Dependent Model for Flame Initiation, Propagation and Quenching," Naval Research Laboratory, Washington, D.C., Rept. 4910, February, 1982.

[10]Dixon-Lewis, G., "Chemical Mechanism and Properties of Freely Propagating Hydrogen-Oxygen Supported Flames," Archivum Combustionis, 4(4), 279 (1984).

[11]Smooke, M.D., Puri, I.K., and Seshadri,K., "A comparison Between Numerical Calculations and Experimental Measurements of the Structure of a Counterflow Diffusion Flame Burning Diluted Methane in Diluted Air," Proceedings of the Twenty-First Symposium (International) on Combustion, The Combustion Institute, Pittsburgh, PA, 1986, p. 1783.

[12]Puri, I.K., Seshadri, K., Smooke, M.D., and Keyes, D.E.,"A comparison Between Numerical Calculations and Experimental Measurements of the Structure of a Counterflow Methane Air Diffusion Flame," Combust. Sci. Tech., vol. 56, 1987, p. 1.

[13]Stull, D.R., and Prophet, H., Janaf Thermochemical Tables, 2nd ed., NSRDS-NBS 37, U. S. Department of Commerce, 1971.

[14]Peters, N., Prog. Energy Combust. Sci., vol. 10, 1984, p.319.

[15]Bilger, R.W., "Turbulent Jet Diffusion Flames," Prog. Energy Combust. Sci., vol. 1, 1976, p. 87.

[16]Linan, A., "The Asymptotic Stucture of Counterflow Diffusion Flames for Large Activation Energies," Acta Astronautica, vol. 1, 1974, p.1007.

[17]Williams, F.A., Combustion Theory, 2nd ed., Benjamin/Cummings, San Fransisco, 1985.

[18]Peters, N., "Premixed Burning in Diffusion Flames - The Flame Zone Model of Libby and Economos," Int. J. Heat Mass Transfer, vol. 22, 1979, p. 691.

[19]Chung, S.H., and Williams, F.A., "Asymptotic Structure ana Extinction of CO/H2 Diffusion Flames with Reduced Kinetic Mechanisms," Combust. Flame, (submitted for publication).

Flame Structure of Axisymmetric Hydrogen-Air Diffusion Flames

Seishiro Fukutani,* N. Kunioshi,† and Hiroshi Jinno‡
Kyoto University, Yoshida-honmachi, Sakyo-ku, Kyoto, Japan

Abstract

A detailed model that contains both a full scheme of combustion reactions and the governing equations of fluid mechanics was made for hydrogen–oxygen systems, and applied to a Burke-Schumann type hydrogen–air diffusion flame to elucidate their flame structures and combustion reactions. The air flowing outside the burner was controlled for the overall equivalence ratio to be fixed at 0.5 and 1.25; these two combustion conditions were experimentally ascertained to give stable flames. The predicted results were as follows: both conditions give well-developed diffusion flames above the height of about 5 mm from the burner mouth. In the well-developed regions, the flame structures and the reaction mechanisms in both flames are not remarkably different at any heights; the chemical reactions and the release of heat are confined into annular zones about 5 mm thick, in which there is a point where H_2 and O_2 are simultaneously exhausted, and in whose vicinity the temperature becomes maximum. The Burke-Schumann theory was proved to be valid in the well-developed regions, although there were some differences between the flame structures assumed by them and those predicted by the present simulation. At the lowest part of the flames, the fluxing air and the H_2 diffusing downward meet and react like premixed gas mixtures; a large amount of thermal energy is released through the reactions during short time intervals and it holds the whole flames at the burner rim.

Introduction

Burke and Schumann's flame-sheet model[1] has been applied in many theoretical researches on diffusion flames, but recent improvements in numerical methods as well in the computational facilities have stimulated the use of detailed models[2,3].

A precise model including a full reaction scheme and, at the same time, the governing equations of fluid mechanics are expected to ascertain the limits of validity of the Burke-Schumann's theory and to yield

* Associate Professor, Faculty of Engineering, Dept. of Industrial Chemistry
† Doctor's Course Student
‡ Professor

detailed information about the structure of a diffusion flame such as the flow field and the distributions of temperature, species concentrations, and chemical reaction rates.

Another point of interest to be investigated is the mechanism of flame holding. The holding of a diffusion flame is, usually, simply explained as resulting from premixed gas produced by the interdiffusion of fuel and air at the base of the flame or as recirculating combustion gas above the burner nozzle. Premixed gases burn in a fairly short time and pull down the whole flame toward a burner rim.[4] The precise model can allow a detailed understanding of that mechanism.

In this investigation, the flame structure, the chemical reactions, and the mechanism of flame holding in hydrogen–air diffusion flames were investigated through a simulation model constructed for axisymmetric diffusion flames. A hydrogen–air diffusion flame is expected to exhibit the typical properties of diffusion flames, since it contains not only the fuel with a large diffusion coefficient but also hydrogen atoms, which are the most diffusive intermediate.

Simulation Model

A model for computer simulation of diffusion flames was formulated on the basis of the following assumptions:

1) The flames are axially symmetric.

Table 1 H_2/O_2 reaction scheme
$k = AT^n \exp(-E/T)$

No.	Reaction	A	n	E	Ref.
R1	$H_2 + O_2 \rightarrow OH + OH$	2.50E06	0.0	19600.0	5
R2	$H + O_2 \rightarrow OH + O$	1.20E11	−0.9	8310.0	6
R3	$O + H_2 \rightarrow OH + H$	1.80E04	1.0	4480.0	5
R4	$OH + OH \rightarrow O + H_2O$	6.30E06	0.0	550.0	5
R5	$OH + H_2 \rightarrow H + H_2O$	2.20E07	0.0	2590.0	5
R6	$H + H + M \rightarrow H_2 + M$	2.60E06	−1.0	0.0	5
R7	$O + O + M \rightarrow O_2 + M$	1.90E01	0.0	−900.0	5
R8	$H + O + M \rightarrow OH + M$	3.60E06	−1.0	0.0	7
R9	$OH + H + M \rightarrow H_2O + M$	4.06E10	−2.0	0.0	5
R10	$H + O_2 + M \rightarrow HO_2 + M$	5.00E03	0.0	−500.0	5
R11	$H + HO_2 \rightarrow H_2 + O_2$	2.50E07	0.0	350.0	5
R12	$H + HO_2 \rightarrow OH + OH$	2.50E08	0.0	950.0	5
R13	$H + HO_2 \rightarrow O + H_2O$	9.00E05	0.5	2000.0	5
R14	$OH + HO_2 \rightarrow H_2O + O_2$	5.00E07	0.0	500.0	8
R15	$O + HO_2 \rightarrow OH + O_2$	6.30E07	0.0	350.0	9
R16	$HO_2 + H_2 \rightarrow H + H_2O_2$	7.30E05	0.0	9400.0	5
R17	$HO_2 + HO_2 \rightarrow H_2O_2 + O_2$	8.50E06	0.0	500.0	5
R18	$OH + H_2O_2 \rightarrow HO_2 + H_2O$	1.00E07	0.0	910.0	5
R19	$H + H_2O_2 \rightarrow OH + H_2O$	2.20E09	0.0	5900.0	5
R20	$O + H_2O_2 \rightarrow OH + HO_2$	2.80E07	0.0	3200.0	5
R21	$H_2O_2 + M \rightarrow OH + OH + M$	1.20E11	0.0	22900.0	5

The rate constants are expressed in m·mol·s units.

2) The flames under consideration contain nine species: OH, H, O, HO_2, H_2O_2, H_2O, H_2, O_2, and N_2. Twenty-one pairs of elementary reactions listed in Table 1 may occur among these species. Nitrogen is inert and acts only as a diluent and a third body in recombination reactions.

3) Heat transfer from the flames to their surroundings and radiant heat transport inside the flames are neglected.

The governing equations are expressed by Eqs.(1–6) in cylindrical coordinates according to the above assumptions.

$$\frac{\partial \rho}{\partial t} + \frac{1}{r}\frac{\partial}{\partial r}(\rho u r) + \frac{\partial}{\partial z}(\rho v) = 0 \tag{1}$$

$$\begin{aligned}\rho\left(\frac{\partial u}{\partial t} + u\frac{\partial u}{\partial r} + v\frac{\partial u}{\partial z}\right) &= \frac{\partial}{\partial r}\left[\mu\left\{2\frac{\partial u}{\partial r} - \frac{2}{3}\left(\frac{1}{r}\frac{\partial}{\partial r}(ur) + \frac{\partial v}{\partial z}\right)\right\}\right] \\ &+ \frac{\partial}{\partial z}\left[\mu\left(\frac{\partial u}{\partial z} + \frac{\partial v}{\partial r}\right)\right] + \frac{2\mu}{r}\left(\frac{\partial u}{\partial r} - \frac{u}{r}\right) - \frac{\partial p}{\partial r}\end{aligned} \tag{2}$$

$$\begin{aligned}\rho\left(\frac{\partial v}{\partial t} + u\frac{\partial v}{\partial r} + v\frac{\partial v}{\partial z}\right) &= \frac{\partial}{\partial z}\left[\mu\left\{2\frac{\partial v}{\partial z} - \frac{2}{3}\left(\frac{1}{r}\frac{\partial}{\partial r}(ur) + \frac{\partial v}{\partial z}\right)\right\}\right] \\ &+ \frac{1}{r}\frac{\partial}{\partial r}\left[\mu r\left(\frac{\partial u}{\partial z} + \frac{\partial v}{\partial r}\right)\right] - \frac{\partial p}{\partial z} - \rho g\end{aligned} \tag{3}$$

$$\begin{aligned}c_p\rho\left(\frac{\partial T}{\partial t} + u\frac{\partial T}{\partial r} + v\frac{\partial T}{\partial z}\right) &= \frac{1}{r}\frac{\partial}{\partial r}\left(\lambda r\frac{\partial T}{\partial r}\right) + \frac{\partial}{\partial z}\left(\lambda\frac{\partial T}{\partial z}\right) \\ &+ \sum_i D_i\rho\frac{\partial h_i}{\partial r}\frac{\partial \omega_i}{\partial r} + \sum_i D_i\rho\frac{\partial h_i}{\partial z}\frac{\partial \omega_i}{\partial z} - \sum_i h_i\phi_i\end{aligned} \tag{4}$$

$$\begin{aligned}\rho\left(\frac{\partial \omega_i}{\partial t} + u\frac{\partial \omega_i}{\partial r} + v\frac{\partial \omega_i}{\partial z}\right) &= \frac{1}{r}\frac{\partial}{\partial r}\left(D_i\rho r\frac{\partial \omega_i}{\partial r}\right) \\ &+ \frac{\partial}{\partial z}\left(D_i\rho\frac{\partial \omega_i}{\partial z}\right) + \phi_i\end{aligned} \tag{5}$$

$$p = \rho RT\sum_i\frac{\omega_i}{m_i} \tag{6}$$

where r is the radius; z, the axial distance; u, the radial flow velocity; v, the axial flow velocity; h, the enthalpy; ω, the mass fraction; ϕ, the production rate due to chemical reactions; m, the molecular weight; the subscript i denotes the i-th species.

The viscosity, the thermal conductivity, and the diffusion coefficient of the components were estimated using Hirschfelder's, Eucken's, and Hirschfelder's approximate equations, respectively.[10] The other thermodynamic data, including the equilibrium constants, which were used to evaluate the rate constants of the reverse reactions, were obtained from JANAF data.[11]

The set of partial differential equations were transformed into the corresponding finite difference equations based on the control-volume method.[12] The resulting finite difference equations were solved iteratively until time-independent solutions were obtained.

The boundary conditions were as follows:

1) The burner set in this simulation consists of two cylinders concentrically disposed: an inner tube of 10 mm in diameter with an infinitely thin wall and an outer cylinder of 40 mm in diameter. Hydrogen and air issue as well-developed flows with given velocity distributions[13] and at 298 K.

2) The gradients of all the time-dependent variables except the radial velocity are zero at the upper and the radial boundaries. Temperature at the cylinder walls was not allowed to raise above maximum limits.

Results and Discussion

Hydrogen issued from the inner tube with a fixed volumetric flow rate of 30 ml/s, and air flew into the annular space with two different rates, 143 ml/s and 57 ml/s, corresponding to the overall equivalence ratios of 0.5 and 1.25. Experimentally, stable laminar flames were obtainded under both conditions.

Figure 1 shows the predicted temperature distribution of the overventilated flame. Both this and the underventilated flames hang down and cling to the burner wall at 2–3 mm below the burner tip; the hanging down is remarkable particularly in the underventilated flame because of the slow air flow outside the burner tube and the consequent easier diffusion of the fuel downward along the burner tube surface. In addition, the high temperature region spreads out toward the outer cylinder in the underventilated flame. These phenomena were experimentally confirmed.

At the base of both flames, the contour lines, laying close together, indicate that large temperature gradients take place there. In that region, the maximum temperatures through the whole flames were found: 2480 K in the overventilated flame and 2370 K in the underventilated flame. The maximum temperature in the radial profiles decreases with increasing height. The hanging down of the flames and the temperature distributions indicate that the combustion reactions are activated intensely at low positions, and this fact is interesting in connection with the flame holding mechanism of hydrogen diffusion flames.

Another remarkable point shown in the temperature distributions in both flames is that the regions with temperature higher than 2200 K,

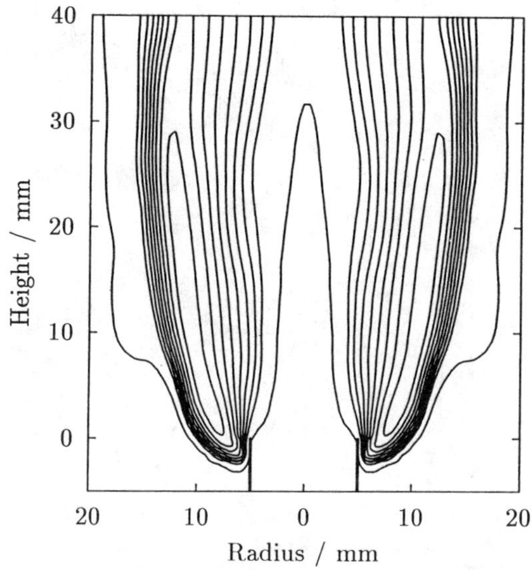

Fig. 1 Temperature distribution in the overventilated flame expressed with the contour lines for 500 K and for every 200 K between 1000 K and 2400 K. The two thick lines show the burner walls.

for instance, are about 5 mm thick independently of the axial position and, consequently, it is supposed that well-developed diffusion flames are obtained from fairly low heights.

The velocity distribution in the overventilated flame is shown in Fig. 2. This figure and that for the underventilated flame show that the environmental air is drawn toward the high-temperature regions at the base of the flames.

Figure 3 gives the distributions of three active species, OH radicals and H and O atoms, in the overventilated flame. The distributions of OH radicals, which is the main fuel oxidizer, are almost overlapped with the regions with local equivalence ratio ranging from 0.01 to 100. Hydrogen atoms are distributed in a region covering a wider range of equivalence ratios due to their mobility; in particular, they diffuse toward the center line and are present even in regions of about 1200 K in temperature in the overventilated flame. On the contrary, oxygen atoms are present in narrow regions of equivalence ratio around unity. Although the concentrations of the three active species in the underventilated flame quantitatively were appreciably smaller than those in the overventilated flame, they showed qualitatively similar distributions in both flames.

The flame front is experimentally defined mainly by the distributions of temperature and chemical species, particularly of active species. According to this definition, our simulation gives the result that the flame fronts of these two hydrogen diffusion flames are located in fairly nar-

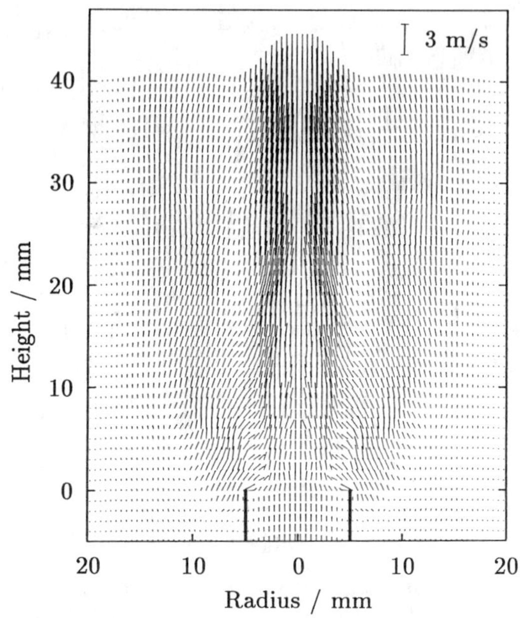

Fig. 2 Flow velocity distribution in the overventilated flame.

row radial ranges, but these extends to heights below the burner mouth in both flames.

The distribution of heat-release rate in the overventilated flame is shown in Fig. 4. In both flames, the predominant heat-releasing regions, where the rates exceed 100 $MJ/m^3 \cdot s$, similarly to the distributions of temperature and active species, extend to heights lower than the burner mouth and, in addition, attach to the burner wall.

The above described distributions of temperature, active species, and heat-release rates in the overventilated and underventilated flames show that both flames apparently have similar flame structures.

The radial profiles of temperature and concentration of stable species at 20 mm from the burner mouth in the overventilated flame are given in Fig. 5, and those of active species in Fig. 6. The temperature gradient in the air side is much larger than that in the fuel side. Hydrogen and oxygen are simultaneously exhausted at about 12 mm in radius. The maximum values of temperature and H_2O concentration occur almost at the same radius, although the position of the latter shifts a little inside. Hydrogen atoms peak at the same position as H_2O, while both OH radicals and O atoms have their maximum concentrations a little outside. The combustion reactions, therefore, are not necessarily taking place within an infinitely thin zone. This structure of the flame front stand for heights higher than 5 mm.

At the height of 20 mm in the underventilated flame, the flame front is located at about 16 mm in radius, and the width of the region of tem-

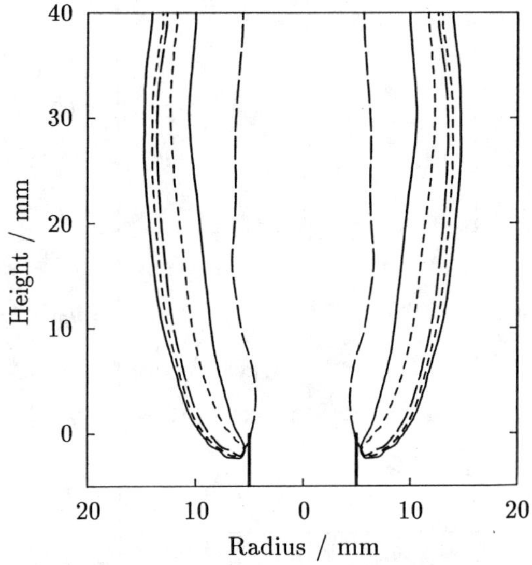

Fig. 3 Distributions of OH radicals (solid lines), H atoms (long dashed lines), and O atoms (short dashed lines) in the overventilated flame.

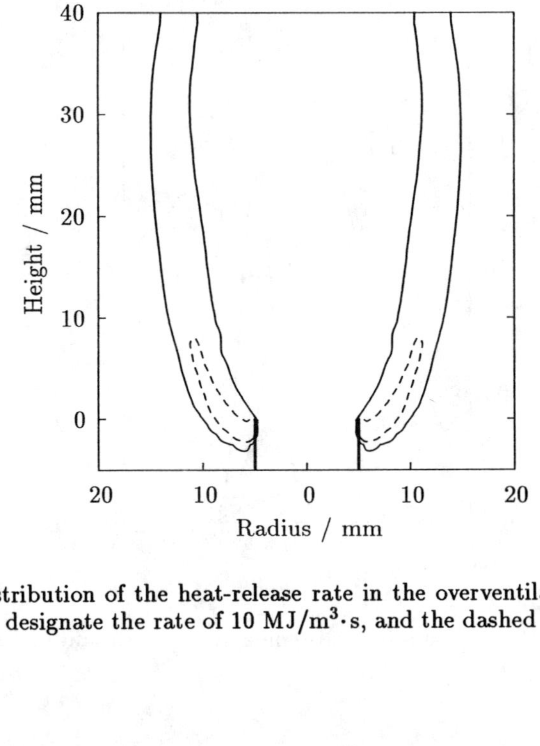

Fig. 4 Distribution of the heat-release rate in the overventilated flame. The solid lines designate the rate of 10 $MJ/m^3 \cdot s$, and the dashed lines that of 100 $MJ/m^3 \cdot s$.

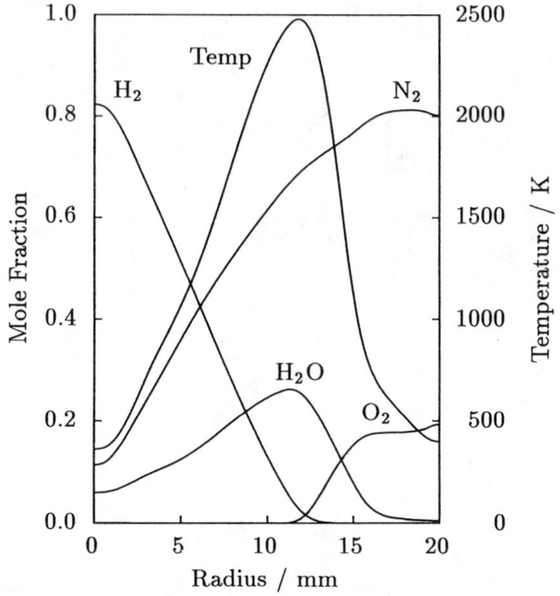

Fig. 5 Profiles of stable species and temperature at 20 mm in height in the overventilated flame.

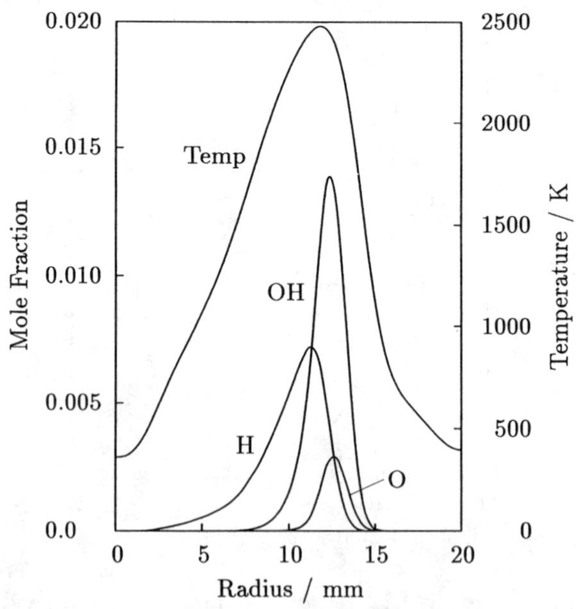

Fig. 6 Profiles of active species and temperature at 20 mm in height in the overventilated flame.

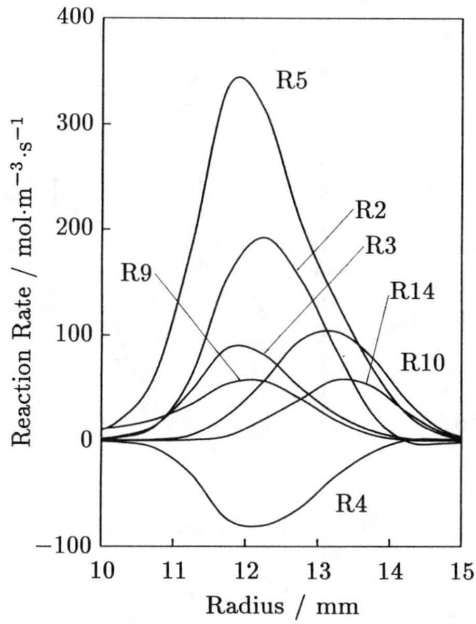

Fig. 7 Profiles of the rates of the predominant reactions and temperature at 20 mm in height in the overventilated flame. The reaction numbers in the figure correspond to those in Table 1.

perature higher than 1500 K, for example, is about 13 mm whereas that in the overventilated flame is only about 7 mm; and the active species concentrations are in general half of those in the overventilated flame. However, as a whole, the underventilated flame has almost the same flame front structure and both flames are well-developed at positions more than 5 mm above the burner.

As shown in Fig. 7, the predominant reactions in the overventilated flame at 20 mm in height are:

$$OH + H_2 \rightarrow H + H_2O \qquad (R5)$$
$$H + O_2 \rightarrow OH + O \qquad (R2)$$
$$H + O_2 + M \rightarrow HO_2 + M \qquad (R10)$$

$$O + H_2 \rightarrow OH + H \qquad (R3)$$
$$OH + H + M \rightarrow H_2O + M \qquad (R9)$$
$$OH + HO_2 \rightarrow H_2O + O_2 \qquad (R14)$$

and

$$OH + OH \rightarrow O + H_2O \qquad (R4)$$

The last reaction proceeds in the reverse direction. For convenience, the reaction zone can be considered as composed of two regions separated at the position where the local equivalence ratio equals unity: the air side

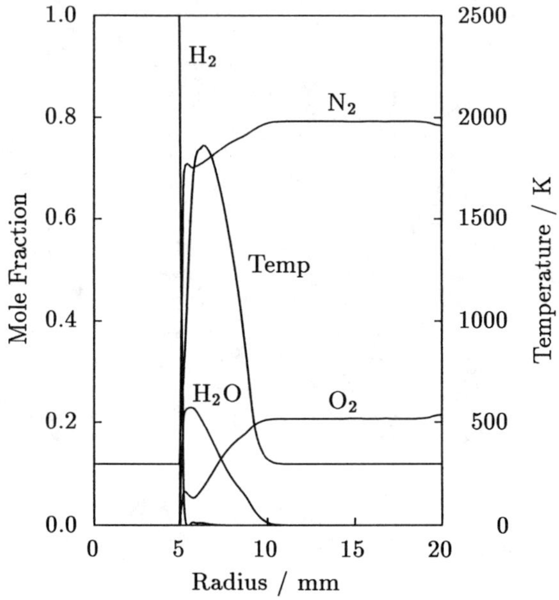

Fig. 8 Profiles of stable species and temperature at −2 mm in height in the overventilated flame.

and the fuel side. The separating position approximately corresponds to the position where reaction (R5) has maximum rate.

All the reaction rates in the hydrogen diffusion flames are lower than those in premixed hydrogen flames. In the air side, the temperature gradient is larger than that in the fuel side, and therefore, reaction (R5) has larger rate there when compared with the rate at the same temperatures in the fuel side.[14] The distributions of OH radicals and the rate of reaction (R5) suggest that the hydrogen combustion reaction proceeds effectively even at relatively low temperatures in the air side.

In the fuel side, OH radicals is supplied, in almost equal contributions, by diffusion and by reactions (R2) and (R3). Atomic and molecular oxygen must be present to activate the above two reactions and both species are supplied mostly by diffusion to this side. Consequently, the production of OH radicals and the supply of atomic and molecular oxygen into the fuel side become a necessary condition for the fuel oxidation.

On the other hand, reaction (R5) has large rates even at lower temperatures in the air side and OH radicals consumed by this reaction are produced particularly by reaction (R2). Therefore, H atoms must diffuse from the fuel side to activate the production of OH radicals, promoting also the production of H_2O.

The maximum heat-release rate at 20 mm in height is given at almost the same position of the maximum temperature and is about 15

$MJ/m^3 \cdot s$. Thermal energy is released, for the most part, within a narrow region which is almost overlapped with the region where the rate of reaction (R5) is large. The same situation was observed at the same height in the underventilated flame. The rates of the chemical reactions in the underventilated flame are smaller than those in the overventilated flame by a factor of about 0.6. However, their relative rates show almost the same profiles. Consequently, the flame fronts of both flames embody the characteristics of a typical diffusion flame at this height. Here, the Burke-Schumann model is valid although the flame fronts are not so thin and the chemical reactions take place in both sides of the position of the maximum temperature.

Temperature at the burner rim must be raised up within a small time interval to hold the flame. This seems to be the most strict requirement for steady combustion of a flame. One of the main roles of the analyses of stable combustion mechanisms is to explain how the temperature rises in that region.

In both overventilated and underventilated flames, hanging down is observed as shown in Fig. 1, for instance. Figures 8 and 9 show, respectively, the radial profiles of temperature and stable species concentrations, and those of active species at −2 mm in height in the overventilated flame; at this height, the positions of radius less than 5 mm correspond to the inside of the burner tube. These profiles are of course confined into a region narrower than that in the well-developed flame.

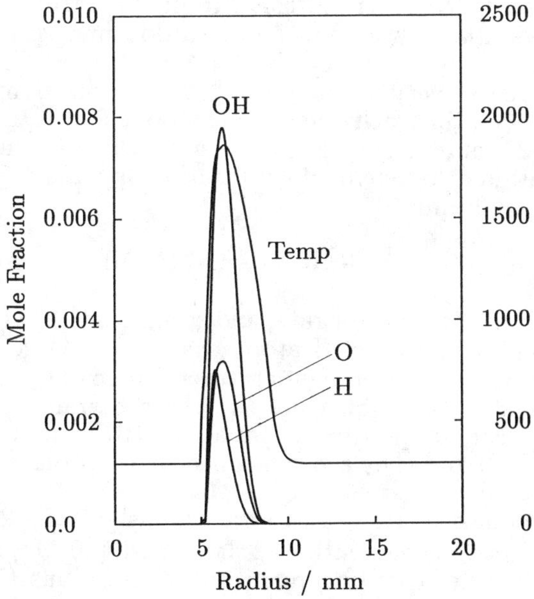

Fig. 9 Profiles of active species and temperature at −2 mm in height in the overventilated flame.

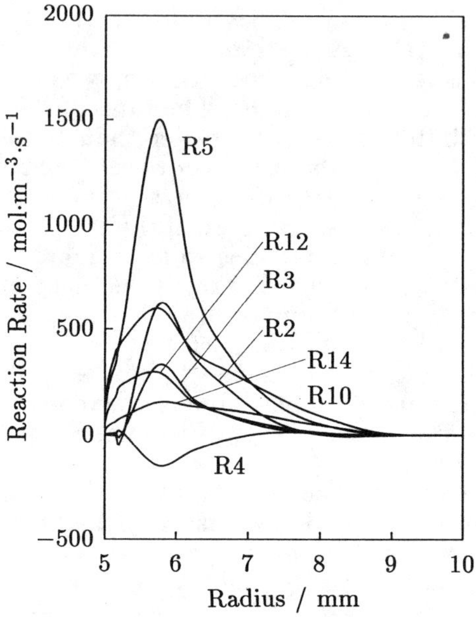

Fig. 10 Profiles of the rates of the predominant reactions and temperature at −2 mm in height in the overventilated flame. The reaction numbers in the figure correspond to those in Table 1.

Small amounts of hydrogen diffusing from the burner mouth down to this height meet the oxygen flowing upward forming premixed gas mixtures that can react in a very short time interval. The temperature, therefore, increases very sharply, so that the axial temperature gradient in the overventilated flame reaches about 950 K/mm at 6 mm in radius and −2 mm in height. Figure 10 gives the predominant reactions at that height; another elementary reaction appears in addition to the reaction set at 20 mm:

$$H + HO_2 \rightarrow OH + OH \qquad (R12)$$

Reaction (R5) has the largest rate, and reactions (R2) and (R10), which are reactions of oxidation of H atoms, have fairly large rates. Reaction (R10) is activated at lower temperatures particularly near the burner wall. Hydrogen atoms, which are the chain carriers for the reactions taking place there, are produced by reaction (R5) and, in a smaller part, by reaction (R3); and they are consumed by reactions (R2), (R10), and (R12).

The maximum heat-release rates at the height of −2 mm are about 380 $MJ/m^3 \cdot s$ in the overventilated flame and 160 $MJ/m^3 \cdot s$ in the underventilated flame, resulting mainly from reactions (R5) and (R10). These heat-release rates, being the largest through the whole flames, can hold the flames at the burner rim even under unfavorable conditions.

Concluding Remarks

A simulation model including a detailed reaction scheme for the combustion of hydrogen was constructed and applied to coaxial hydrogen–air diffusion flames. The simulation using this model gave the following conclusions on the flame structures and the reaction mechanisms of the overventilated and the underventilated flames:

1) The structures and the reaction mechanisms of the two flames do not have remarkable differences at any heights in the well-developed regions, which are attained at heights above 5 mm in both flames. The chemical reactions and the heat release are confined into narrow zones about 5 mm thick. At each transversal section of the flame, both fuel and oxidizer are consumed simultaneously at a single point, where temperature has maximum value. The Burke-Schumann theory is proved to be valid in these heights, although chemical reactions were find to take place within regions of considerable thickness.

2) At the lowest part of both flames, the fluxing air and the diffusing hydrogen, traveling in opposite directions, meet and react like premixed gas mixtures. A large amount of thermal energy is released in very short time intervals and it holds the whole flames at the burner rim.

References

[1]Burke, S. P., and Schumann, T. E. W., "Diffusion Flames," Industrial and Engineering Chemistry, Vol. 20, Oct. 1928, pp. 998-1004.

[2]Miller, J. A., and Kee, R. J., "Chemical Nonequilibrium Effects in Hydrogen–Air Laminar Jet Diffusion Flames," Journal of Physical Chemistry, Vol. 81, Dec. 1977, No. 25, pp. 2534-2542.

[3]Fukutani, S., and Jinno, H., "Simulation of Hydrogen/Air Laminar Diffusion Flame," Memoirs of the Faculty of Engineering, Kyoto University, Vol. 50, Pt. 3, July 1988, pp. 201-214.

[4]Gaydon, A. G., and Wolfhard, H. G., Flames, 4th ed., Chapman and Hall, London, 1979, Chap. 6, pp. 129-157.

[5]Baulch, D. L., Drysdale, D. D., Horne, D. G., and Lloyd, A. C., Evaluated Kinetic Data for High Temperature Reactions, Vol. 1, Butterworths, London, 1972.

[6]Warnatz, J., Combustion Chemistry, edited by W. C. Gardiner, Jr., Springer-Verlag, New York, 1984, pp. 197-360.

[7]Jensen, D. E., and Jones, G. A., "Reaction Rate Coefficients for Flame Calculations," Combustion and Flame, Vol. 32, No. 1, May 1978, pp. 1-34.

[8]Heap, M. P., Tyson, T. J., Cichanowicz, J. E., Gershman, R., and Kau, C. J., "Environmental Aspects of Low Btu Gas Combustion," Sixteenth Symposium(International) on Combustion, The Combustion Institute, Pittsburgh, 1976, pp. 535-545.

[9]Baulch, D. L., Drysdale, D. D., Duxbury, J., and Grant, S., Evaluated Kinetic Data for High Temperature Reactions, Vol. 3, Butterworths, London, 1976.

[10]Perry, R. H., and Chilton, C. H., Chemical Engineer's Handbook, 5th ed., McGraw-Hill, New York, 1973.

[11]Stull, D. R., and Prophet, H., JANAF Thermodynamical Tables, 2nd ed., U.S. Dept. of Commerce, Washington, 1971.

[12]Patankar, S. V., Numerical Heat Transfer and Fluid Flow, Hemisphere Publishing Co., New York, 1980, Chap. 3, pp. 25-40.

[13]Bird, R. B., Stewart, W. E., and Lightfoot, E. N., Transport Phenomena, John Wiley & Sons, New York, 1960, Chap. 2, p. 51.

[14]Fukutani, S., and Jinno, H., "Simulation of Hydrogen/Air Diffusion Flame," Nensho Kenkyu, Vol. 75, No. 75, Aug. 1987, pp. 1-16.

Numerical Modeling of the Structure and Properties of Tubular Strained Laminar Premixed Flames

G. Dixon-Lewis*
University of Leeds, Leeds, England, United Kingdom
and
V. Giovangigli†
Ecole Polytechnique et CNRS, Palaiseau, France
and
R. J. Kee‡ and J. A. Miller§
Sandia National Laboratories, Livermore, California
and
B. Rogg¶
University of Cambridge, Cambridge, England, United Kingdom
and
M. D. Smooke**
Yale University, New Haven, Connecticut
and
G. Stahl†† and J. Warnatz‡‡
Universität Stuttgart, Stuttgart, Germany

Abstract

The paper addresses itself to the numerical representation of the structure of tubular counterflow laminar premixed flames, with the use of complex chemistry and detailed formulation of the transport fluxes. The approach flow in such systems is radially inwards from the wall of a porous tube, and the exit flow is along the tube axis. A stoichio-

---

* Emeritus Professor, Department of Fuel and Energy.
† Research Scientist, Centre de Mathématiques Appliquées.
‡ Supervisor, Computational Mechanics Division.
§ Distinguished Member of Staff, Combustion Chemistry Division.
¶ University Lecturer, Department of Engineering.
**Professor, Department of Mechanical Engineering.
††Research Assistant, Institut für Technische Verbrennung.
‡‡Professor, Institut für Technische Verbrennung.

metric methane-air flame is chosen for investigation. The equations describing the system are reduced to a boundary value problem in one spatial dimension, and the solutions obtained by discretization and solution of the resulting algebraic equations. Both the positions and structures of the flames are calculated. Five independent groups participated in the computations, with some differences between groups in matters of detail of method of solution, and in matters of reaction mechanism and transport flux formulation. The radius of a stable tubular flame under prescribed strain conditions showed a significant, though not large, sensitivity to the reaction mechanism employed. The strain rates at extinction showed a larger sensitivity. For a fixed reaction mechanism and rate parameters, the internal heat release rate structure of a less strongly strained flame (radius 4 mm in the configuration employed) was almost indistinguishable from that of a planar counterflow flame at similar strain rate. On the other hand, the strain rates at extinction were considerably different. A second computation related to a 5.59% methane-air mixture flowing radially inwards from a porous tube of 15 mm diameter. The computed structure of this flame is compared with experiment.

Aspects of the fluid mechanics of both flames are discussed.

Introduction

Laminar flamelet models of turbulent combustion have received considerable attention recently because of their power to simulate chemical kinetic and molecular transport effects in turbulent flames.[1] These models are appealing because they decouple the turbulent flowfield problems from the detailed flame structure computation: the only way the turbulent flowfield affects the flame structure is through parameters such as the scalar dissipation rate, or mixture fraction, or geometric effects like flame curvature. Central to this approach is the compilation of "libraries" of flame structures in well-defined laminar flame configurations, which must introduce the effects of these parameters in ways that can be clearly and unambiguously defined. In the case of planar flames a series of studies has already been made. Planar counterflow diffusion flames, and more recently counterflow premixed flames, have been studied both experimentally[2,3] and computationally.[4-8]

An alternative configuration is the tubular flame. Here the premixed reactants flow radially inward through the porous wall of a long tube. An annular flame may be stabilized in the tube, and the products of reaction flow radially out at the ends. The overall flow configuration is shown in Fig. 1. The configuration introduces both the

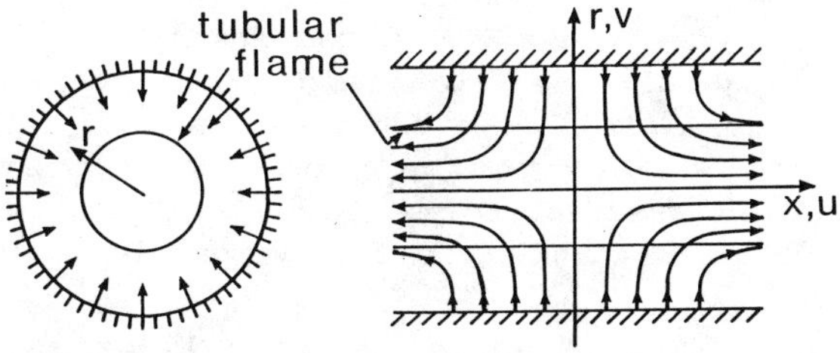

Fig. 1 Flow configuration.

effects of stretch and flame curvature. Ishizuka[9] and Kitano and Kobayashi[10] have determined extinction limits for such flames in methane-air mixtures, and Takeno et al.[11-13] have employed simplified kinetic models to analyze the flames. The present paper is concerned with the detailed numerical representation of two such methane-air flames, with realistic chemical kinetics and molecular transport formulation. The paper summarizes results obtained by several groups working independently. The major part of the results was initially discussed at a workshop entitled "Numerical Calculation of the Structure and Properties of Tubular Strained Laminar Premixed Flames," held at the University of Leeds in April, 1988.

Formulation of Problem

In the steady state, axisymmetric flow of Fig. 1, the continuity condition

$$\frac{\partial}{\partial x}(r\rho u) + \frac{\partial}{\partial r}(r\rho v) = 0 \tag{1}$$

is satisfied exactly if a stream function $\psi(x,r) = xV(r)$ is defined so that

$$-\frac{\partial \psi}{\partial r} = r\rho u = -x\frac{dV}{dr} \tag{2}$$

$$\frac{\partial \psi}{\partial x} = r\rho v = V \tag{3}$$

It follows that the radial velocity v is a function only of r, and the axial velocity u varies linearly with x. We seek solutions such that the temperature T, the species mass fractions Y_k, and the density ρ are functions of r alone, whence it follows that the axial transport fluxes are zero, and that the transport coefficients (viscosity μ, thermal conductivity λ, and diffusion coefficients D_{ik}) are also functions only of r. We also make a small Mach number approximation, in which the thermodynamic pressure p remains constant throughout the flowfield, but pressure gradient terms still remain in the momentum equation. Finally, body forces are neglected. In the end we encounter no contradictions to our assumed forms for T, Y_k, and ρ, so we may regard the solutions as exact for compatible sets of boundary conditions.

The Navier-Stokes equations in cylindrical coordinates are stated as

$$\rho u\frac{\partial u}{\partial x} + \rho v\frac{\partial u}{\partial r} = -\frac{\partial p}{\partial x} + \frac{\partial}{\partial x}\left[\mu\left\{\frac{4}{3}\frac{\partial u}{\partial x} - \frac{2}{3r}\frac{\partial}{\partial r}(rv)\right\}\right] + \frac{1}{r}\frac{\partial}{\partial r}\left[\mu r\left(\frac{\partial v}{\partial x} + \frac{\partial u}{\partial r}\right)\right] \tag{4}$$

$$\rho u\frac{\partial v}{\partial x} + \rho v\frac{\partial v}{\partial r} = -\frac{\partial p}{\partial r} + \frac{\partial}{\partial r}\left[\mu\left\{2\frac{\partial v}{\partial r} - \frac{2}{3}\frac{\partial u}{\partial x} - \frac{2}{3r}\frac{\partial}{\partial r}(rv)\right\}\right] + \frac{\partial}{\partial x}\left[\mu\left(\frac{\partial v}{\partial x} + \frac{\partial u}{\partial r}\right)\right] + \frac{2\mu}{r}\left(\frac{\partial v}{\partial r} - \frac{v}{r}\right) \tag{5}$$

Introducing the above assumptions and stream function, and then isolating the pressure gradients, Eqs. (4) and (5) become

$$\frac{1}{x}\frac{\partial p}{\partial x} = -\frac{1}{\rho r^2}\left(\frac{dV}{dr}\right)^2 + \frac{V}{r}\frac{d}{dr}\left(\frac{1}{\rho r}\frac{dV}{dr}\right) - \frac{1}{r}\frac{d}{dr}\left[\mu r\frac{d}{dr}\left(\frac{1}{\rho r}\frac{dV}{dr}\right)\right] \tag{6}$$

$$\frac{\partial p}{\partial r} = f(r) \tag{7}$$

where $f(r)$ is a long expression analogous to the r.h.s. of Eq. (6). From Eq. (6), the pressure gradient term $(1/x)(\partial p/\partial x)$ is also a function only of r. Since Eq. (7) implies that the further derivatives $\partial^2 p/\partial r\partial x$ and $\partial/\partial r\{(1/x)\partial p/\partial x\}$ are both zero, the result is obtained that, for specific boundary conditions, $(1/x)\partial p/\partial x$ is a constant eigenvalue of the problem. If we denote this eigenvalue by J, then after multiplying Eq. (6) by ρr^2, the u-momentum equation becomes

$$\left(\frac{dV}{dr}\right)^2 - \rho r V\frac{d}{dr}\left(\frac{1}{\rho r}\frac{dV}{dr}\right) + \rho r\frac{d}{dr}\left[\mu r\frac{d}{dr}\left(\frac{1}{\rho r}\frac{dV}{dr}\right)\right] + \rho r^2 J = 0 \tag{8}$$

The problem has thus been reduced to a single third-order differential equation boundary value problem. For numerical implementation, it is convenient additionally to convert Eq. (8) into a system of two equations by defining a new variable F, where

$$F(r) = -\frac{1}{\rho r}\frac{dV}{dr} \tag{9}$$

The additional substitution leads to Eqs. (17) and (18) below.

With parallel assumptions to the above, the energy equation, neglecting viscous dissipation and pressure work, is

$$\rho v\frac{dh}{dr} = -\frac{1}{r}\frac{d}{dr}\{r\,(Q_D + Q_T)\} \tag{10}$$

and the species conservation equations are

$$\rho v\frac{d\sigma_i}{dr} = -\frac{1}{r}\frac{d}{dr}\left(\frac{rj_i}{m_i}\right) + R_i \qquad (i = 1, 2, \ldots, N) \tag{11}$$

Here ρ is the density, h is the specific enthalpy of the mixture, $\sigma_i = Y_i/m_i$ where m_i is the molecular mass of the i-th component, and R_i is the net molar rate of formation of the species, presumed to result from a system of elementary reactions that proceed according to the law of mass action. The enthalpy h, the pressure p, the diffusive fluxes j_i, and the energy transport fluxes Q_D and Q_T are given by

$$h = \sum_{i=1}^{N}(\sigma_i H_i) \tag{12}$$

$$p = \rho RT\sum_{i=1}^{N}(\sigma_i) \tag{13}$$

$$j_i = \frac{\rho m_i}{M^2}\sum_{k=1}^{N} m_k D_{ik}\frac{dX_k}{dr} - \frac{D_i^T}{T}\frac{dT}{dr} \tag{14}$$

$$Q_D = \sum_{i=1}^{N}\frac{j_i H_i}{m_i} \tag{15}$$

$$Q_T = -\lambda\frac{dT}{dr} - \sum_{i=1}^{N}\frac{RT}{m_i X_i}D_i^T\frac{dX_i}{dr} \tag{16}$$

where the molar enthalpies H_i are polynomial functions of temperature, X denotes mole fraction, and M is the mean molecular mass.

The complete set of equations to be solved becomes

$$\frac{dV}{dr} + \rho r F = 0 \tag{17}$$

$$\rho r V \frac{dF}{dr} = \rho r \frac{d}{dr}\left(\mu r \frac{dF}{dr}\right) - \rho r^2 J - (\rho r F)^2 \tag{18}$$

$$V \frac{dh}{dr} = - \frac{d}{dr}\{r\,(Q_D + Q_T)\} \tag{19}$$

$$\frac{V}{r}\frac{d\sigma_i}{dr} = -\frac{1}{r}\frac{d}{dr}\left(\frac{rj_i}{m_i}\right) + R_i \qquad (i = 1, 2, \ldots, N) \tag{20}$$

with boundary conditions (22) at the wall of the tube (radius R) and the symmetry conditions (21) at the tube axis

$$r = 0: V = 0,\ dF/dr = 0,\ d\sigma_i/dr = 0,\ dh/dr = 0 \tag{21}$$

$$r = R: V_R = \rho_R R v_R,\ F = 0,\ V_R\,(G_i/m_i - \sigma_i) = R\,(j_i/m_i)_R,\ T = T_R \tag{22}$$

Here G_i denotes the mass fraction of the i-th component in the input stream. In most cases of practical importance the diffusive fluxes $(j_i/m_i)_R$ at the wall are negligible, and $\sigma_{i,R} = G_i/m_i$.

Methods of Solution and Input Data

The system of differential Eqs. (17-20) is solved by an implicit finite difference or finite domain procedure. The resulting set of algebraic equations is solved either directly by a steady state, modified, damped Newton algorithm, or by a time dependent approach. Further, if in the former case the Newton algorithm fails to converge, the solution estimate is conditioned by a period of time integration. This provides a new starting point for the Newton algorithm that is closer to the solution and, thus, more likely to be in the region of convergence. The initial discretization is on a coarse grid. After obtaining a solution on the coarse grid, new grid points are added adaptively in regions of high spatial activity. In addition, one grid point is placed at a location very close to the cylinder axis, so as to meet the symmetry conditions at the axis without forcing the flame. Interpolation from coarse grids on to the finer ones gives the initial estimates for the solutions on the fine grids. Grid points are added until the solution is resolved to sufficient accuracy.

The specific approaches employed by the various groups were as follow:

1) G. Dixon-Lewis (Leeds): Solution by finite domain method,[14] with micro-integration over computational cells on

the assumption of linear variation of dependent variables with ξ between nodes, where $\delta\xi = \rho r \delta r$. Enthalpy and species by block implicit solution of time-dependent equations, F by Newton method. Eighteen species, 45 reaction mechanism,[15,16] with C_2 hydrocarbons truncated at ethylene stage. Multicomponent diffusion and thermal diffusion coefficients, and thermal conductivity, by detailed calculation.[17,18] Viscosity of mixtures by Buddenberg and Wilke[19] formula.

2) R.J. Kee and J.A. Miller (Sandia): Second order, conservative finite difference approximations to discretize the conservation equations. Solution by combination of Newton and time-dependent approaches.[20-23] Extinction limit by arc-length continuation method.[24-27] Reaction mechanism involves 134 reactions and 29 species.[28] Multicomponent diffusion velocities, with thermal diffusion included.[17,18] Mixture viscosities by Wilke's formula.

3) B. Rogg (Cambridge): Hybrid finite difference approach[29,30] with adaptive gridding. Newton's method for solution of steady state equations, preceded by use of an extrapolation method for the time-dependent problem (necessary initially to bring the solution into the domain of convergence of Newton). Reaction mechanism involved C_1 chemistry with 16 reacting species plus N_2.[see Ref. 31] Detailed transport model, with thermal diffusion included.[32]

4) M.D. Smooke (Yale) and V. Giovangigli (Palaiseau): Initial calculation used a finite difference method in which adaptive time steps were used to bring solution into the convergence domain of Newton's method. Subsequent calculations used a phase-space, pseudoarclength continuation method coupled with Newton-like iterations and global adaptive gridding.[33] Method allows any solution component to be used in the reparameterization. Global adaptive gridding procedure enables the mesh to follow the flame front as it moves in the computational domain. All thermodynamic and transport quantities, and chemical production rates, were computed with use of highly vectorized subroutine libraries.[34] Reaction mechanism included 79 reactions with 26 chemical species.[35] All the computations except one set of those reported employed the forward rate expression $k_2 = 2.0\times10^{14}\exp(-8400/T)$ cm^3 mol^{-1} s^{-1} for the reaction (*ii*) $H + O_2 = OH + O$. The remaining set (see Table 2 and Fig. 3) employed $k_2 = 5.13\times10^{16}\,T^{-0.816}\exp(-8253/T)$.

5) G. Stahl and J. Warnatz (Stuttgart): Solution of differential/algebraic system by the package LIMEX due to Deuflhard and Nowak.[36] Uses extrapolation method for solution of time-dependent problem, internal time step determination, and semi-implicit Euler discretization. Three chemical reaction mechanisms, comprising 1) a full mechanism

with 228 reactions of 31 species, including C_3 species[37]; 2) a mechanism of 103 reactions of 19 species (C_1 species only); and 3) a mechanism as in (1), but with a changed rate coefficient for the reaction $CHO + M \rightarrow CO + H + M$. Binary mixture approximation for transport model. Thermal diffusion included.

Results

Tubular Stoichiometric Methane-Air Flame

Extinction limits for tubular strained methane-air flames have been measured by Kobayashi and Kitano.[10] The flame radius decreases with increasing radial injection velocity v_R, until extinction is observed. In a tube of radius 8 mm, with tube surface temperature T_R = 300 K, extinction occurs for the stoichiometric mixture at an inlet velocity v_R = -470 cm s^{-1}.

Initial computations for this system were performed for a pressure curvature eigenvalue J = -0.339 MN m^{-4}. Figure 2 compares the radial velocity profiles and temperature profiles computed by groups A, D and E. Table 1

Table 1 Computed inlet velocities, maximum temperatures, flame positions, and maximum species mole fractions in tubular strained stoichiometric methane-air flame with J = -0.339 MN m^{-4}, in tube of radius 8 mm.

	Group				
	A	B	C	D	E
v_R/ (cm s^{-1})	-110	-107	-120	-105	-101
$\|v\|_{max}$/ (cm s^{-1})	236	-	-	241	253
T_{max}/ (K)	2170	2129	2153	2156	2181
Radius of $CH_{3,max}$/ (mm)	4.35	3.75	3.27	4.51	4.75
Maximum mole fractions:					
10^2 CO	5.24	5.12	4.34	4.95	4.68
10^2 H_2	2.02	2.10	1.57	1.98	1.82
10^3 H	8.64	7.65	6.21	7.72	7.58
10^3 O	3.92	3.38	2.93	3.71	3.55
10^3 OH	8.06	7.59	7.21	7.39	7.73
10^3 CH_3	3.38	2.39	4.54	1.98	3.15
10^3 CH_2O	1.41	1.04	1.96	2.74	1.75
10^4 HO_2	1.29	1.54	1.28	2.06	1.97
10^5 H_2O_2	-	3.64	1.21	-	2.80
10^5 CHO	11.9	5.54	2.11	8.40	2.31
10^5 CH_2	-	6.43	1.63	0.32	1.62
10^6 CH	-	1.78	7.89	21.5	9.43
10^6 C_2H	-	-	-	0.80	1.19

records, for all groups, the inlet velocities at the tube wall, the maximum temperatures achieved, the flame positions as given by the CH_3 maximum, and the peak mole fractions of all intermediate species. In both Fig. 2 and Table 1, the results of group E are for their mechanism 1. Agreement between the groups is satisfactory, particularly in respect to the results shown in Fig. 2. However, it should be added that the results obtained by group E with use of their mechanisms 2 and 3 showed flame positions 0.85 mm and 1.57 mm closer to the tube axis than with mechanism 1.

Augmentation of the injection velocity v_R increases the strain rate in the system. When this occurs, the flame moves towards the tube axis, the maximum temperature falls, and eventually the flame is extinguished at sufficiently high velocities. The computed injection velocities and the corresponding eigenvalues J at extinction are recorded in Table 2; and the upper forks of curves 1 and 2 in Fig. 3 show the computed maximum temperatures corresponding with the two reaction kinetic conditions studied by group D. The pairs of results obtained by both groups D and E show a very appreciable sensitivity of the extinction limit to the assumed reaction mechanism and rates. The mechanisms that

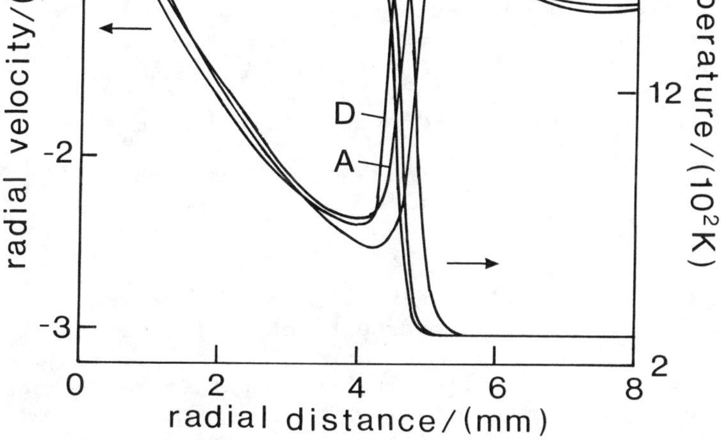

Fig. 2 Temperature profiles and radial velocity profiles for flame of Table 1, as computed by groups A, D, and E.

Table 2 Computed injection velocities and "pressure curvature" eigenvalues J at extinction, for tubular strained stoichiometric methane-air flame in tube of radius 8 mm.

	Group				
	A	B	C	D	E
J_{ext}/ (MN m^{-4})	-	-1.68	-	-2.1[a]	-3.5[b]
$v_{R,ext}$/ (cm s^{-1})	-	-285	-	-	-435

[a] Value obtained with use of $k_2 = 2.0\times10^{14} \exp(-8400/T)$ cm^3 mol^{-1} s^{-1}, where reaction *(ii)* is $H + O_2 = OH + O$. With $k_2 = 5.13\times10^{16} T^{-0.816} \exp(-8253/T)$, J_{ext} = -0.9 MN m^{-4}.

[b] Value quoted is for mechanism 1. For mechanism 2, J_{ext} = -2.5 MM m^{-4}.

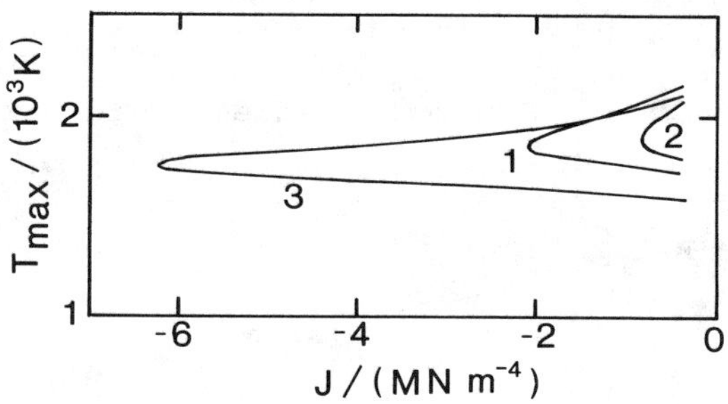

Fig. 3 Computed maximum temperatures in strained stoichiometric methane-air flames (group D).
Curve 1: Tubular flame, $k_2 = 2.0\times10^{14} \exp(-8400/T)$.
Curve 2: Tubular flame, $k_2 = 5.13\times10^{16} T^{-0.816} \exp(-8253/T)$.
Curve 3: Planar flame, k_2 as for curve 1.
Reaction *(ii)* is $H + O_2 = OH + O$. Units of k_2 are cm^3 mol^{-1} s^{-1}.

lead to smaller flame radii at the fixed eigenvalue of Table 1 also give the lower extinction limits.

Also shown in Fig. 3 (curve 3) are the computed maximum temperatures achieved in the planar Cartesian configuration arising from initially nondivergent discharges from nozzles separated by a distance of 16 mm. Extinction occurs only at a considerably higher strain rate than that computed for the tubular flame.

Tubular Strained Methane-Air Flame Near the Lean Flammability Limit

Sakai and Ishizuka[38] have examined the location and structure of a strained flame supported by a 5.59% methane-air mixture in a tube of 15 mm radius. The normal injection velocity at the tube wall, v_R, was -10 cm s^{-1}. The flame itself was not perfectly cylindrical, but had a somewhat triangular cross section; and to enhance the stability and symmetry of the flame, a small disc with a hole in its centre was placed on one end of the cylinder. Species profiles were measured by sampling and gas chromatography, and the temperature profile by use of a fine silica-coated Pt/Pt-13%Rh thermocouple (wire diameter 0.1 mm). The measured temperatures are shown by the points in Fig. 4.

Computations for these conditions were performed by groups C and D. Both agree on the computed position of the flame, the calculated temperature profile of which is shown by the line 1 in Fig. 4. Repetition of the computation with an inlet velocity v_R = -25 cm s^{-1} led to the temperature profile shown by line 2. The "luminous zone" of the latter flame moves to approximately the measured position, but the computed profile is much steeper than the measured one. Indeed, the two computed temperature profiles are virtually

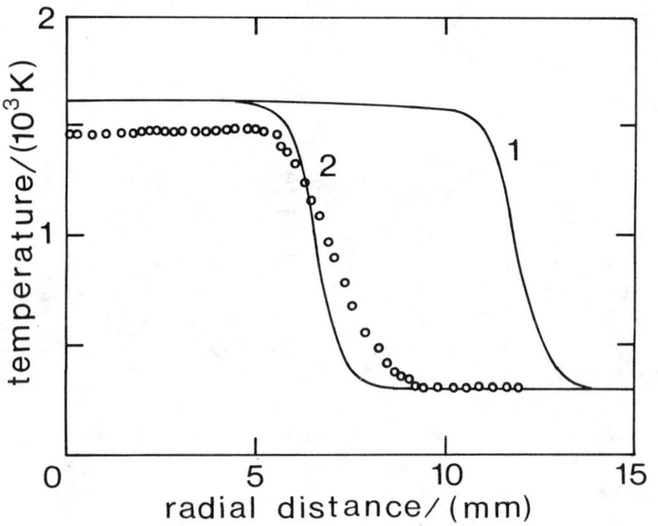

Fig. 4 Temperature profiles for tubular 5.59% methane-air flame in tube of radius 15 mm. Points represent measurements of Sakai and Ishizuka[38] for injection velocity v_R = -10 cm s^{-1}. Line 1: computation for v_R = -10 cm s^{-1}; line 2: computation for v_R = -25 cm s^{-1}.

identical in shape, suggesting that under the conditions of their comparison, and in the absence of heat losses, the flame position is principally determined by an almost curvature-independent "burning velocity" just short of 10 cm s^{-1}. The measured temperature profile corresponds with a "burning velocity" less than this.

It is noteworthy in this context that the "uncorrected" measured burnt gas temperature is some 170 K below the computed value. At the temperature in question it is unlikely that the whole of this shortfall can be attributed to radiative losses from the fine thermocouple; and any remaining shortfall after a radiation correction has been applied will lead to a reduction of the effective "burning velocity." The measurements are to this extent consistent with the computation; but the explanation for such a further shortfall remains open to question. Unaccounted for heat losses, or maybe incorrectly accounted for selective diffusion effects, are possibilities.

Discussion

Additional Features of the Counterflow Geometry

As a preliminary to discussion we note, first, that the transformation of Eqs. (17) and (18) to the density-weighted coordinate $\xi = \int_0^r \rho r dr$ leads to

$$\frac{dV}{d\xi} + F = 0 \tag{23}$$

$$V\frac{dF}{d\xi} = \frac{d}{d\xi}\left(\rho\mu r^2 \frac{dF}{d\xi}\right) - \frac{J}{\rho} - F^2 \tag{24}$$

and secondly that, by Eq. (2), the definition of F implies $F = u/x$, that is, F is a local axial velocity gradient which may vary with radial position in the flowfield. Next, the formulation of the boundary value problem, as outlined in "Formulation of Problem", is exactly analogous to that used by Kee et al.[39] for the planar axisymmetric strained flame. In the latter case the flows of Fig. 1 are reversed in direction, and the proous tube is, of course, absent. A stream function $\psi(x,r) = r^2 U(x)$ is defined so that

$$\frac{\partial\psi}{\partial r} = r\rho u = 2rU \tag{25}$$

$$-\frac{\partial\psi}{\partial x} = r\rho v = -r^2 U(x) \tag{26}$$

and the analogues of Eqs. (23) and (24) become

$$\frac{dU}{d\zeta} + \overline{F} = 0 \tag{27}$$

$$2U\frac{d\overline{F}}{d\zeta} = \frac{d}{d\zeta}\left(\rho\mu\frac{d\overline{F}}{d\zeta}\right) - \frac{\overline{J}}{\rho} - \overline{F}^2 \tag{28}$$

where now $\zeta = \int_0^x \rho dx$. Here $\overline{J}$ is an eigenvalue of $(1/r)(\partial p/\partial r)$, and the definition of $\overline{F}$ implies $\overline{F} = v/r$, a local radial velocity gradient.

In parallel with the local velocity gradients, the "pressure curvature" eigenvalues J or $\overline{J}$ can be associated at density ρ with "fully developed" potential flow velocity gradients $a = (-J/\rho)^{1/2}$ or $\overline{a} = (-\overline{J}/\rho)^{1/2}$. We note 1) that these gradients also are functions of the local conditions; and 2) that equality of a with F, or $\overline{a}$ with $\overline{F}$, implies zero overall contribution of the derivative terms in Eqs. (24) or (28). As has already been pointed out by Kee et al.,[39] the proper characterization of a counterflow flame system requires

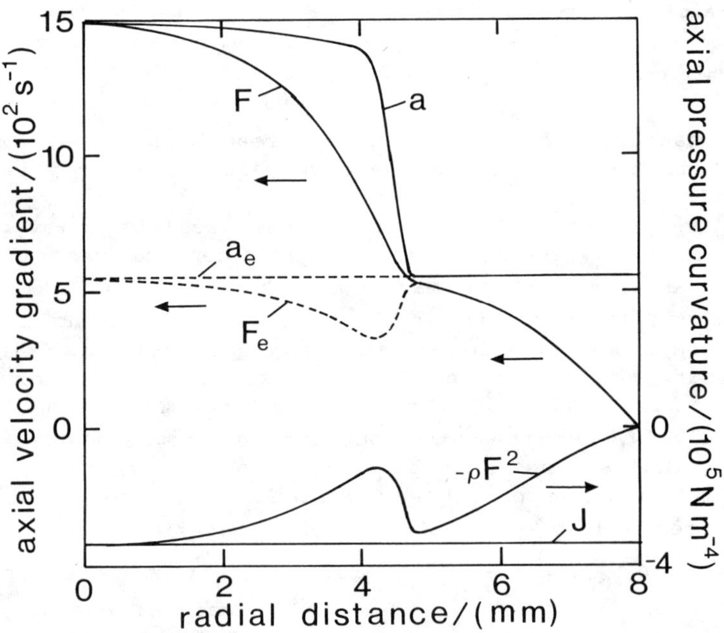

Fig. 5 Graphic representation of several fluid mechanical properties associated with flame of Table 1. For explanation of symbols, see first paragraph under "Discussion" in text.

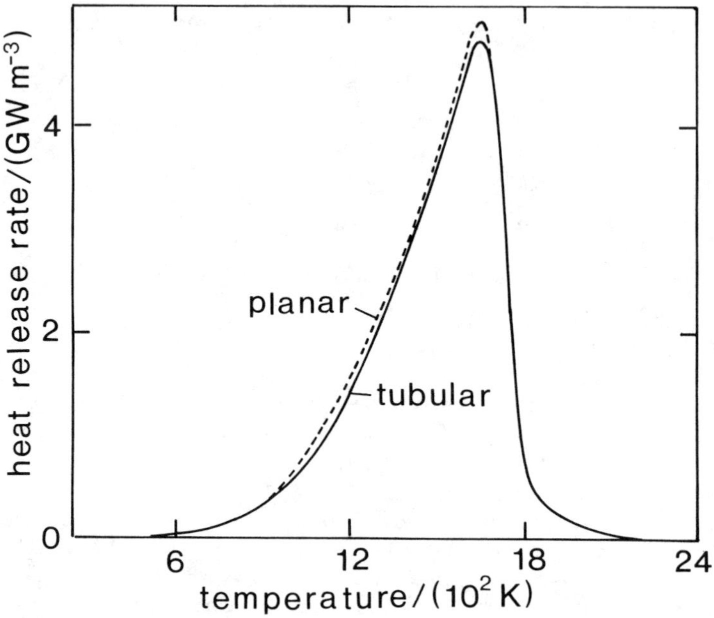

Fig. 6 Heat release rates in counterflow, laminar, stoichiometric methane-air flames. Continuous line: tubular flame of Table 1, with eigenvalue velocity gradient a_e = 549 s^{-1}; dotted curve: planar counterflow flame with $\bar{a}_e$ = 500 s^{-1} and free stream outer boundary condition $\bar{F}_e = \bar{a}_e$.

specification of F or $\bar{F}$ at its upstream boundaries, as well as information from which the eigenvalue J or $\bar{J}$ may be calculated.

As a converse to the above, the local velocity gradient F or $\bar{F}$ may be considered in the general case to contribute a component of magnitude $-\rho F^2$ to J, or $-\rho\bar{F}^2$ to $\bar{J}$, with the difference $(J+\rho F^2)$ or $(\bar{J}+\rho\bar{F}^2)$ providing for the "acceleration" and viscous terms in Eqs. (24) and (28). To illustrate the situation for the case of the tubular flame of Fig. 2 and Table 1, the velocity gradients a and F, the eigenvalue J, and the contributions $-\rho F^2$ to J, are shown by the continuous lines in Fig. 5.

Yet another possible formulation replaces J or $\bar{J}$ by the corresponding eigenvalue velocity gradient $a_e = (-J/\rho_e)^{1/2}$ or $\bar{a}_e = (-\bar{J}/\rho_e)^{1/2}$, in each of which ρ_e is the density of the unburnt gas entering the flame. The partial characterization of strained laminar flames in terms of an eigenvalue velocity gradient under such standard density conditions has been widely used, particularly in discussions relating to turbulent combustion. It implies $a_e = a(\rho/\rho_e)^{1/2}$, or

$\bar{a}_e = \bar{a}(\rho/\rho_e)^{1/2}$. The value of a_e is shown as the upper dotted line in Fig. 5. The parallel transformation $F_e = F(\rho/\rho_e)^{1/2}$ leads to the lower dotted line. The distinction between F and a, and between F_e and a_e, is apparent.

Comparison Between Properties of Tubular and Planar Flames

A major objective of the study of the tubular configuration is an examination of the effect of curvature on the flame structure and properties. We specifically examine the following two features.

Heat Release Rate Profiles

The tubular flame of Fig. 2 and Table 1 has an eigenvalue velocity gradient a_e = 549 s^{-1}, for ρ_e = 1.142 kg m^{-3}. Its velocity gradient profiles are shown in Fig. 5, for the outer (wall) boundary condition F = 0.

The continuous line in Fig. 6 shows the heat release rates for this flame, plotted agaonst temperature. The dotted line shows the similar profile for the symmetric back-to-back planar flame in a Cartesian geometry, with a_e = 500 s^{-1}, but with the free stream potential flow boundary condition $\bar{F} = \bar{a}_e$. Apart from an increasing truncation at the hot end of flames with higher velocity gradients, the heat release rate profiles for such planar flames are themselves remarkably independent of the velocity gradient and the outer boundary condition. The curves of Fig. 6 show that, at least for flame radii of 4 mm or more under the present conditions, the curvature also has little effect.

In the same context, it is interesting to note further from Fig. 2 that the *radial* velocity gradient in the cold gas just ahead of the tubular flame reaches a value of about 500 s^{-1}, not far short of the local *axial* gradient of 520 s^{-1} (Fig. 5). Equality of the two gradients is a characteristic of the Cartesian rather than the planar axisymmetric counterflow geometry. The tubular flame results again exhibit a local fluid mechanical behaviour reminiscent of the planar Cartesian flame.

On the hot sides of the planar and tubular flames, the maximum temperatures are 2170 and 2119 K, at distances of 4.3 and 2.1 mm respectively from the positions of maximum heat release rate (group A).

Extinction Limits of Stoichiometric Methane-Air Flame

The behaviour at the extinction limit contrasts strongly with that just described. Thus, the limits computed by group D for the tubular and planar flames under otherwise identical conditions are at J = -2.1 and -6.2 MN m^{-4} (curves 1 and 3 of Fig. 3). This rather dramatic change has still to be investigated. It may be caused by changes in the flame profiles themselves, or it may be due to fluid mechanical effects such as, for example, an enhanced sensitivity to viscous effects in the tubular as compared with the planar flame boundary layer.

Conclusions

The more important conclusions from the present study are thus

1) The equations describing the tubular counterflow premixed flame may be reduced to a one-dimensional boundary value problem, and a stationary solution obtained for realistic chemical kinetic and transport flux formulations. The systems are characterized by an eigenvalue pressure curvature in the axial (exit) direction, the magnitude of which depends on the outer wall boundary conditions. For purposes of characterization, the pressure curvature may be replaced by an eigenvalue velocity gradient. However, the latter is not the same as the (measurable) local velocity gradient, which varies with radial position.

2) The computed extinction limits for the stoichiometric methane-air flame show a marked sensitivity to details of the reaction kinetic scheme. The radial position of a less strongly strained flame shows a significant but reduced sensitivity. However, a full sensitivity analysis has not yet been undertaken.

3) For the same flame, and for otherwise identical conditions, the computed extinction eigenvalue pressure curvature for the tubular counterflow geometry is only some 35% of that for a planar geometry.

4) At a radius of approximately 4 mm, the heat release rate profile of the tubular stoichiometric methane-air flame with an eigenvalue pressure curvature J = -0.339 MN m^{-4} (eigenvalue velocity gradient 550 s^{-1}) is almost indistinguishable from that of the corresponding counterflow planar flame at a strain rate of 500 s^{-1}, both profiles plotted against temperature as the independent variable.

Acknowledgments

We thank Professor T. Takeno for his advice and encouragement in setting up the workshop problem; Dr. Y. Sakai and Professor S. Ishizuka for permission to use their tubular flame results prior to publication; and Professor S. Fukutani, Dr. H. Kobayashi, and Professor M. Kitano for further valuable contributions and discussion.

References

[1]Peters, N., "Laminar Flamelet Concepts in Turbulent Combustion," *Proceedings of the Twenty-First Symposium (International) on Combustion*, The Combustion Institute, Pittsburgh, PA, 1988, pp. 1231-1250.

[2]Tsuji, H., and Yamaoka, I., "The Counterflow Diffusion Flame in the Forward Stagnation Region of a Porous Cylinder," *Proceedings of the Eleventh Symposium (International) on Combustion*, The Combustion Institute, Pittsburgh, PA, 1967, pp. 979-984.

[3]Wu, C.K., and Law, C.K., "On the Determination of Laminar Flame Speeds from Stretched Flames," *Proceedings of the Twentieth Symposiun (International) on Combustion*, The Combustion Institute, Pittsburgh, PA, 1985, pp. 1941-1949.

[4]Dixon-Lewis, G., David, T., Gaskell, P.H., Fukutani, S., Jinno, H., Miller, J.A., Kee, R.J., Smooke, M.D., Peters, N., Effelsberg, E., Warnatz, J., and Behrendt, F., "Calculation of the Structure and Extinction Limit of a Methane-Air Counterflow Diffusion Flame in the Forward Stagnation Region of a Porous Cylinder," *Proceedings of the Twentieth Symposium (International) on Combustion*, The Combustion Institute, Pittsburgh, PA, 1985, pp. 1893-1904.

[5]Giovangigli, V., and Smooke, M.D., "Calculation of Extinction Limits for Premixed Laminar Flames in a Stagnation Point Flow," *Journal of Computational Physics*, Vol. 68, February 1987, pp. 327-345.

[6]Stahl, G., Warnatz, J., and Rogg, B., "Simulation of Stretched Premixed CH_4-Air and C_3H_8-Air Flames with Detailed Chemistry," *Dynamics of Reactive Systems, Part I*, edited by A.L. Kuhl, J.R. Bowen, J.-C. Leyer, and A. Borisov, *(Progress in Astronautics and Aeronautics, Vol. 113)*, A.I.A.A., Washington, D.C., 1988, pp. 195-214.

[7]Dixon-Lewis, G., "Structure and Extinction Limits of Some Strained Premixed Flames," *Dynamics of Reactive Systems, Part I*, edited by A.L. Kuhl, J.R. Bowen, J.-C. Leyer, and A. Borisov, *(Progress in Astronautics and Aeronautics, Vol. 113)*, A.I.A.A., Washington, D.C., 1988, pp. 166-183.

[8]Dixon-Lewis, G., and Missaghi, M., "Structure and Extinction Limits of Counterflow Diffusion Flames of Hydrogen-Nitrogen Mixtures in Air," *Proceedings of the Twenty-Second Symposium (International) on Combustion*, The Combustion Institute, Pittsburgh, PA, 1989, pp. 1461-1470.

[9]Ishizuka, S., "On the Behaviour of Premixed Flames in a Rotating Flow Field: Establishment of Tubular Flames," *Proceedings of the Twentieth Symposium (International) on Combustion*, The Combustion Institute, Pittsburgh, PA, 1985, pp. 287-294.

[10]Kobayashi, H., and Kitano, M., "Extinction Characteristics of a Stretched Cylindrical Premixed Flame," *Combustion and Flame*, Vol. 76(3), 1989, pp. 285-295.

[11]Takeno, T., and Ishizuka, S., "A Tubular Flame Theory," *Combustion and Flame*, Vol. 64(1), 1986, pp. 83-98.

[12]Takeno, T., Nishioka, M., and Ishizuka, S., "A Theoretical Study of Extinction of a Tubular Flame," *Combustion and Flame*, Vol. 66(3), 1986, pp. 271-283.

[13]Nishioka, M., Takeno, T., and Ishizuka, S., "Effects of Variable Density on a Tubular Flame," *Combustion and Flame*, Vol. 73(3), 1988, pp. 287-301.

[14]Dixon-Lewis, G., "Computer Modelling of Combustion Reactions in Flowing Systems with Transport," *Combustion Chemistry*, edited by W.C. Gardiner, Jr., Springer-Verlag, New York, 1984, Chapter 2.

[15]Dixon-Lewis, G., "Aspects of the Kinetic Modelling of Methane Oxidation in Flames," *Proceedings of the First Specialists' Meeting (International) of the Combustion Institute (University of Bordeaux)*, Section Française du Combustion Institut, 1981, pp. 284-289.

[16]Dixon-Lewis, G., and Islam, S.M., "Flame Modelling and Burning Velocity Measurement," *Proceedings of the Nineteenth Symposium (International) on Combustion*, The Combustion Institute, Pittsburgh, PA, 1983, pp. 283-291.

[17]Dixon-Lewis, G., "Flame Structure and Flame Reaction Kinetics II: Transport Phenomena in Multicomponent Systems," *Proceedings of The Royal Society of London*, Vol. A 307, October 1968, pp. 111-135.

[18]Kee, R.J., Dixon-Lewis, G., Warnatz, J., Coltrin, M.E., and Miller, J.A., "A Fortran Computer Package for the Evaluation of Gas-Phase Multicomponent Transport Properties," Sandia National Laboratories, Livermore, CA, Rept. SAND86-8246, December 1986.

[19]Buddenberg, J.W., and Wilke, C.R., "Calculation of Gas Mixture Viscosities," *Industrial and Engineering Chemistry*, Vol. 41, July 1949, pp. 1345-1347.

[20]Smooke, M.D., "Solution of Burner-Stabilized Premixed Flames by Boundary Value Methods," *Journal of Computational Physics*, Vol. 48, October 1982, pp. 72-105.

[21]Smooke, M.D., Miller, J.A., and Kee, R.J., "Determination of Adiabatic Flame Speeds by Boundary Value Methods," *Combustion Science and Technology*, Vol. 34(1-6), 1983, pp. 79-90.

[22]Kee, R.J., Grcar, J.F., Smooke, M.D., and Miller, J.A., "A Fortran Program for Modelling Steady Laminar One-Dimensional Premixed Flames," Sandia National Laboratories, Livermore, CA, Rept. SAND85-8240, December 1985.

[23]Grcar, J.F., Kee, R.J., Smooke, M.D., and Miller, J.A., "A Hybrid Newton/Time-Integration Procedure for the Solution of Steady, Laminar, One-Dimensional, Premixed Flames," *Proceedings of the Twenty-First Symposium (International) on Combustion*, The Combustion Institute, Pittsburgh, PA, 1988, pp. 1773-1782.

[24]Keller, H.B., "Numerical Solution of Bifurcation and Nonlinear Eigenvalue Problems," *Applications of Bifurcation Theory*, edited by P Rabinowitz, Academic Press, New York, 1977, pp. 359-384.

[25]Keller, H.B., "The Bordering Algorithm and Path Following Near Singular Points of Higher Nullity," *SIAM Journal on Scientific and Statistical Computing*, Vol. 4(4), December 1983, pp. 573-582.

[26]Heinemann, R.F., Overholser, K.A., and Reddien, G.W., "Multiplicity and Stability of the Hydrogen-Oxygen-Nitrogen Flame: The Influence of Chemical Pathways and Kinetics on Transitions Between Steady States," *American Institute of Chemical Engineers Journal*, Vol. 26, September 1980, pp. 725-734.

[27]Giovangigli, V., and Smooke, M.D., "Extinction of Strained Premixed Laminar Flames with Complex Chemistry," *Combustion Science and Technology*, Vol. 53(1), 1987, pp. 23-49.

[28]Glarborg, P., Miller, J.A., and Kee, R.J., "Kinetic Modelling and Sensitivity Analysis of Nitrogen Oxide Formation in Well-Stirred Reactors," *Combustion and Flame*, Vol. 65(2), 1986, pp. 177-202.

[29]Rogg, B., "Numerical Analysis of Premixed CH_4-Air Flames with Detailed Chemistry," *Numerical and Applied Mathematics*, edited by W.F. Ames, Baltzer-Verlag, Switzerland, 1989, pp. 159-167.

[30]Rogg, B., "Numerical Modelling and Computation of Reactive Stagnation Point Flows," *Computers and Experiments in Fluid Flow*, edited by G.M. Carlomagno and C.A. Brebbia, Springer-Verlag, Berlin - Heidelberg, 1989, pp. 85-85.

[31]Warnatz, J,, "Rate Coefficients in the *C/H/O* System," *Combustion Chemistry*, edited by W.C. Gardiner, Jr., Springer-Verlag, New York, 1984, Chapter 5.

[32]Rogg, B., "Response and Flamelet Structure of Stretched Premixed Methane-Air Flames," *Combustion and Flame*, Vol. 73(1), 1988, pp. 45-65.

[33]Giovangigli, V., and Smooke, M.D., "Adaptive Continuation Algorithms with Application to Combustion Problems," *Applied Numerical Mathematics*, Vol. 5, 1989, pp. 305-331.

[34]Giovangigli, V., and Darasbiha, N., "Vector Computers and Complex Chemistry Combustion," *Proceedings of the Conference on Mathematical*

Modelling in Combustion, edited by C.-M. Brauner and C. Schmidt-Lainé, N.A.T.O., A.S.I. Series E, No. 140, 1987, p. 491.

[35]Puri, I.K., Seshadri, K., Smooke, M.D., and Keyes, D.E., "A Comparison Between Numerical Calculations and Experimental Measurements of the Structure of a Counterflow Methane-Air Diffusion Flame," *Combustion Science and Technology*, Vol. 56(1), 1987, pp. 1-22.

[36]Deuflhard, P., and Nowak, U., "Extrapolation Integrators for Quasilinear Implicit O.D.E's," University of Heidelberg, Sonderforschungsbereich 123, Rept. 332, 1985.

[37]Frenklach, M., and Warnatz, J., "Detailed Modelling of PAH Profiles in a Sooting Low-Pressure Acetylene Flame," *Combustion Science and Technology*, Vol. 51(4-6), 1987, pp. 265-283.

[38]Sakai, Y., and Ishizuka, S., private communication, 1988.

[39]Kee, R.J., Miller, J.A., Evans, G.H., and Dixon-Lewis, G., "A Computational Model of the Structure and Extinction of Strained, Opposed Flow, Premixed Methane-Air Flames," *Proceedings of the Twenty-Second Symposium (International) on Combustion*, The Combustion Institute, Pittsburgh, PA, 1989, pp. 1479-1494.

Simulation of Partially Premixed Methane-Air Counterflow-Diffusion Flames and Comparison with Experimental Results

Frank Behrendt* and Jürgen Warnatz†
Universität Stuttgart, Stuttgart, Germany

Abstract

The main reason for the strong interest in laminar counterflow diffusion flames is their key role in modelling of turbulent diffusion flames. The basic concept of this model is the treatment of turbulent flames as an ensemble of thin laminar diffusion flames, the so-called laminar diffusion "flamelets". This concept is based on the availability of libraries consisting of results of laminar diffusion flame calculations. To allow for a realistic simulation of turbulent diffusion flames, these libraries must include partially premixed as well as pure diffusion flames. Counterflow flames of the Tsuji type have been proven as a valuable tool for experimental and numerical studies of laminar diffusion flame structures. According to this, results of a theoretical study of counterflow diffusion flames consisting of two methane-air streams of various composition are presented here. The governing boundary-layer equations are transformed into an one-dimensional form, and are integrated by an implicit finite-differences method. An improved mathematical model for the flow field description of this configuration is discussed. A detailed reaction mechanism is used, and a semiempirical multicomponent transport model is applied. Profiles for temperature and species mole fractions are shown for different degrees of premixedness. The structure of these profiles is discussed and compared with experimental results.

* Postdoctoral Research Fellow, Institute of Combustion Technology

† Professor, Institute of Combustion Technology

I. Introduction

The laminar flamelet concept of turbulent diffusion flames[1–3] strongly depends on the understanding and availability of libraries of laminar diffusion flames at various conditions. For a detailed prediction of temperature and species distributions as well as formation of pollutants within turbulent diffusion flames, numerical simulations of laminar flamelets using a detailed reaction mechanism with elementary reactions and realistic transport models are necessary.

In turbulent diffusion flames quenching and re-ignition of some areas leads to an appearence of partially premixed reaction zones. For the example of a CO-air flame the flamelet library has been extended by partially premixed flamelets. This library has been proven as a valuable tool for achievement of better agreement between experimental data and numerical results.[4]

Before including partially premixed diffusion flames in flamelet libraries, the numerical results must be checked against reliable experimental data. Compared with pure diffusion flames, the flame structure of partially premixed flames is, due to the presence of a premixed subflame, more complicated. In contrast to the transport-controlled flame structure of pure diffusion flames, the location and structure of the premixed subflame depends strongly on the reaction mechanism.

The laminar flamelet computations presented in this paper are based on a counterflow geometry described first by Tsuji and Yamaoka.[5–7] The overall flow configuration is given in Fig. 1. The location of

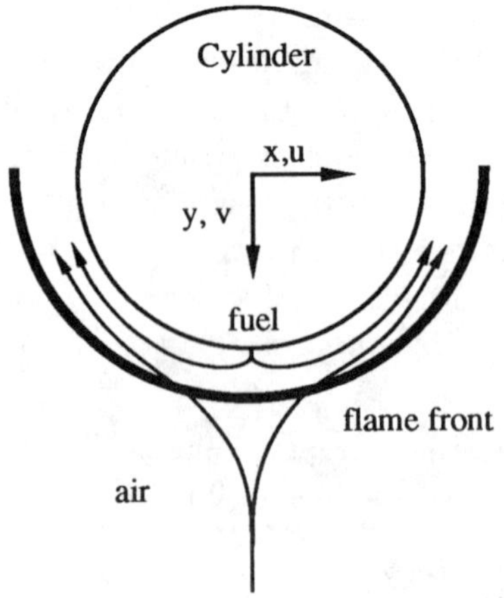

Fig. 1 Counterflow geometry introduced by Yamaoka and Tsuji.

the flame zone is within the boundary layer but outside of the free stagnation line.

These counterflow flames allow the examination of partially premixed flames of stoichiometries near and beyond their upper and lower limits of flammability. Profiles of temperature and stable species for two classes of partially premixed CH_4-air diffusion flames were measured by Yamaoka and Tsuji. First, CH_4-air mixtures with stoichiometries around the upper flammability limit were blown out of the cylinder against up-streaming air.[8] Second, CH_4-air mixtures below and above the lower flammability limit were blown against a rich CH_4-air mixture with $\phi = 3$.[9]

The choice of the experimental data of Yamaoka and Tsuji is guided by the long experience of this group in running counterflow diffusion flames. Their data have been proven as reliable in the past. In this paper, their experimental results are compared with numerical calculations.

For a realistic numerical treatment of the flame structure of counterflow diffusion flames, a correct description of the flowfield is essential. Earlier models assumed an outer potential flow surrounding the boundary layer.[10] The pressure gradient within the boundary layer was stated to be constant and identical with the one of the outer flow. Moreover, a single parameter, the velocity gradient a, was used for characterization of the strain field to which the flame is exposed. Application of experimental boundary conditions for both gas streams was not possible.

The new formulation presented in this paper estimates the pressure gradient within the boundary layer as an eigenvalue. No a priori assumptions are made for the outer flow. Different types of boundary conditions can be specified at the free stream edge of the boundary layer. This allows for proper application of experimental boundary conditions to the calculations. As an outcome calculated and measured position and structure of the flame within the boundary layer will show better agreement compared with the results of the traditional approach.

II. Mathematical Formulation

For a quantitative treatment of the flame structure, the governing two-dimensional Navier-Stokes equations are simplified traditionally by applying the boundary-layer assumptions and by assuming an outer potential flow. As an outcome of this procedure the problem is reduced spatially to one dimension, and the strain field to which the flame is exposed can be specified by the potential flow velocity gradient a. The pressure gradient within the boundary layer is assumed to be constant and is derived from the outer potential flow.[10]

To overcome certain deficiencies of the assumption of an outer potential flow, an alternate formulation is used. This formulation is based on ideas of Chapman and Bauer [11] and Seshadri and Williams.[12] An adaption to plane counterflow geometry was presented by Kee et al.[13]

The general description of the laminar reacting flow around a cylinder is given by the governing two-dimensional equations for overall continuity, momentum, enthalpy, and mass of species i. To reduce the spatial order of the problem a streamfunction

$$\Psi(x, y) \equiv xF(y) \tag{1}$$

identically satifying the continuity equation is introduced. The definition of Ψ is chosen so that

$$\begin{aligned} -\frac{\partial\Psi}{\partial x} &= \rho v = -F(y) \\ \frac{\partial\Psi}{\partial y} &= \rho u = x\frac{\partial F}{\partial y} \end{aligned} \tag{2}$$

is fulfilled.

The boundary-layer assumption allows treating density ρ, mass fractions of species i w_i, temperature T, and the transport coefficients heat conductivity λ, diffusion coefficents $\mathcal{D}_{ij}$, and viscosity μ as functions of the radial coordinate y only. The tangential velocity u varies linearly with x, and the radial velocity v is a function of y only, too. The thermodynamic pressure p is assumed to be constant within the flowfield, with the pressure gradient still appearing in the momentum equations (small Mach number approximation).

Introduction of the stream function, application of the previously described assumptions to the two-dimensional governing equations, and resolving for the pressure gradients shows that the gradients $\partial p/\partial x$ and $\partial p/\partial y$ are functions of y only. Pressure p is a state variable which implicates the mixed second derivatives are equal. Because both gradients are functions of y only, the mixed second derivatives must be equal to zero. So $\partial p/\partial y$ must be a constant and can be estimated as an eigenvalue of the system. In the following this eigenvalue is abbreviated by H. For technical reasons a new variable G with $G = \partial F / \partial y$ is introduced.

The modified boundary-layer equations used for the description of the Tsuji counterflow configuration are:

Continuity:

$$\rho v = (\rho v)_w - \int_0^y \rho G dy \tag{3}$$

Momentum:

$$\frac{\partial G}{\partial t} = \frac{1}{\rho} \frac{\partial}{\partial y} \left(\mu \frac{\partial G}{\partial y} \right) - v \frac{\partial G}{\partial y} - G^2 - \frac{H}{\rho} \tag{4}$$

Temperature:

$$\frac{\partial T}{\partial t} = \frac{1}{\rho c_p} \frac{\partial}{\partial y} \left(\lambda \frac{\partial T}{\partial y} \right) - v \frac{\partial T}{\partial y} - \frac{1}{\rho c_p} \sum_i c_{p,i} j_i \frac{\partial T}{\partial y} - \frac{1}{\rho c_p} \sum_i h_i r_i \tag{5}$$

Mass fraction of species i:

$$\frac{\partial w_i}{\partial t} = \frac{1}{\rho} \frac{\partial}{\partial y} \left(D_i^m \rho \frac{\partial w_i}{\partial y} \right) - v \frac{\partial W_i}{\partial y} + \frac{1}{\rho} \frac{\partial}{\partial y} \left(D_i^T \frac{\partial \ln T}{\partial y} \right) + \frac{r_i}{\rho} \tag{6}$$

where ρ = density, t = time, v = flow velocity in y–direction, y = cartesian space coordinates, p = pressure, μ = dynamic viscosity, c_p = specific heat capacity, T = temperature, λ = heat conductivity, r = mass scale chemical rate of formation, w = mass fraction, j_i = diffusive mass flux, D_i^m = modified binary diffusion coefficient, D_i^T = thermo-diffusion and index i stands for the ith chemical species of the system.

The boundary conditions at $y = 0$ are given by:

Mass flow:

$$\rho v = (\rho v)_w \tag{7}$$

Momentum :

$$G = 0 \tag{8}$$

Table 1: Velocities of the upstreaming air and CH_4-air mixtures

Flame no	Φ	v [cm s^{-1}]
1	1.55	10.0
2	1.80	12.0
3	3.00	15.0
4	0.42	10.0
5	0.58	10.0

Temperature :

$$T = T_w \tag{9}$$

Mass fraction of species i :

$$w_i = w_{i,u} - \frac{j_{i,w}}{(\rho v)_w} \qquad \text{i} = 1,2,\ldots,\text{N} \tag{10}$$

Due to the recombination of H-atoms at the cylinder wall surface, mass fractions for these atoms are handled differently:

$$w_H = 0 \qquad \text{and} \qquad w_{H_2} = w_{H_2,u} - \frac{j_{H_2,w}}{(\rho v)_w} - \frac{j_{H,w}}{(\rho v)_w} \tag{11}$$

where index w denotes conditions at the cylinder and index u denotes conditions of the unburnt gas. The respective boundary conditions at the outer edge of the flame y_e are

$$\frac{\partial G}{\partial y} = 0 \tag{12}$$

$$T = T_e \tag{13}$$

$$w_i = w_{i,e} \qquad \text{i} = 1,2,\ldots,\text{N} \tag{14}$$

where index e denotes conditions of the free up-streaming gas.

Equations (4) to (6) are transformed into finite difference expressions by discretizing them parabolically in space and linearly in time.[14] At the beginning of the integration, 40 grid points are nonuniformly spread across the flame front, with a distribution adapted to the temperature profile gradient. For educts cubic polynomals, for

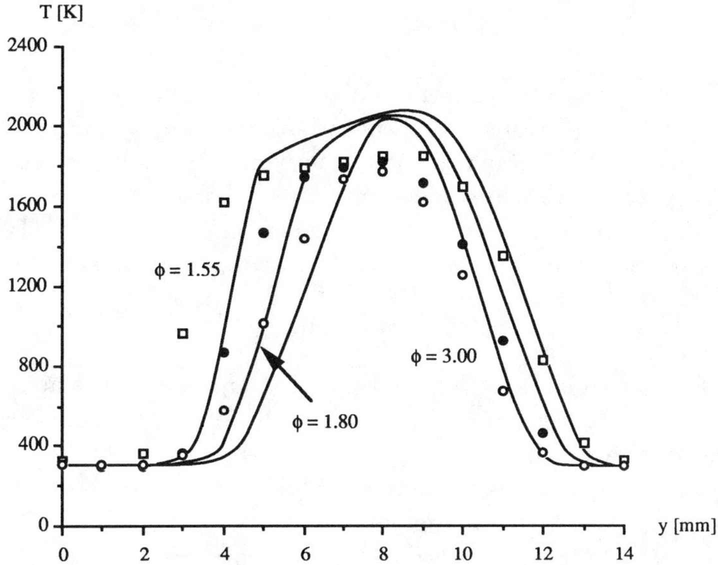

Fig. 2 Measured (symbols, Yamaoka and Tsuji 1975) and calculated (solid lines, this paper) temperature profiles for $\phi = 1.55$, $\phi = 1.8$, and $\phi = 3.0$.

temperature, intermediate species, and products Gaussian profiles are chosen as starting estimates.

Due to stiffness of the system considered, an implicit solution method is used.[15] During the integration the grid point system is statically adapted according to a weighted norm of gradients of temperature and some species profiles.[16] Interpolation onto the new mesh is performed by piece-wise monotonic cubic Hermite interpolation.[17]

III. Transport Processes, Reaction Mechanism and Thermochemistry

For the calculation of diffusion coefficients D_i^M, thermal diffusion coefficients D_i^T, heat conductivities λ, and dynamic viscosities μ, data are taken from molecular parameters. To calculate the transport parameters, formulas given by Kee et al. are used.[18] A simplified transport model is used, because comparison with multicomponent transport models leads to errors relatively small if compared to those caused by uncertainties in the reaction mechanism.[19]

The reaction mechanism consists of 188 elementary steps (including reverse reactions), and of 28 chemical species. The major components of this mechanism include the H_2-O_2-system, CO oxidation, degradation of methane to CO, alkyl recombination to ethane, and its consumption.

Kinetic data are taken from Warnatz [20] with revisions because of new experimental results, and due to outcomings of a new evalution of data available.[21] The rates for the reverse reactions are calculated with help of equilibrium constants derived from thermodynamical data. Thermochemical data are taken from the JANAF Tables [22] and addenda, processed by Burcat to sets of polynomal coefficients.[23]

IV. Results

The data given by Yamaoka and Tsuji are the results of two different sets of experiments. In the first set CH_4-air mixtures with equivalence ratios of $\phi = 1.55$, $\phi = 1.8$, and $\phi = 3.0$ were ejected from the cylinder against upstreaming air.[8] The second set consists of two flames with lean CH_4-air mixtures of $\phi = 0.42$ and $\phi = 0.58$ ejected from the cylinder against a rich CH_4-air-mixture with $\phi = 3.00$.[9]

In all cases the ejection velocity from the cylinder was 15.6 cm/s. The velocity of the upstreaming air- or fuel-airstream at a distance of $y = 14$ mm from the cylinder was estimated from the velocity profiles given by Yamaoka and Tsuji. Table 1 lists the velocities used for the calculations.

Table 1: Velocities of the upstreaming air and CH_4-air mixtures

The upper limit of flammability for CH_4-air mixtures lies near $\phi =$ 1.6 Flame 2 and 3 would not burn in a purely premixed configuration. However, in the counterflow configuration the premixed subflame is stabilized by the coexisting diffusion flame.

Similary, flame 4 lies below the lower limit of flammability of CH_4-air mixtures ($\phi = 0.45$), and becomes stabilized by burned gases of the upstreaming rich ($\phi = 3.0$) mixtures. Again, this upstreaming mixture itself could not burn in a pure premixed configuration. The lean premixed subflame in flame 5 would burn without stabilization. In comparision with flame 4 the double structure of the flame is more distinct.

In Fig. 2 experimental and calculated temperature profiles for flames 1 - 3 are compared. The general shape of the calculated profiles agrees well with the experimental results. The agreement of flame position between experiment and calculation is very good. The deviation at the fuel side must be attributed to limitations of the

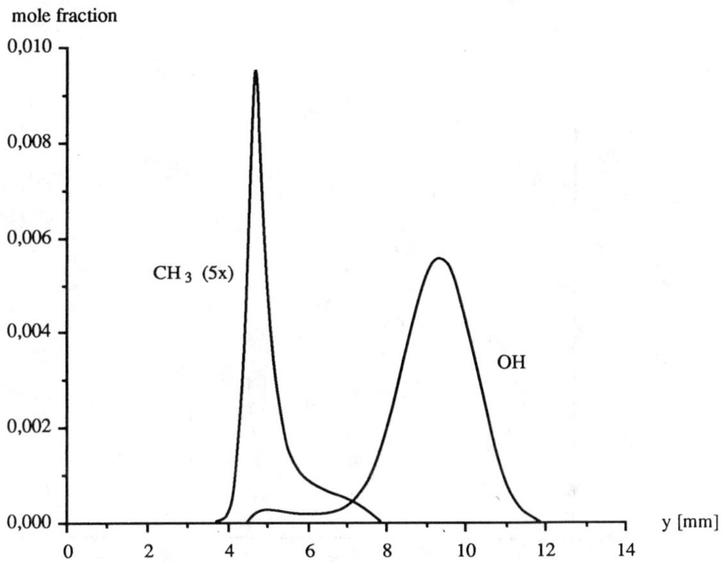

Fig. 3a Calculated profiles OH and CH_3 radicals for $\phi = 1.55$.

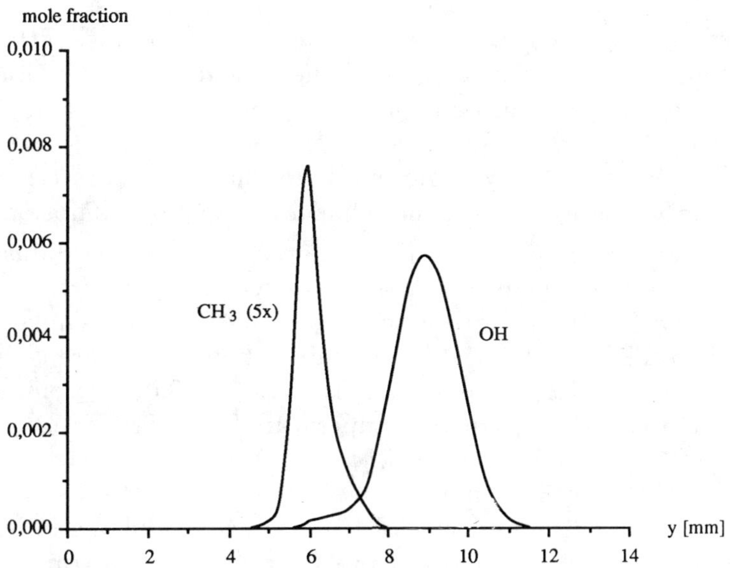

Fig. 3b Calculated profiles OH and CH_3 radicals for $\phi = 1.8$.

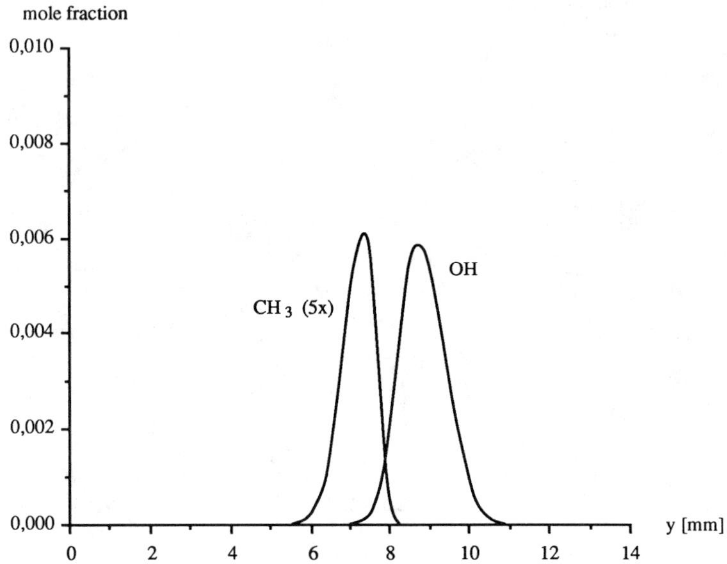

Fig. 3c Calculated profiles OH and CH_3 radicals for $\phi = 3.0$.

reaction mechanism in a very fuel rich regime, because of the exclusion of higher hydrocarbons than ethane. The deviating maximum values of temperature can be explained by experimental errors caused by using of uncorrected thermo-couples.

Flame 1 clearly exhibits a twin flame structure consisting of a premixed subflame near the cylinder followed by a diffusion subflame. In flame 2 the premixed flame moves closer to the diffusion flame, because only there it can be sustained from extinction by radical supply. The temperature profile of flame 3 is very similar to that of a pure diffusion flame. The premixed flame structure is not present any longer. The increase of ϕ from 1.55 to 3.0 is accompanied by a decrease of the maximum temperature from 2082 K ($\phi = 1.55$) over 2062 K ($\phi = 1.8$) to 2040 K ($\phi = 3.0$).

The analysis of intermediate species profiles allows a more precise look at the disappearance of the premixed subflame in increasingly rich systems. CH_3 and OH profiles are given in Figure 3. In flame 1 (Fig. 3a) two seperated flame fronts can clearly be identified by two maxima of both profiles. Increase of the stoichiometry above the upper limit of flammability (Fig. 3b) leads to a reduction of the second

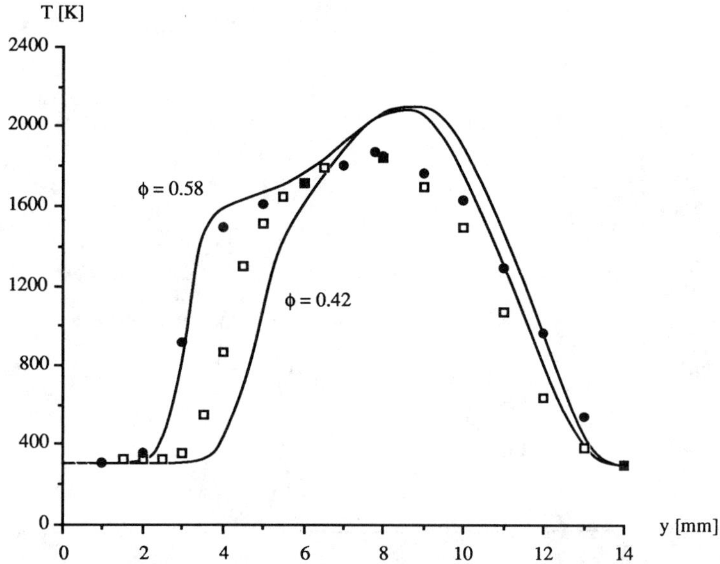

Fig. 4 Measured (symbols, Yamaoka and Tsuji 1977) and calculated (solid lines, this paper) temperature profiles for $\phi = 0.42$, and $\phi = 0.58$.

maximum, while flame 3 exhibits the structure of a pure diffusion flame (Fig. 3c).

Flames 4 and 5 are distinct from the first three flames. They consist of two premixed gas streams of different composition. Besides two premixed outer subflames, a diffusion flames is established in between, stabilizing both outer flames. While in flame 4 the lean premixed subflame causes a small shoulder only on the temperature profile only, the temperature profile of flame 5 shows clearly the existence of at least two subflames. The profile is broadened due to the premixed subflame; the maximum temperature is increased from 2085 K to 2105 K (Fig. 4).

The general agreement between calculations and experimental results is very well validating both experimental and calculated results. The difference between calculated and measured temperature maxima must be explained the uncorrected thermocouple measurements, as can be confirmed by comparison of similar thermocouple measurements with CARS measurements.[24]

Again, a view at the profiles of the intermediate species CH_3 and OH allows a closer look into the internal structure of flame system.

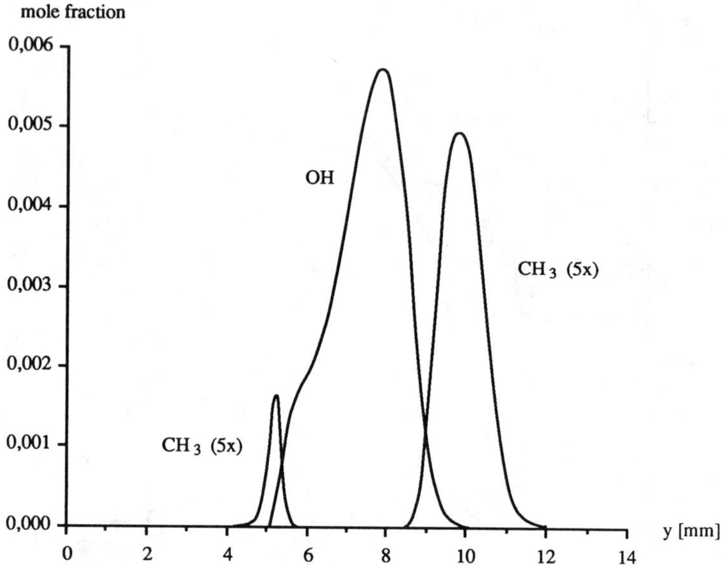

Fig. 5a Calculated profiles OH and CH_3 radicals for $\phi = 0.42$.

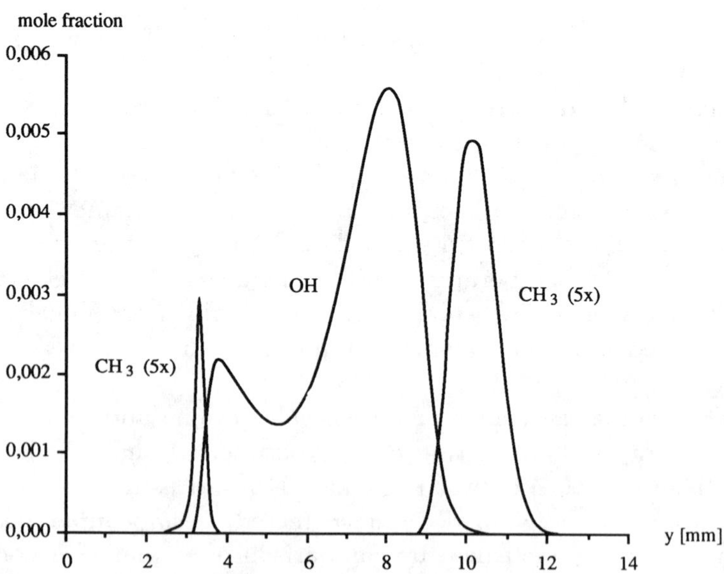

Fig. 5b Calculated profiles OH and CH_3 radicals for $\phi = 0.58$.

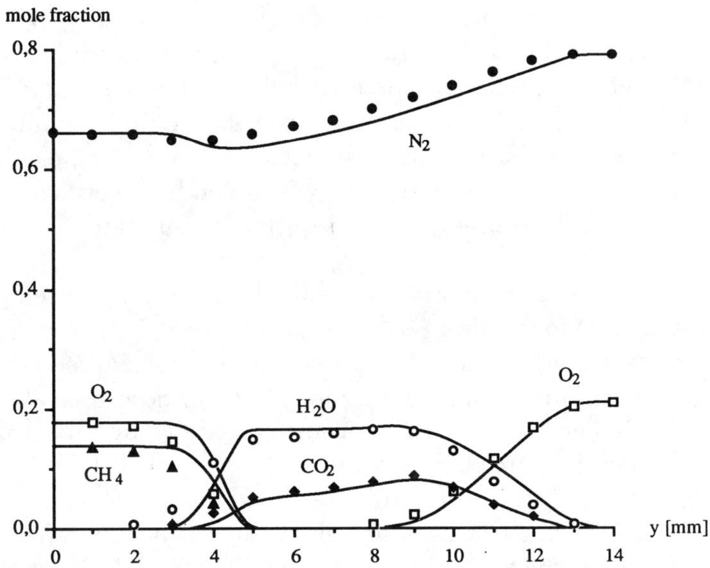

Fig. 6a: Measured (symbols, Yamaoka and Tsuji 1977) and calculated (solid lines, this paper) profiles of CO_2 and H_2O for $\phi = 1.55$.

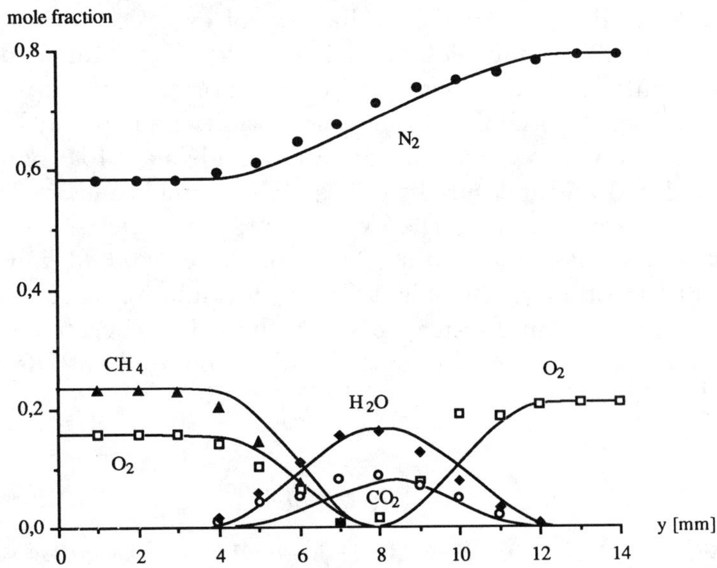

Fig. 6b: Measured (symbols, Yamaoka and Tsuji 1977) and calculated (solid lines, this paper) profiles of CO_2 and H_2O for $\phi = 3.0$.

The outer rich premixed sub-flame nearly is not visible, because it merges with the diffusion flame. This situation is similiar to that of flame 3. On the cylinder side of the flame the increase of ϕ from 0.42 to 0.58, i.e. the move across the lower flammability limit, causes a significant change of the flame structure. The CH_3 mole fraction shows a clear double peak structure, which can be identified with the reaction zones of the premixed and diffussion subflame (Figures 5a and 5b).

Figures 6a and 6b compare the calculated and measured profiles of CO_2 and H_2O for flame 1 ($\phi = 1.55$) and 3 ($\phi = 3.0$). Again, shape and absolute values of the calculated mole fractions agree very good with the measurement. Similiar to the behaviour described above, the structural change of the profiles from a partially premixed doublet flame to a nearly pure diffusion flame is obvious. While in flame 1 the premixed and diffusion subflames lead to two local maxima of the CO_2 profile, flame 3 does not show such a behaviour. Flame 3 resembles the product distribution of a similiar pure diffusion flame.[10]

V. Conclusions

The description of turbulent diffusion flames by the flamelet model strongly depends on the availability of extended libraries, with partially premixed flamelets included. Calculated laminar diffusion flame profiles should be verified by comparision with experimental results. The potential flow formulation traditionally applied to simulation of counterflow diffusion flames is inadequate for comparision of calculated and measured profiles. The formulation used instead allows the application of the experimental boundary conditions to the flowfield. Position and structure of the calculated flame profiles that are obtained by this alternate formulation agree very well with the experimental results. Flamelet libraries generated using this formulation offer a reliable database for enhanced turbulent flame calculations.

References

[1] Williams, F.A., "Turbulent Mixing in Non-Reactive and Reactive Flows", Turbulent Mixing in Non-Reactive and Reactive Flows, edited by S.N.B. Murphy, Plenum, New York, 1975, p. 189.

[2] Peters, N., Combust. Sci. Technol., Vol. 30, 1983, p. 1.

[3] Peters, N., "Laminar Diffusion Flamelet Models in Non-Premixed Turbulent Combustion", Prog. Energy Combust. Sci., Vol. 10, 1984, p. 319.

[4] Rogg, B., Behrendt, F., Warnatz, J., "Turbulent Non-Premixed Combustion in Partially Premixed Diffusion Flamelets with Detailed Chemistry", Proceedings of the 21st Symposium (International) on Combustion, The CombustionInstitute, Pittsburgh, PA, 1988, p. 1533.

[5] Tsuji, H., Yamaoka, I., "The Counterflow Diffusion Flame in the Forward Stagnation Region of a Porous Cylinder", Proceedings of the 11th Symposium (International) on Combustion, The Combustion Institute, Pittsburgh, PA, 1967, p. 979.

[6] Tsuji, H., Yamaoka, I., "The Structure of Counterflow Diffusion Flames in the Forward Stagnation Region of a Porous Cylinder", Proceedings of the 12th Symposium (International) on Combustion, The Combustion Institute, Pittsburgh, PA, 1969, p. 997.

[7] Tsuji, H., Yamaoka, I., "Structure Analysis of Counterflow Diffusion Flames in the Forward Stagnation Region of a Porous Cylinder", Proceedings of the 13th Symposium (International) on Combustion, The Combustion Institute, Pittsburgh, PA, 1971, p. 723.

[8] Yamaoka, I., Tsuji, H., "Structure Analysis of Rich Fuel-Air Flames in the Forward Stagnation Region of a Porous Cylinder", Proceedings of the 16th Symposium (International) on Combustion, The Combustion Institute, Pittsburgh, PA, 1977, p. 1145.

[9] Yamaoka, I., Tsuji, H., "An Experimental Study of Flammability Limits Using Counterflow Flames", Proceedings of the 17th Symposium (International) on Combustion, The Combustion Institute, Pittsburgh, PA, 1979, p. 843.

[10] Dixon-Lewis, G., David, T., Gaskell, P.H., Fukutani, S., Jinno, H., Miller, J.A., Kee, R.J., Smooke, M.D., Peters, N., Effelsberg, E., Warnatz, J., Behrendt, F., "Calculation of the Structure and Extinction Limits of a Methane-Air Counterflow Diffusion Flame in the Forward Stagnation Region of a Porous Cylinder", Proceedings of the 20th Symposium (International) on Combustion, The Combustion Institute, Pittsburgh, PA, 1984, p. 1893.

[11] Chapman, T.W., Bauer, G.L., Appl. Sci. Res. Vol. 31, 1975, p. 223.

[12] Seshadri, K., Williams, F.A., Int. J. Heat Mass Transfer, Vol. 21, 1978, p. 251.

[13] Kee, R.J., Miller, J.A., Evans, G.H., Dixon-Lewis, G., "A Computational Model of the Structure and Extinction of Strained, Opposed Flow, Premixed Methane-Air Flames", Proceedings of the 22nd Symposium (International) on Combustion, The Combustion Institute, Pittsburgh, PA, 1989, p. 1479.

[14] Warnatz, J., "Calculation of the Structure of Laminar Flat Flames I: Flame Velocity of Freely Propagating Ozone Decomposition Flames", Ber. Bunsenges. Phys. Chem., Vol. 82, 1981, p. 193.

[15] Forsythe, G.E., Wasow, W.R., Finite-Differences-Methods for Partial Differential Equations, Wiley, New York, 1960.

[16] Kautsky, J., Nichols, N.K., Equidistributing Meshes with Constraints, SIAM J. Sci. Stat. Comput., Vol. 1, 1980, p. 449.

[17] Fritsch, F.N., Carlson, R.E., SIAM J. Numer. Anal., Vol. 17, 1980, p. 238.

[18] Kee, R.J., Warnatz, J., Miller, J.A., "A FORTRAN Computer Code Package for the Evaluation of Gas-Phase Viscosities, Conductivities, and Diffusion Coefficients", Sandia National Laboratories Report SAND83-8209, Livermore, CA, 1983.

[19] Warnatz, J., "Influence of Transport Models and Boundary Conditions on Flame Structure", Numerical Methods in Laminar Flame Propagation, edited by N. Peters, J. Warnatz, Friedr. Vieweg & Sohn, Braunschweig/Wiesbaden, 1982, p. 87.

[20] Warnatz, J., "Rate Coefficients in the $C/H/O$-System", Combustion Chemistry, edited by W.C. Gardiner, Jr. Springer, New York/Berlin/Heidelberg/ Tokyo, 1985, p. 197.

[21] Baulch, D.L., Just, Th., Kerr, J.A., Pilling, M., Troe, J., Walker, R.W., Warnatz, J., "Compilation of Rate Data on C_1/C_2 Species Oxidation", J. Phys. Chem. Ref. Data, publication in preparation (1990)

[22] Stull, D.R., Prophet, H., JANAF Thermochemical Tables, 2nd Ed., National Bureau of Standards, Washington, D.C., 1971.

[23] Burcat, A., "Thermochemical Data for Combustion Calculations", Combustion Chemistry, edited by W.C. Gardiner, Jr., Springer, New York/Berlin/Heidelberg/ Tokyo, 1985, p. 455.

[24] Dreier, Th., Lange, B., Wolfrum, J., Zahn, M., Behrendt, F., Warnatz, J., "CARS Measurements and Computations of the Structure of Laminar Stagnation-Point Methane-Air Counterflow Diffusion Flames", Proceedings of the 21st Symposium (International) on Combustion, The Combustion Institute, Pittsburgh, PA, 1988, p. 1729.

Extinction of Premixed Curved Flames Stabilized in a Stagnation Flow

K. Asato,* H. Nagata,† and T. Kawamura†
Gifu University, Gifu, Japan
and
K. Mabuchi‡
Sanyo Electric Company, Ltd., Osaka, Japan
and
T. Ban‡
Nippondenso Company, Ltd., Kariya, Japan

Abstract

The characteristics of laminar premixed flames convex and concave to the stream of an unburned mixture (convex flame and concave flame), which are established in a stagnation region of potential flow with a point source in the uniform stream, are studied numerically. The effects of curvature on the extinction are discussed. First, since convex flames are stabilized in a location closer to the stagnation surface compared with concave flames under the conditions of the same first Damköhler number (D_I), the reaction zones of the convex flames are pressed more on the stagnation surface than those of the concave flames; thus, the incomplete combustion of the convex flame compared with the concave flame proceeds with a decrease in D_I. Therefore, the flame temperatures of the convex flame decrease rapidly compared with those of the concave flame, and the convex flames are extinguished at the large value of D_I compared with the concave flames. Second, in the case of the convex flames, the heat flow by conduction related to curvature is so largely divergent from the reaction zone to the unburned mixture that curvature further promotes flame extinction. In the case of the concave flames, however, the heat flow by conduction related to curvature is so largely

---

*Associate Professor, Department of Mechanical Engineering.
†Professor, Department of Mechanical Engineering.
‡Research Engineer, Department of Engineering.

convergent from the reaction zone to the unburned mixture that curvature retards flame extinction.

Introduction

Premixed flames stabilized in shear flow are exposed to stretching and are given curvature simultaneously. Examples of this include the tip of Bunsen flames, cellular flames, the flame base of inverted flames, a minute tubular flame suggested as an element of the turbulent combustion, and the initial flames propagated in the swirl flow in internal combustion engines. Examination of the effects of curvature and stretch on flame characteristics is important to elucidate not only the laminar flame characteristics but also the minute structures of turbulent combustion.[1,2]

Theoretical and experimental studies on the structure of the Bunsen cone,[3-13] on the cylindrical stretched flames,[14,15] on the tubular flames,[16-18] and on the flame base of inverted flames[19,20] have been carried out. It has also been pointed out that the blowoff of inverted flames occurs when the radius of curvature of the flame base approaches the thickness of the preheat zone[21] and that the burning velocity is affected significantly by the combined effects of curvature and stretch.[22-25] It seems, however, that the effects of the curvature on the flame characteristics have not been elucidated sufficiently.

In a previous study,[26] laminar premixed double flames with a small radius of curvature were stabilized in the stagnation region established by a two-dimensional jet of premixed gases opposed to a uniform stream of premixed gases. The characteristics of the flame convex to the uniform stream (convex flame, one of the double flames) were examined experimentally. The effects of the curvature and the combined effects of the curvature with the Lewis number on the extinction of the convex flames were discussed.

In the present work, the characteristics of laminar premixed flames convex and concave to the stream of the unburned mixture, which were established in the stagnation regions of potential flow with a point source in the uniform stream, are studied numerically. The effects of curvature on the extinction of the flame are discussed independently of the effects of the flame stretch.

Model and Governing Equations

Figure 1 shows the schematic diagram of the flame model adopted herein. The point source of the unburned mixture is in the uniform stream (U) of the unburned mixture. The mixture supplied through the point source is of the same

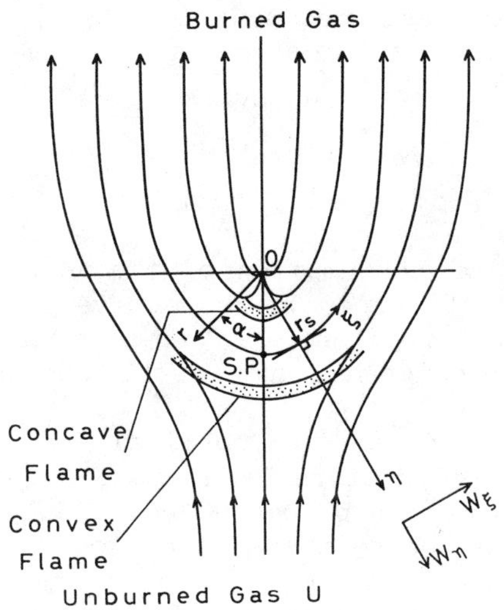

Fig. 1 Schematic diagram of adopted flame model.

composition as the uniform stream. Laminar premixed double flames convex and concave to the stream of the unburned mixture (the convex flame and the concave flame) are established at either side of the stagnation point (S.P.), as shown in Fig. 1. In Fig. 1, r and α represent, respectively, the radial distance from the point source and the azimuthal angle between a streamline along the center-line passing through the stagnation point and the r axis. Under the assumption of potential flow, the flowfield is expressed as

$$W_\alpha = U \sin\alpha \tag{1}$$

$$W_r = -U \cos\alpha + \frac{m}{2\pi}\frac{1}{r} \tag{2}$$

Here m, U, W_α, and W_r are the flow rate from the point source, the velocity of the uniform stream, and the velocity components of the α and r directions, respectively. The following assumptions will be adopted in the analysis.

1) The streamline that passes through the stagnation point and is along the flame (the stagnation surface) is regarded as an adiabatic and noncatalytic surface.

2) In the energy equation, the work done by pressure, the dissipation due to viscosity, and radiation can be neglected.

3) The density ρ, the specific heat at constant pressure c_p, the thermal conductivity λ, the diffusion

coefficient D, and the frequency factor B are constant. The Lewis number is unity.

4) Two species, reactant and product, exist in the flame; a unimolecular reaction is considered.

5) The reaction rate is given by a first-order Arrhenius kinetic expression,

$$\dot{\omega}_R = -\dot{\omega}_P = -B\rho Y_R \exp(-E/RT)$$

where $\dot{\omega}_R$ and $\dot{\omega}_P$ are the rate of consumption of the reactant and the rate of production of the product, respectively. Y_R is the mass fraction of the reactant, E the activation energy, R the universal gas constant, and T the absolute temperature.

6) The flame is regarded to be parallel to the stagnation surface, and the temperature gradient along the flame near the stagnation point can be neglected. Analysis is confined to the region near the stagnation point ($\alpha \rightarrow 0$).

As shown in Fig. 1, ξ and η represent the coordinates along and normal to the stagnation surface, respectively. Since analysis is confined to the region near the stagnation point ($\alpha \rightarrow 0$), η approximately coincides with r. Therefore, the flowfield described by Eqs. (1) and (2) can be transformed into the following equations near the stagnation point:

$$W_\xi = U\alpha \tag{3}$$

$$W_\eta = -U + \frac{m}{2\pi(r_s + \eta)} \tag{4}$$

Here W_ξ, W_η, and r_s are the velocity components of the ξ and η directions, and the radius of curvature of the stagnation surface near the stagnation point, respectively. Using the preceding assumptions, the continuity equation and the energy and species conservation equations written in the curvilinear orthogonal coordinate system (ξ and η)[27] result as follows:

Continuity:

$$\frac{\partial(\rho W_\eta)}{\partial \eta} + \frac{1}{r_s + \eta}\rho W_\eta + \frac{\partial(\rho W_\xi)}{\partial \xi} = 0 \tag{5}$$

Energy:

$$\lambda \frac{d^2 T}{d\eta^2} + \frac{\lambda}{r_s + \eta}\frac{dT}{d\eta} - \rho c_p W_\eta \frac{dT}{d\eta} + \dot{\omega}_p q = 0 \tag{6}$$

Reactant species conservation:

$$\rho D \frac{d^2 Y_R}{d\eta^2} + \frac{\rho D}{r_s + \eta}\frac{dY_R}{d\eta} - \rho W\eta \frac{dY_R}{d\eta} + \dot{\omega}_R = 0 \tag{7}$$

Here q is the heat of reaction. Following assumption 1 outlined earlier and room temperature T_e at $\eta \to \infty$ and $\eta = -r_s$, the boundary conditions of Eqs. (6) and (7) for the convex and concave flames are

$$\frac{dT}{d\eta} = 0, \quad \frac{dY_R}{d\eta} = 0 \quad \text{for both flames at } \eta = 0 \tag{8a}$$

$$T \to T_e, \quad Y_R \to 1 \quad \text{for convex flame as } \eta \to \infty \tag{8b}$$

$$T = T_e, \quad Y_R = 1 \quad \text{for concave flame at } \eta = -r_s \tag{8c}$$

If we introduce the following dimensionless variables and parameters,

$$\eta^* = \frac{\eta}{r_s}, \quad \theta = \frac{T}{T_e}, \quad m^* = \frac{\rho c_p m}{2\pi\lambda}$$

$$E^* = \frac{E}{RT_e}, \quad Le = \frac{\rho c_p D}{\lambda}, \quad D_I = \frac{B r_f}{U} = \frac{S r_f}{r_s}$$

$$Q = \frac{q}{c_p T_e}, \quad S = \frac{Bm}{2\pi U^2} = \frac{B r_s}{U} \tag{9}$$

Eqs. (6) and (7) can be transformed into

$$\frac{d^2\theta}{d\eta^{*2}} + \left[\frac{1 + m^*\eta^*}{1 + \eta^*}\right]\frac{d\theta}{d\eta} + Q Y_R m^* S \exp\left[-\frac{E^*}{\theta}\right] = 0 \tag{10}$$

$$Le \frac{dY_R}{d\eta^{*2}} + \left[\frac{Le + m^*\eta^*}{1 + \eta^*}\right]\frac{dY_R}{d\eta^*} - Y_R m^* S \exp\left[-\frac{E^*}{\theta}\right] = 0 \tag{11}$$

and the boundary conditions become

$$\frac{d\theta}{d\eta^*} = 0, \quad \frac{dY_R}{d\eta^*} = 0 \quad \text{for both flames at } \eta^* = 0 \tag{12a}$$

$$\theta \to 1, \quad Y_R \to 1 \quad \text{for convex flame as } \eta^* \to \infty \tag{12b}$$

$$\theta = 1, \quad Y_R = 1 \quad \text{for concave flame at } \eta^* = -1 \tag{12c}$$

Here r_f is the distance from the point source to the flame, η^* the dimensionless distance from the stagnation surface, θ the dimensionless temperature, m* the dimensionless flow rate from the point source, E* the dimensionless activation energy, D_I the first Damköhler number, Le the Lewis number, Q the dimensionless heat of reaction, and S the parameter that determines the shape of the stagnation surface.

The numerical solutions of the preceding governing equations were obtained using the Runge-Kutta-Gill method. There are five parameters in Eqs. (10) and (11): E*, Q, Le, D_I (or S), and m*. The objective of this study is to examine the effects of curvature on the extinction of the convex and concave flames. Therefore, numerical integration of the governing equations was carried out only for m* = 10 and 1000 at E* = 50, Q = 5, and Le = 1.0. The flames of m* = 10 and 1000 correspond to the flames of the small and large radii of curvature (namely, the flat flame established in the stagnation region), respectively.

Results and Discussions

Flame Temperature and Flame Structure

Figure 2 shows the relation between the flame temperature θ_f and the first Damköhler number D_I. The slid and dashed lines in Fig. 2 represent the results for the convex

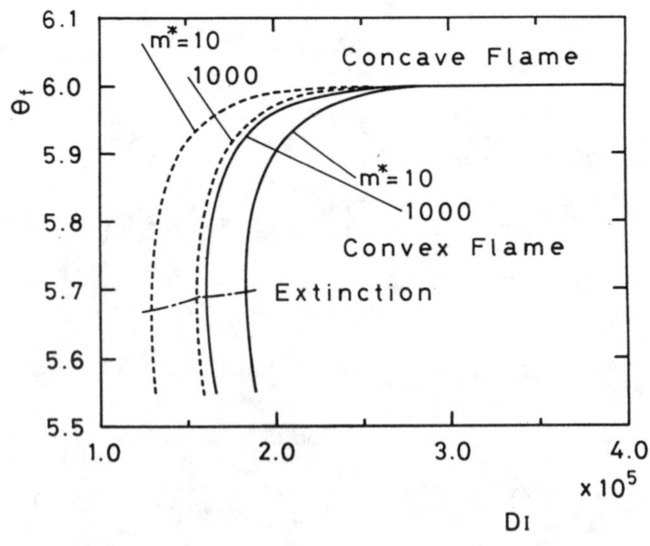

Fig. 2 Variation of flame temperatures with first Damköhler number.

and concave flames, respectively. The well-known S-curves are obtained. Since we are interested in the intense combustion region, the upper branch and a part of the middle branch are illustrated in this figure. As D_I is decreased by increasing the velocity of the uniform stream, the flame temperature θ_f decreases according to its value on the upper branch. As is well known, the flame is finally extinguished at the minimum value ($D_{I\,cr}$) of D_I, which is shown by chain lines in Fig. 2. When D_I approaches infinity, the flame temperature θ_f tends toward the one-dimensional adiabatic flame temperature θ_b (in this case, $\theta_b = 6$) for all flames, and curvature has no effect on the flame temperature. As D_I is decreased from infinity, however, the curvature manifests a marked effect on flame behavior. In the vicinity of $D_{I\,cr}$, θ_f of the convex flame decreases rapidly with a decrease in D_I compared with that of the concave flame. Therefore, the flame temperatures of the near-limit convex flames are lower than those of the concave flames at the same first Damköhler number. The convex flames compared with the concave flames are extinguished at the large value of $D_{I\,cr}$. Thus, the concave flames, rather than the convex flames, are able to withstand the extinction.

The flame temperatures of the near-limit convex flames of m* = 10 are lower than those of m* = 1000 at the same first Damköhler number (the same flame stretch factor), and the extinction of the convex flame of m* = 10 occurs at the larger first Damköhler number compared with that of m* = 1000. The results of the convex flames of m* = 10 and 1000

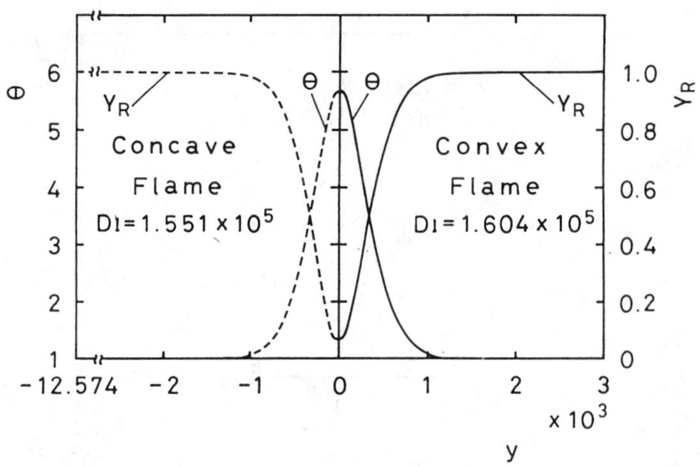

Fig. 3 Flame structures at extinction limits for flames of m* = 1000.

are in qualitative agreement with the experimental results on the lean methane convex flames and the flat flames of the Lewis number of unity.[26] Conversely, the flame temperatures of the near-limit concave flames of m* = 10 are higher than those of m* = 1000 at the same first Damköhler number, and the extinction of the concave flame of m* = 10 occurs at the smaller first Damköhler number compared with that of m* = 1000. It appears that the concave flame of m* = 10 is more stable than that of m* = 1000. The result of the premixed flat flame formed in the stagnation flow by Sato and Tsuji[28] must exist between the results of the convex and concave flames of m* = 1000. Therefore, the flames of m* = 1000 correspond to the flat flames stabilized in the stagnation flow.

Figure 3 shows the flame structures of the convex and concave flames of m* = 1000 at the extinction limit. The abscissa y in Fig. 3 indicates the dimensionless distance from the stagnation surface that is proportional to the physical distance, given as the following equation:

$$y = \eta\sqrt{B\rho c_P/\lambda} = \eta^*\sqrt{r_s D_I m^*/r_f} = \eta^*\sqrt{S m^*} \qquad (13)$$

The structures of the convex and concave flames of m* = 1000 are symmetric with respect to the axis of y = 0, namely, the stagnation surface, and the flame temperature of both flames at the extinction limit is the same. In the cases of m* = 1000 (namely, the large radius of curvature), the flame structure of the convex flame is nearly similar to that of the concave flame, and curvature has little effect on the flame structure.

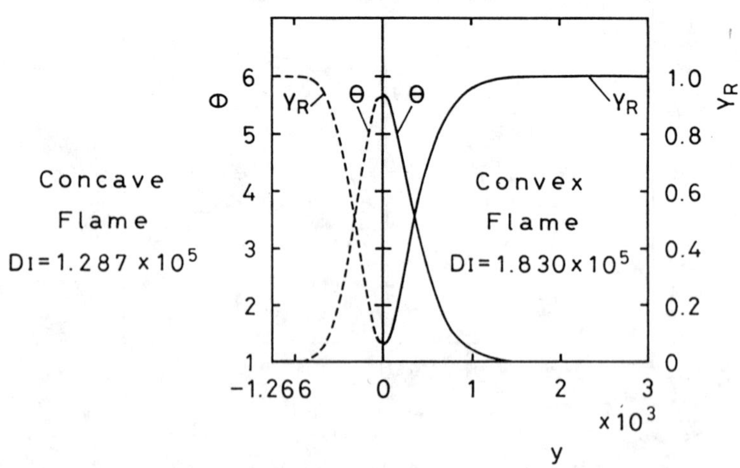

Fig. 4 Flame structures at extinction limits for flames of m* = 10.

Figure 4 shows the flame structures of m* = 10 at the extinction limit. The concentration of the reactant at the stagnation surface (y = 0) is not zero, and incomplete combustion occurs. Proportions of the incomplete combustion and a decrease in the flame temperature for the convex flame are the same as those of the concave flame. The gradients of the temperatures and the concentrations for the convex flame, however, are slightly smaller than those of the concave flame.

Radius of Curvature and Flame Location

Figure 5 gives the relation between the radius of curvature and the first Damköhler number D_I. The ordinate r* in Fig. 5 indicates the dimensionless distance from the point source that is proportional to the physical distance, given as the following equation. Here the flame front is defined as the deflection point of the temperature profiles, as shown in Figs. 3 and 4.

$$r^* = (r_s + \eta)\sqrt{B\rho c_P/\lambda} = (1+\eta^*)\sqrt{r_s D_I m^*/r_f}$$
$$= (1+\eta^*)\sqrt{S m^*} \qquad (14)$$

The radius of curvature of m* = 1000 for the convex flame is nearly equal to that of the concave flame, but the radius of curvature r_f* of m* = 1000 for both flames is much larger than that of m* = 10. We can see that the flames of m* = 10

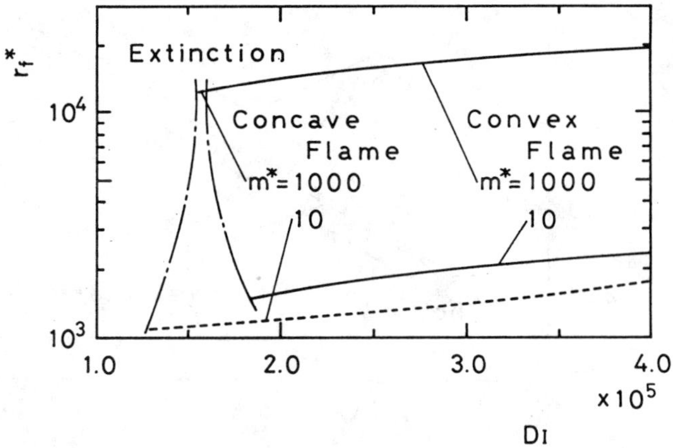

Fig. 5 Variation of radius of curvature for convex and concave flames.

and 1000 correspond to the flames with small and large radii of curvature in Fig. 5, respectively. The radius of curvature of the flame r_f* monotonously decreases with a decrease in D_I for all cases, and the flames are extinguished at the critical value of r_f*. At m* = 10, r_f* of the convex flame at the extinction limit is slightly larger than that of the concave flame. The variation of the radius of curvature with D_I in the case of the convex flames is in qualitative agreement with the experimental results obtained previously.[26]

Figure 6 shows the relation between the distance from the stagnation surface to the flame (flame location) and the first Damköhler number D_I. The ordinate in Fig.6 is shown by the absolute value of y_f to compare the convex and concave flames. The flame locations y_f monotonously decrease with D_I for all cases, and finally the flames are extinguished. The flame locations are the same for all cases at each of the extinction limits.

The locations of the convex flames, however, are smaller than those of the concave flame at the same value of D_I. Since the convex flames are stabilized in the location closer to the stagnation surface compared with the concave flame at the same value of D_I (Fig. 6), the reaction zone of the convex flame is pressed more on the stagnation surface than that of the concave flame; thus, the incomplete combustion of the convex flame compared with the concave

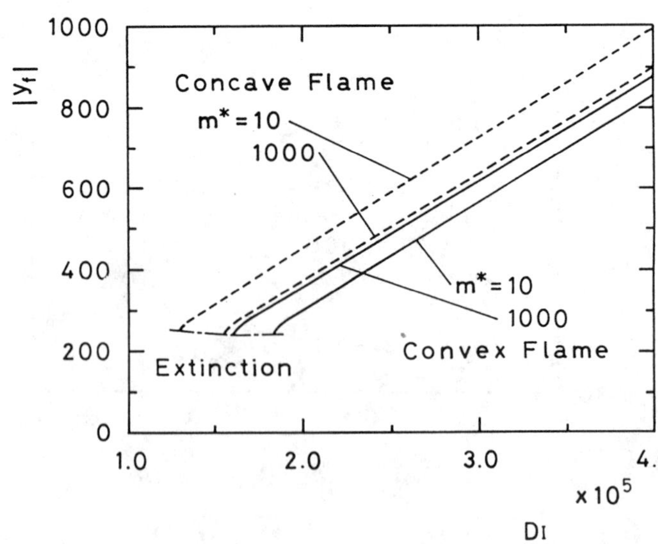

Fig. 6 Variation of flame locations for convex and concave flames.

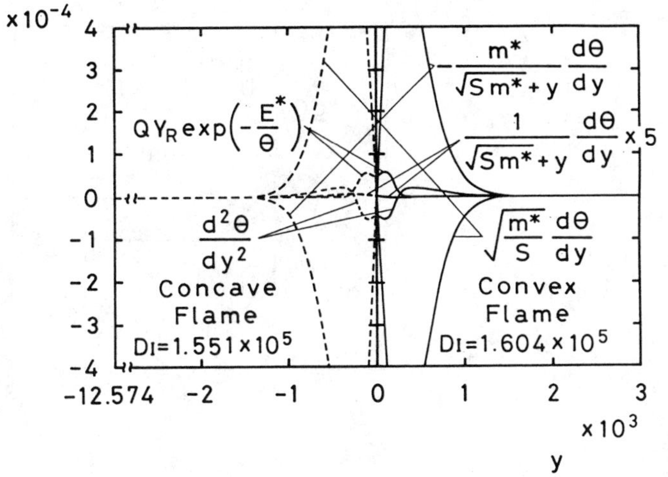

Fig. 7 Profiles of individual terms of the energy equation at extinction limits for flames of m* = 1000.

flame proceeds with a decrease in D_I (Fig. 4). Therefore, the flame temperature of the convex flame decreases rapidly compared with that of the concave flame, and the convex flames are extinguished at the large value of D_I compared with the concave flames (Fig. 2).

Effects of Individual Terms of the Energy Equation on Extinction

To discuss the influence of individual terms of the energy equation on the extinction, Eq. (10) can be transformed using the dimensionless distance given by Eq. (13), which is proportional to the physical distance, into

$$\frac{d^2\theta}{dy^2}+\frac{1}{\sqrt{Sm^*}+y}\frac{d\theta}{dy}+\sqrt{\frac{m^*}{S}}\frac{d\theta}{dy}-\frac{m^*}{\sqrt{Sm^*}+y}\frac{d\theta}{dy}$$

$$+QY_R\exp\left[-\frac{E^*}{\theta}\right]=0 \qquad (15)$$

The first, second, third, fourth, and fifth terms on the left-hand side of Eq. (15) are considered to be normal conduction, conduction related to curvature, convection due to the uniform stream, convection related to the point source, and the reaction term, respectively.

Figure 7 shows the profiles of the individual terms in Eq. (15) at the extinction limit for the convex and concave

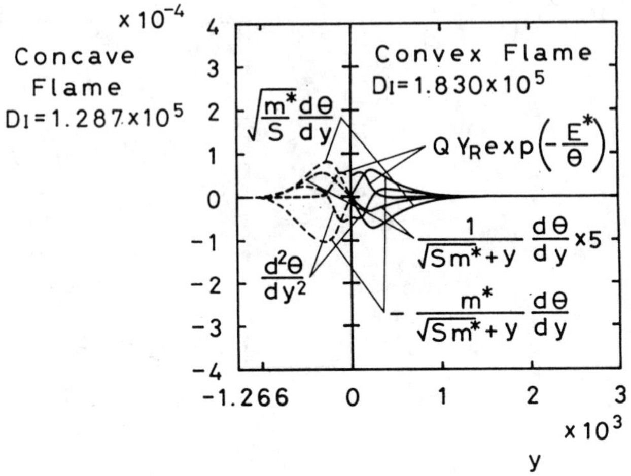

Fig. 8 Profiles of individual terms of the energy equation at extinction limits for flames of m* = 10.

flames of m* = 1000. The value of the conduction term related to curvature (second term) for both flames is nearly zero in all positions of the flame zone. These results indicate the characteristics of the flat flame established in the stagnation flow. Figure 8 indicates the results of the convex and concave flames of m* = 10. The value of the conduction term related to curvature becomes large compared with that of m* = 1000. The effect of curvature on extinction is intensified with decreasing of the radius of curvature of the flame.

Since the signs of the third and fourth terms in Eq. (15) are opposite, the third and fourth terms due to convection cancel each other, as shown in Figs. 7 and 8. In the case of m* = 1000, namely, the flat flame established in the stagnation flow (Fig. 7), the value of the conduction term related to curvature (second term) is nearly zero in all positions of the flame zone. Therefore, the imbalance between the normal conduction term (first term) and the reaction term (fifth term) causes flame extinction. However, in the case of m* = 10, namely, the small radius of curvature (Fig. 8), the conduction term related to curvature (second term) of the convex flame shows a negative value; thus, the heat flow by the conduction related to curvature is so largely divergent from the reaction zone to the unburned mixture that curvature of the convex flame further promotes flame extinction. The conduction term related to curvature of the concave flame, however, shows a positive value; thus, the heat flow by the conduction related to curvature is so largely convergent from the reaction zone to the unburned mixture that curvature of the concave flame retards flame extinction.

Conclusions

The main conclusions obtained are summarized as follows.

1) Since convex flames are stabilized in the location closer to the stagnation surface compared with concave flames under the conditions of the same first Damköhler number (D_I), the reaction zones of the convex flames are pressed more on the stagnation surface than those of the concave flames; thus, the incomplete combustion of the convex flame compared with the concave flame proceeds with a decrease in D_I. Therefore, the flame temperatures of the convex flame decrease rapidly compared with those of the concave flame, and the convex flames are extinguished at the large value of D_I compared with the concave flames.

2) In the case of the convex flames, the heat flow by conduction related to curvature is so largely divergent from the reaction zone to the unburned mixture that curvature further promotes flame extinction. In the case of the concave flames, however, the heat flow by conduction related to curvature is so largely convergent from the reaction zone to the unburned mixture that curvature retards flame extinction.

References

[1]Chomiak, J., "Dissipation Fluctuations and the Structure and Propagation of Turbulent Flames in Premixed Gases at High Reynolds Numbers," Proceedings of the Sixteenth Symposium (International) on Combustion, The Combustion Institute, Pittsburgh, PA, 1977, pp. 1665-1673.

[2]Peters, N., "Laminar Flamelet Concepts in Turbulent Combustion," Proceedings of the Twenty-First Symposium (International) on Combustion, The Combustion Institute, Pittsburgh, PA, 1988, pp. 1231-1250.

[3]Buckmaster, J., "A Mathematical Description of Open and Closed Flame Tips," Combustion Science and Technology, Vol. 20, 1979, pp. 33-40.

[4]Buckmaster, J. and Crowley, A. B., "The Fluid Mechanics of Flame Tips," Journal of Fluid Mechanics, Vol. 131, 1983, pp. 341-361.

[5]Frankel, M. L. and Sivashinsky, G. I., "On Effects Due to Thermal Expansion and Lewis Number in Spherical Flame Propagation," Combustion Science and Technology, Vol. 31, 1983, pp. 131-138.

[6]Frankel, M. L. and Sivashinsky, G. I., "On Quenching of Curved Flames," Combustion Science and Technology, Vol. 40, 1984, pp. 257-268.

[7]Law, C. K., Ishizuka, S. and Cho, P., "On the Opening of Premixed Bunsen Flames Tips," Combustion Science and Technology, Vol. 28, 1982, pp. 89-96.

[8]Law, C. K., Cho, P., Mizomoto, M. and Yoshida, H., "Flame Curvature and Preferential Diffusion in the Burning Intensity of Bunsen Flames," Proceedings of the Twenty-First Symposium (International) on Combustion, The Combustion Institute; Pittsburgh, PA, 1988, pp. 1803-1809.

[9]Mizomoto, M., Asaka, Y., Ikai, S. and Law, C. K., "Effects of Preferential Diffusion on the Burning Intensity of Curved Flames," Proceedings of the Twentieth Symposium (International) on Combustion, The Combustion Institute, Pittsburgh, PA, 1985, pp. 1933-1939.

[10]Mizomoto, M. and Yoshida, H., "Effects of Lewis Number on the Burning Intensity of Bunsen Flames," Combustion and Flame, Vol. 70, 1987, pp. 47-60.

[11]Sivashinsky, G. I., "The Diffusion Stratification Effect in Bunsen Flames," Journal of Heat Transfer, Transactions of the ASME, Vol. 11, 1974, pp. 530-535.

[12]Sivashinsky, G. I., "Structure of Bunsen Flames," Journal of Chemical Physics, Vol. 62, 1975, pp. 638-643.

[13]Sivashinsky, G. I., "On a Distorted Flame Front as a Hydrodynamic Discontinuity," Acta Astronautica, Vol. 3. 1976, pp. 889-918.

[14]Kobayashi, H., Kitano, M. and Otsuka, Y., "A Study on the Extinction of a Stretched Cylindrical Premixed Flame," Transactions of JSME, Ser. B, Vol. 52, 1986, pp. 3811-3817.

[15]Kobayashi, H., Kitano, M. and Otsuka, Y., "An Analysis of a Stretched Cylindrical Premixed Flame," Combustion Science and Technology, Vol. 57, 1988, pp. 17-36.

[16]Ishizuka, S., "On the Behavior of Premixed Flames in a Rotating Flow Field: Establishment of Tubular Flames," Proceedings of the Twentieth Symposium (International) on Combustion, The Combustion Institute, Pittsburgh, PA, 1985, pp. 287-294.

[17]Takeno, T. and Ishizuka, S., "A Tubular Flame Theory," Combustion and Flame, Vol. 64, 1986, pp. 83-98.

[18]Takeno, T., Nishioka, M. and Ishizuka, S., "A Theoretical Study of Extinction of a Tubular Flame," Combustion and Flame, Vol. 66, 1986, pp. 271-283.

[19]Kawamura, T., Asato, K., and Mazaki, T., "Reexamination of the Blowoff Mechanism of Premixed Flames--Inverted Flames," Combustion and Flame, vol. 45, 1982, pp. 225-233.

[20]Mikolaitis, D. W. and Buckmaster, J., "Flame Stabilization in a Rear Stagnation Point Flow," Combustion Science and Technology, Vol. 27, 1981, pp. 55-68.

[21]Kawamura, T., Asato, K., Mazaki, T., Hamaguchi, T. and Kayahara, H., "Explanation of the Blowoff of Inverted Flames by the Area-Increase Concept," Combustion and Flame, Vol. 35, 1979, pp. 109-116.

[22]Mikolaitis, D. W., "The Cylindrical Stretched Flame," Combustion and Flame, Vol. 56, 1984, pp. 327-335.

[23]Mikolaitis, D. W., "The Interaction of Flame Curvature and Stretch, Part 1: The Concave Premixed Flame," Combustion and Flame, Vol. 57, 1984, pp. 25-31.

[24]Mikolaitis, D. W., "The Interaction of Flame Curvature and Stretch, Part 2: The Convex Premixed Flame," Combustion and Flame, Vol. 58, 1984, pp. 23-29.

[25]Mikolaitis, D. W., "Stretched Spherical Cap Flames," Combustion and Flame, Vol. 63, 1986, pp. 95-111.

[26]Asato, K., Kawamura, T., Ban, T., "Effects of Curvature on Extinction of Premixed Flames Stabilized in Stagnation Flow," Proceedings of the Twenty-Second Symposium (International) on Combustion, The Combustion Institute, Pittsburgh, PA, 1989, pp. 1509-1515.

[27]Tollmien, W., Handb. d. Exper.-Pysik, Band 4, Pt. 1, 248, Akdemische Verlargsgeseiischaft, 1931.

[28]Sato, J. and Tsuji, H., "Extinction of Premixed Flames in a Stagnation Flow Considering General Lewis Number," Combustion Science and Technology, Vol. 33, 1983, pp. 193-205.

Chapter III. Dynamics of Flames and Shear Layers

Dynamics of an Unsteady Diffusion Flame: Effects of Heat Release and Viscosity

J. L. Ellzey,* K. J. Laskey,† and E. S. Oran‡
Naval Research Laboratory, Washington, DC

Abstract

Experiments in jet diffusion flames show that the flowfield consists of two types of instabilities. Low-frequency instabilities form in the outer region of the flowfield and high-frequency structures form at the interface between the high-velocity and low-velocity fluid. In this paper, we investigate the effects of heat release and viscosity on the development of the high-frequency structures. We solve the conservation equations for mass, momentum, energy, and individual species densities using the BIC-FCT algorithm and assuming an ideal-gas equation of state. The viscosity, mass diffusion, and conduction terms are solved independently, and a simplified model is used to represent the reaction of fuel and oxidizer. Results are presented for computations of an axisymmetric H_2-N_2 fuel jet surrounded by coflowing air, with jet velocity of 10 m/s and air velocity of 15 cm/s. The results indicate that heat release delays the formation of large-scale instabilities and once formed, the structures are weaker than those in the nonreacting case. When viscosity is included in the calculation of the reacting jet, the fluctuations in the flow field are reduced substantially, and the flame appears almost laminar.

Introduction

Visualizations of jet diffusion flames show that both low-frequency and high-frequency structures develop in the flowfield[1,2]. The low-frequency (10-15 Hz) oscillations appear to be buoyancy-driven and form in the low-speed fluid outside the mixing region because of the temperature gradients associated with heat release. These structures, which have been observed in both experiments[1-3] and computations[4], are only weakly affected by changes in jet velocity or fuel composition.

---

*Berkeley Research Associates, Springfield, VA.

†Grumman Space Station Program Support Division, Reston, VA.

‡Laboratory for Computational Physics and Fluid Dynamics, Naval Research Laboratory.

The high-frequency structures are Kelvin-Helmholtz instabilities that form at the interface between the high-velocity and low-velocity fluid, with typical frequencies of a few hundred hertz. Previous research suggests that these structures are affected by heat release. Mahalingam et al.[5] used an inviscid, linearized stability analysis to show that heat release in a low-speed diffusion flame shifts the most amplified mode to lower frequency and reduces the growth rate of the mixing layer. Experiments on a chemically reacting mixing layer[6] indicate that heat release reduces the vorticity thickness.

In this paper, we present time-dependent simulations of an axisymmetric H_2-N_2 jet with coflowing air. Results for reacting and nonreacting jets with and without viscosity are presented. In these calculations, gravity is not included, and the outer buoyancy-driven structures do not develop. Instead, we focus on the effects of heat release and viscosity on the development of the high-frequency structures.

Numerical Technique

The computer code used for this study is based on that developed by Laskey (7) to simulate jet flames and by Patnaik et al.[8] to simulate low-speed premixed flames. The code solves the following conservation equations for mass, momentum, energy, and species:

$$\frac{\partial \rho}{\partial t} + \nabla \cdot (\rho \mathbf{V}) = 0 \tag{1}$$

$$\frac{\partial \rho \mathbf{V}}{\partial t} + \nabla \cdot (\rho \mathbf{V} \mathbf{V}) = -\nabla P - \nabla \cdot \tau \tag{2}$$

$$\frac{\partial E}{\partial t} + \nabla \cdot (E \mathbf{V}) = -\nabla \cdot P\mathbf{V} + \nabla \cdot (\kappa \nabla T) - \nabla \cdot \sum_{k=1}^{n_{sp}} n_k \mathbf{V_k} h_k + Q \tag{3}$$

$$\frac{\partial n_k}{\partial t} + \nabla \cdot (n_k \mathbf{V}) = -\nabla \cdot (n_k \mathbf{V}_k) + w_k \tag{4}$$

where ρ is the density, $\mathbf{V}$ the velocity vector, t time, P pressure, E total energy density, κ thermal conductivity, T temperature, $\mathbf{V}_k$ the diffusional velocity vector of species k, Q the energy released with reaction, and w_k the number of molecules of species k per unit volume per unit time produced with reaction; n_k and h_k are the number of molecules and enthalpy of species k per unit volume.

This system of equations is closed by the equation of state

$$P_k = n_k k T \tag{5}$$

and the relationship between internal energy e and pressure given by

$$\mathrm{d}e = \mathrm{d}P/(\gamma - 1) \tag{6}$$

where γ is the ratio of specific heats.

The equations are rewritten in terms of finite-difference approximations on an Eulerian mesh and then solved numerically for specified boundary conditions. The accuracy of the solution is determined by the finite-difference algorithm, the spatial resolution set by the computational grid, and the temporal resolution set by the time step. In reacting flow problems, there is a wide range of important spatial and temporal scales. Because it is not possible to resolve phenomena on all of these scales, the smallest scales usually must be modeled. The basic assumption in most flow modeling of diffusion flames is that mixing of fuel and oxidizer takes place much more slowly than the chemical reactions and, therefore, a global reaction mechanism may be used. However, because a simulation that includes a detailed set of elementary reaction rates will be possible in the next few years, this program is designed in a modular form such that more detailed chemistry can be substituted later.

The conservation equations contain terms representing convection, conduction, species diffusion, chemical reaction, and viscous effects. In our algorithm, the separate processes are solved independently. The individual algorithms for the various processes and the coupling mechanism are described in the next section. A more detailed description of the method is presented by Laskey[7].

Convection

The solution to Eqs. (1)-(4) is obtained using the high-order implicit method BIC-FCT. The flux-corrected transport (FCT) itself is an explicit, finite-difference algorithm with fourth-order phase accuracy. Recently, Patnaik et al.[8] developed the barely implicit correction for flux-corrected transport, BIC-FCT, for subsonic flows by including an implicit correction step. This technique is based on the idea proposed by Casulli and Greenspan[9] that only the terms containing the pressure in the momentum equation and the velocity in the energy equation must be treated implicitly in order to avoid the sound-speed limitation on the time step. BIC-FCT has three steps. In the first step, the conservation equations are solved explicitly with FCT using a relatively large time step governed by the Courant condition on the fluid velocity. In the second step, the energy and momentum equations are rewritten in terms of a pressure correction δp. These equations can be manipulated such that only one elliptic equation for δp must be solved. In the third step, final values of momentum and energy are obtained by adding the pressure correction terms. Patnaik et al.[10] have incorporated this algorithm in a two-dimensional flame program to investigate laminar instabilities in premixed flames.

Molecular Diffusion

An algorithm for molecular diffusion has been formulated to estimate the molecular diffusion fluxes without having to solve a full matrix problem. The change in species concentration for each species k due to molecular diffusion only is

$$\frac{\partial n_k}{\partial t} = -\nabla \cdot n_k \mathbf{V}_k \tag{7}$$

where $\mathbf{V}_k$ is the diffusion velocity calculated using Fick's Law and then corrected by a procedure described by Kee et al.[11] to satisfy the requirement that the sum of the diffusion fluxes is zero. This method is algebraically equivalent to the first iteration of the DFLUX algorithm[12], an iterative approach that solves for diffusion velocities.

The change in total energy density due to molecular diffusion alone is

$$\frac{\partial E}{\partial t} = -\nabla \cdot \sum_{k=1}^{n_{sp}} n_k h_k \mathbf{V}_k \tag{8}$$

This energy term is calculated during the diffusion algorithm but is added to the total energy at the end of the time step.

The explicit finite-differencing procedure applied to this term introduces a numerical stability condition, $D_{km}\Delta t/[1/\Delta x^2 + 1/\Delta y^2] < 1/2$, where D_{km} is the diffusion coefficient for species k diffusing into a mixture. To maintain stability, this condition can require a time step smaller than that required by the convection algorithm, which could add substantially to the cost of the calculation. To avoid this problem, the diffusion term is evaluated several times during a convection time step. This is especially important if the elevated temperature of the reacting flow results in higher diffusion coefficients.

Binary diffusion coefficients are calculated from kinetic theory and are in the following form

$$D_{kl} = \frac{A_{kl}}{n} T^{B_{kl}} \tag{9}$$

where A_{kl} and B_{kl} are dependent on species k and l. Values for A_{kl} and B_{kl} have been tabulated by Kailasanath et al.[13]. The diffusion coefficient of species k in a mixture of n_{sp} species is calculated according to

$$D_{km} = \frac{1 - Y_k}{\sum\limits_{\substack{kk=1 \\ kk \neq k}}^{n_{sp}} X_{kk}/D_{k,kk}} \tag{10}$$

where Y_k is the mass fraction of species k, X_k the mole fraction of species k, and $D_{k,kk}$ the diffusion coefficient of species k diffusing into species kk.

Thermal conduction

A two-dimensional model has also been formulated to simulate thermal conduction. Restricting our attention only to the Fourier conduction term, the energy equation appears as

$$\frac{\partial E}{\partial t} = -\nabla \cdot (\kappa \nabla T) \tag{11}$$

As with the molecular diffusion algorithm, the use of explicit finite differencing introduces a stability limit for the thermal conduction calculation, $\kappa \Delta t/\rho c_p[1/\Delta x^2 + 1/\Delta y^2] < 1/2$, where $\kappa/\rho c_p$ is the thermal diffusivity. The thermal conduction term is evaluated several times during each convection time step in the same manner as the molecular diffusion term. Thermal conductivities κ_k for the individual species were calculated from kinetic theory over the temperature range 300 to 3300 K, and these values have been fit to a third-order polynomial. The mixture thermal conductivity is then calculated using the expression from Kee et al.[11]

$$\kappa = \frac{1}{2}\left[\sum_{k=1}^{n_{sp}} X_k\kappa_k + \frac{1}{\sum_{k=1}^{n_{sp}} X_k/\kappa_k}\right] \tag{12}$$

<u>Viscous Stress</u>

An algorithm that calculates the viscous terms in the momentum equation is included in the program. The momentum transport associated with viscosity is

$$\frac{\partial\rho\mathbf{V}}{\partial t} = -\nabla\cdot\tau \tag{13}$$

where

$$\tau = \left(\frac{2}{3}\mu - \zeta\right)(\nabla\cdot\mathbf{V})\mathbf{I} - \mu\left[(\nabla\mathbf{V}) + (\nabla\mathbf{V})^T\right] \tag{14}$$

and ζ is the second coefficient of viscosity and is assumed to equal zero. Equation (15) is rewritten in terms of an explicit finite-difference approximation that introduces a numerical stability condition, $\mu\Delta t/\rho[1/\Delta x^2 + 1/\Delta y^2] < 1/2$.

The values for μ_k were calculated from kinetic theory over the temperature range 300-3000 K and fit to a third-order polynomial[14]. The mixture viscosity was calculated from

$$\mu = \sum_{k=1}^{n_{sp}} \frac{X_k\mu_k}{\sum_{j=1}^{n_{sp}} X_j\Phi_{kj}} \tag{15}$$

where Φ_{kj} is a weighting factor

$$\Phi_{kj} = \frac{1}{\sqrt{8}}\left(1+\frac{M_k}{M_j}\right)^{-\frac{1}{2}}\left(1+\left(\frac{\mu_k}{\mu_j}\right)^{\frac{1}{2}}\left(\frac{M_j}{M_k}\right)^{\frac{1}{4}}\right)^2 \tag{16}$$

and for species k, the molecular weight is M_k.

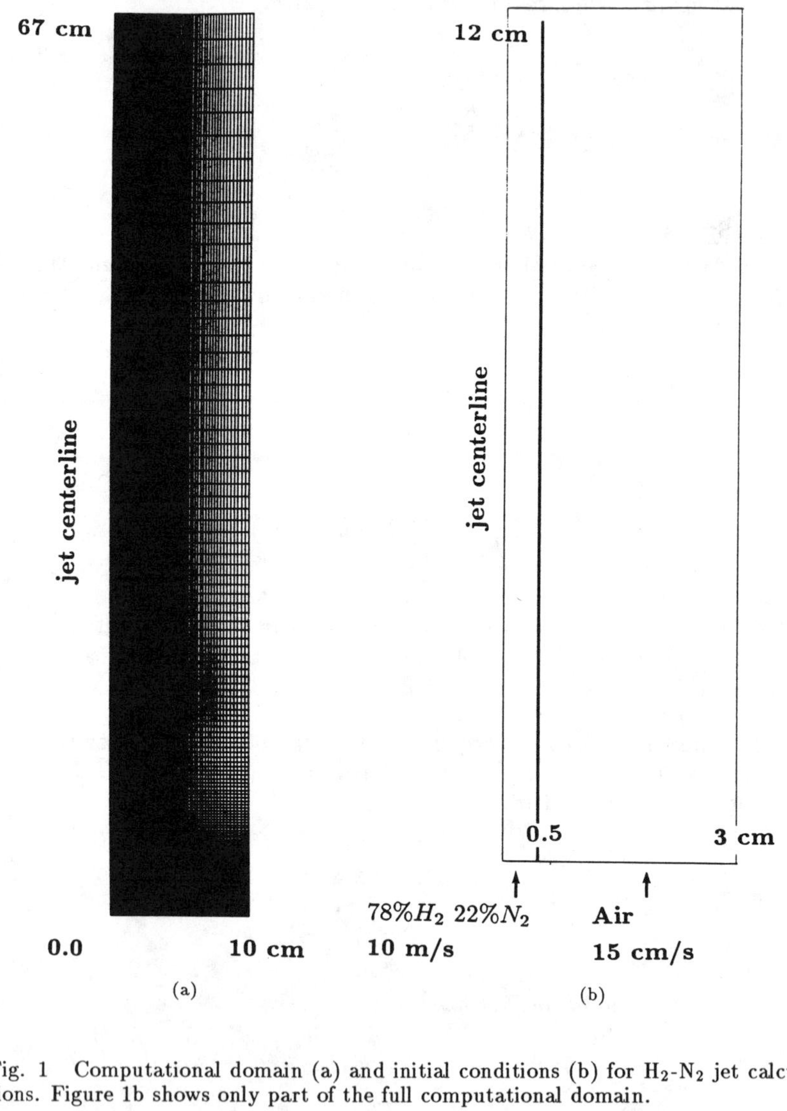

Fig. 1 Computational domain (a) and initial conditions (b) for H_2-N_2 jet calculations. Figure 1b shows only part of the full computational domain.

Model for Chemical Reaction

Ideally, we would like to simulate the chemical reaction by including a detailed set of elementary reactions to describe the production of the individual species and the energy release in the flame. However, the cost of computer time and memory makes this prohibitive for problems in which the flows are complex.

In the original flame sheet model that was proposed by Burke and Schumann[15], the fuel and oxidizer react completely and are not permitted to co-exist. Thus, the flame is an infinitesimal interface between regions of fuel and oxidizer. In the PDR model developed by Laskey[7], a single global reaction mechanism is used, but the fuel and oxidizer do not react instantaneously. Instead, the reaction occurs over a finite time interval. The fraction of fuel that reacts in a particular time step is determined from a separate simulation for a laminar flame. The reaction rate for the laminar flame was adjusted such that the maximum temperature was the adiabatic flame temperature for a stoichiometric mixture of fuel and oxidizer.

Coupling

A complete solution to the governing equations requires solving the terms for individual processes as well as accounting for the interaction among the processes. In the calculations that follow, we use time-step splitting that assumes that the net effect of all the processes is the sum of the solutions to individual processes. This technique is valid if the changes in the dependent variables during a time step are small.

In these computations, the changes in internal energy resulting from the individual processes are not incorporated into the solution as soon as they are computed but instead are accumulated. The entire change in internal energy is then added to the energy equation in the fluid convection step. The coupling technique has been described by Oran and Boris[16], and a modification by Patnaik, et al.[17] has been shown to allow for a greater addition of energy per time step while maintaining numerical stability.

Application of the Model to Jet Diffusion Flames

The computational domain and initial conditions for the jet diffusion flame calculations are shown in Fig. 1. The domain is 10×67 cm and consists of 128×192 cells. Cells of approximately 0.02 cm are concentrated around the jet exit. Beginning at $r = 1$ cm, the size of each cell is increased by 3% over the size of its neighboring cell for both simulations of a nonreacting jet and for the simulation of a reacting jet without viscosity. For the simulation of a reacting jet with viscosity, the cell size is stretched by 5% to increase the size of the domain to 27×67 cm. The cells in the axial direction for all simulations are stretched by 3% starting at $z = 1$ cm.

A fuel mixture consisting of 78% H_2 and 22% N_2 by volume flows through a jet of radius 0.5 cm at 10 m/s at the lower boundary. Air flows through the outer annular region $r \geq 0.5$ at 15 cm/s. The outer boundary is a free-slip wall. The inner boundary at $r = 0.0$ is the jet centerline. An outflow boundary is specified at $z = 67$ cm.

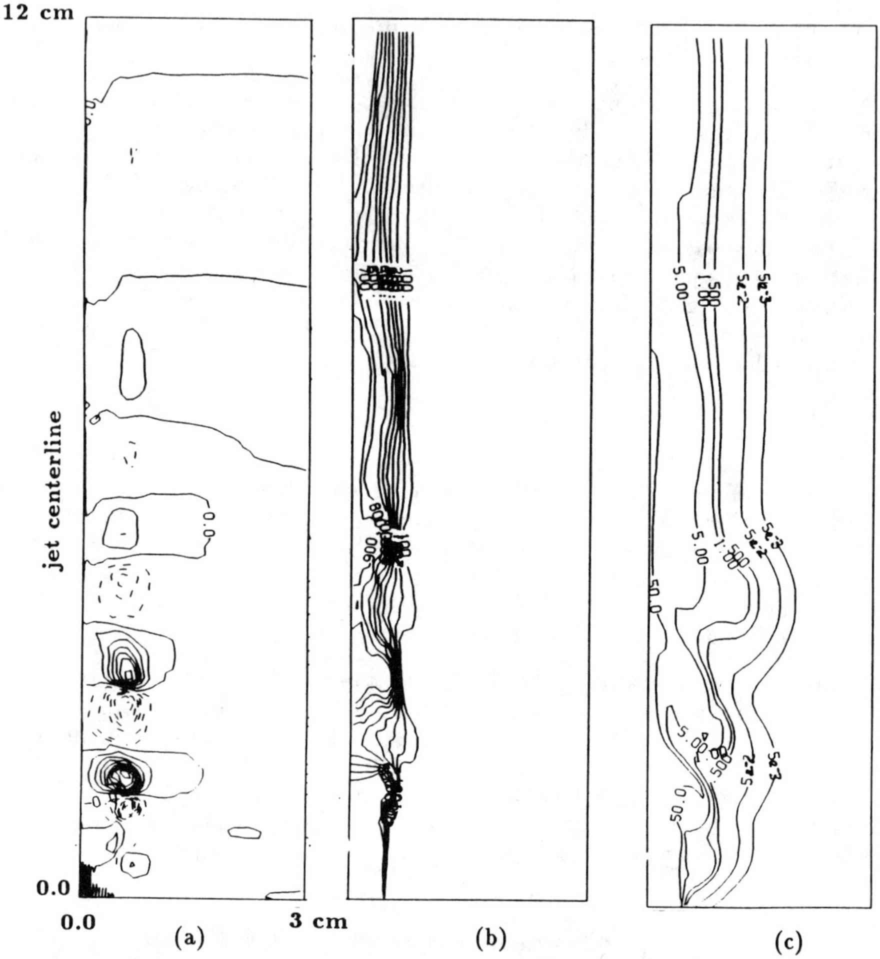

Fig. 2 Contours of (a) radial velocity, (b) axial velocity (cm/s), (c) local equivalence ratio for nonreacting H_2-N_2 jet without viscosity. Radial velocities are nondimensionalized by the value 516 cm/s, and contour values range from -1.0 to 1.0 in 0.1 increments. Dotted lines indicate negative values. Solid lines indicate positive values. Lowest contour value for axial velocity is 100 cm/s; highest is 900 cm/s.

Results

Nonreacting Jet without Viscosity

Figure 2 shows the instantaneous contours for radial velocity, axial velocity, and local equivalence ratio. The instability forms approximately 1.5 diameters downstream of the jet The radial velocity is a maximum (500 cm/s) at this point and decreases downstream. The structures appear fairly symmetrical throughout their development. Although they are not evident in this plot, there are weak instabilities at the top of the figure.

The contours of local equivalence ratio indicate that the concentration field is significantly affected by the rotating structures.

Nonreacting Viscous Jet

Figure 3 shows that the viscosity does not qualitatively alter the flow. The instability forms approximately 1.5 diameters downstream of the jet and weakens rapidly in the downstream direction. The maximum radial velocity is 300 cm/s and occurs in the center of the first recognizable structure. The axial velocity field and local equivalence ratio are similar to the case without viscosity.

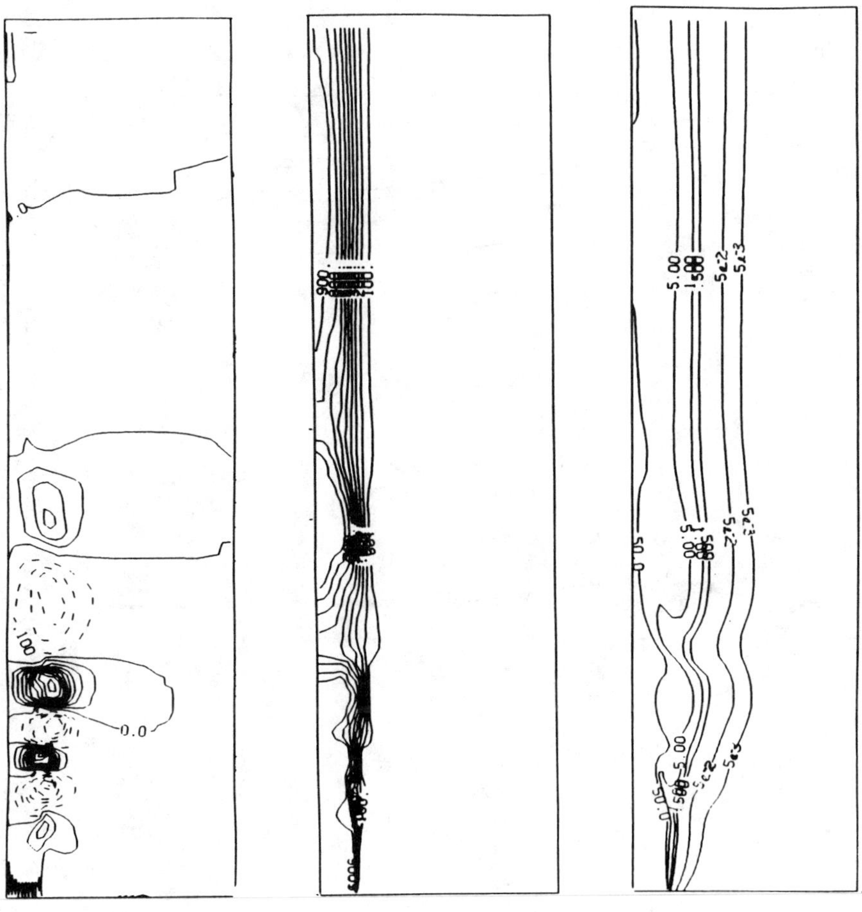

(a) (b) (c)

Fig. 3 Contours of (a) radial velocity, (b) axial velocity, (c) local equivalence ratio for nonreacting H_2-N_2 jet with viscosity. Radial velocities are nondimensionalized by the value 320 cm/s.

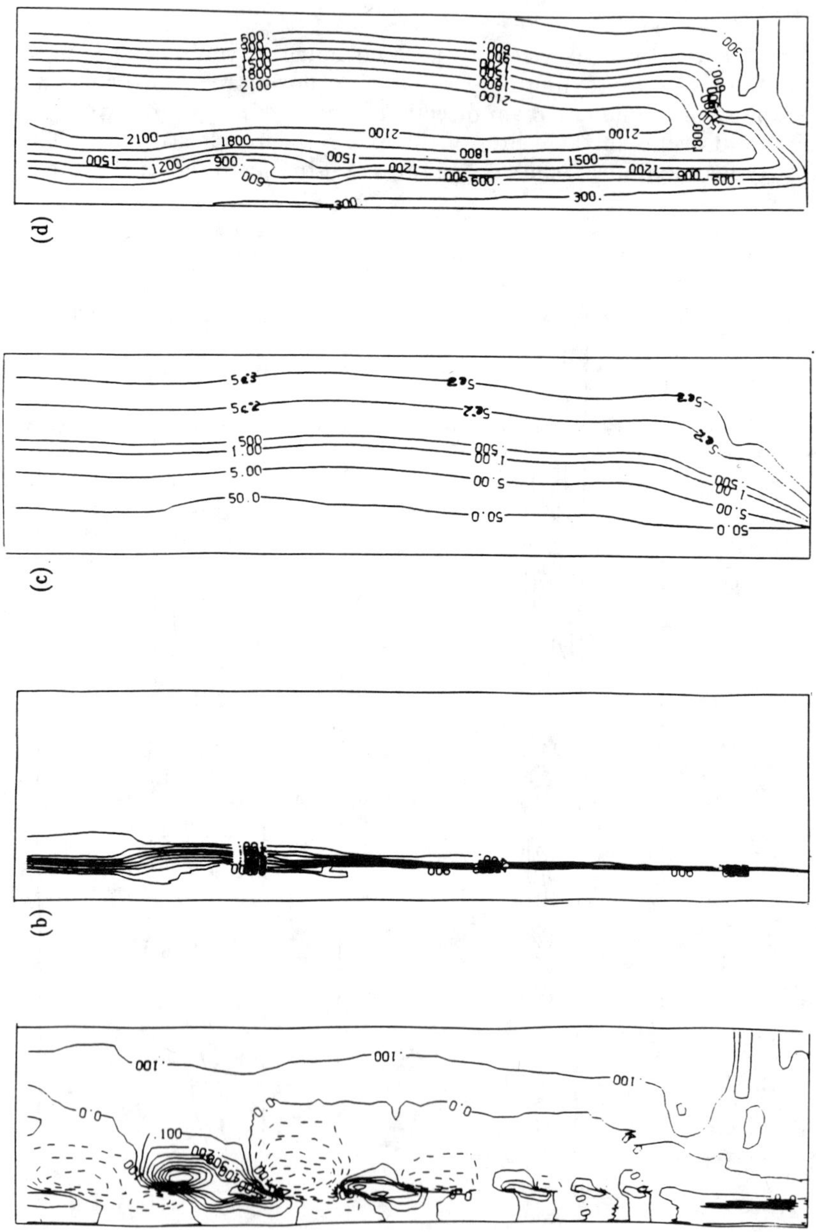

Fig. 4 Contours of (a) radial velocity, (b) axial velocity, (c) local equivalence ratio, (d) temperature for reacting H_2-N_2 jet without viscosity. Radial velocities are nondimensionalized by the value 83 cm/s.

Reacting Jet without Viscosity

When the fuel mixture reacts, the nature of the flowfield changes significantly. The radial-velocity contours in Fig. 4 indicate that the initial instablity still forms close to the jet, approximately three diameters downstream, but it is very weak at this point. The radial velocity increases downstream and is a maximum (80 cm/s) in the center of the structure at about 10 jet diameters. In addition, the structures are elongated as a result of the acceleration at the flame interface.

The lower radial velocity decreases the amount of fluid entrained and results in a lower growth rate of the mixing layer. The axial velocity contours show a narrower mixing layer than in the nonreacting case. The structures have not merged along the centerline, and a large potential core region still exists. The contours of equivalence ratio show that the stoichiometric surface is shifted radially outward and is outside the region of intense mixing.

The maximum temperature occurs along the reaction zone and is approximately 2100 K. A large region around the flame is heated as a result of conduction. The temperature field is distorted slightly at the base of the flame. The reaction in this region is quite intense. As the burned gases are accelerated downstream, coflowing air is entrained. In this simulation, the outside boundary is a free-slip wall, and the entrainment establishes a weak recirculation zone at the bottom of the domain.

Reacting Viscous Jet

Figure 5 shows that viscosity has a significant effect on the reacting jet. The magnitude of the radial velocity has been reduced by viscosity. Weak instabilities are still evident at approximately five diameters downstream, with a maximum velocity of 10 cm/s. Further downstream, the flowfield appears laminar. There is a large region of negative radial velocity indicated by dotted lines. The axial velocity contours show that the mixing region is shifted radially outward by the addition of viscosity. A potential core of undisturbed fluid is evident along the jet centerline. The maximum temperature is approximately 2100 K and occurs along the stoichiometric contour ($\phi = 1$).

Discussion

In these calculations, the Reynolds numbers for the jets without viscosity (Figs. 2 and 4) are poorly defined because there is only numerical viscosity that is dependent on the local cell size. The numerical viscosity is clearly lower than the physical viscosity because the velocity field changes when the viscous terms are included. This implies that the effective Reynolds number for the calculations without viscosity is higher than that for the calculations with viscosity.

The Reynolds number of the nonreacting jet (Fig. 3) is 2100 based on jet diameter and the properties of the fuel jet. At this Reynolds number, viscous effects are relatively small. The instabilities are primarily inviscid, and viscosity reduces the maximum radial velocity in the flow field without altering the nature of the instability. In the reacting jet (Fig. 5), however, the elevated temperature significantly increases the viscosity which reduces

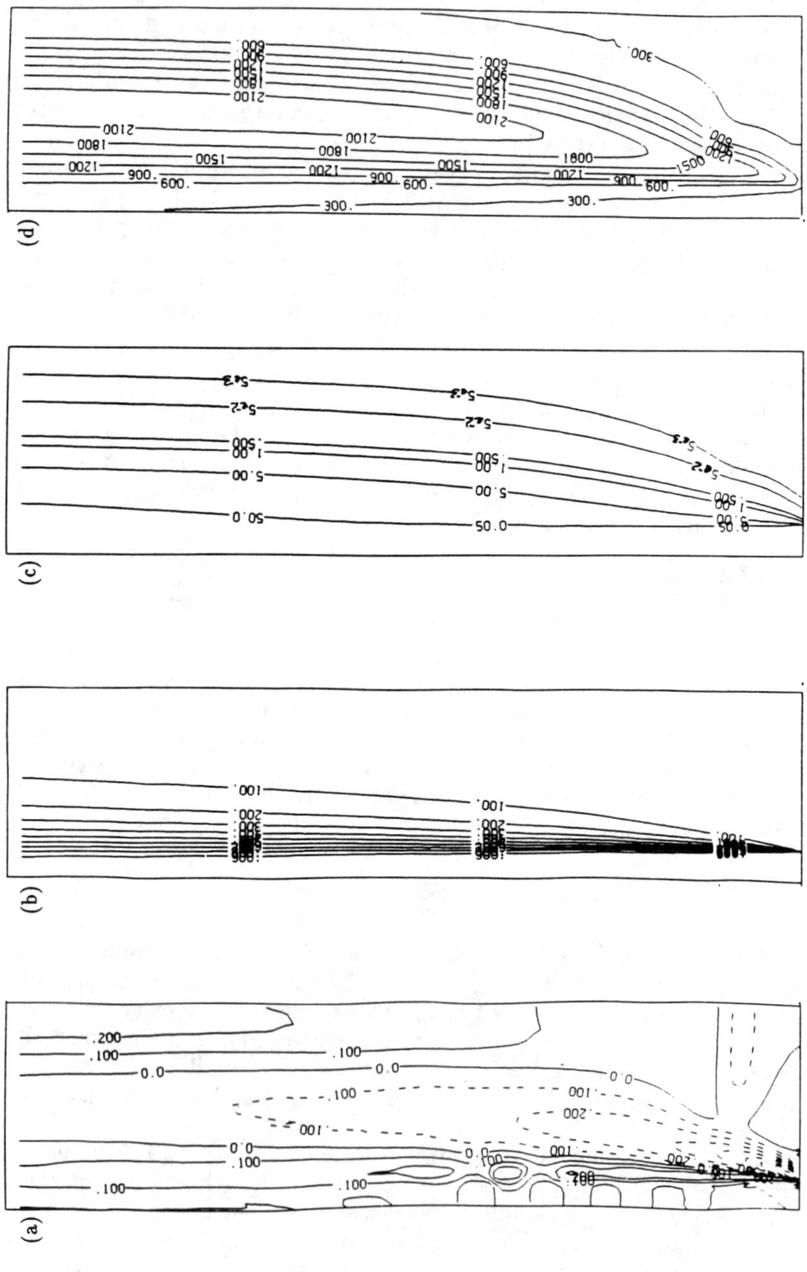

Fig. 5 Contours of (a) radial velocity, (b) axial velocity, (c) local equivalence ratio, (d) temperature for reacting H_2-N_2 jet with viscosity. Radial velocities are nondimensionalized by the value 37 cm/s.

the Reynolds number. At 1000 K, the Reynolds number is 300 and viscous effects are significant.

When reaction is included in the calculations without viscosity (Fig. 4), the heat release decreases the strength of the vortices and delays their formation. This agrees with the inviscid analysis by Mahalingam et al.[5], which shows that heat release stabilizes the mixing layer. Their work also indicates that the stabilizing effect of heat release is greater as the mixing layer spreads. This suggests that the layer becomes more stable further downstream. In our simulations (Fig. 4), the layer is more stable close to the jet and less stable further downstream. The analysis cannot, however, be compared directly because the heat release is not uniform in the axial direction in the calculations. The most intense reaction occurs close to the jet, where pure reactants mix. Further downstream, where the reactants are mixed with products, the heat release decreases significantly.

The calculations that do not include viscosity (Figs. 2 and 4) also agree with experimental data that show that heat release decreases the layer thickness[6] in high Reynolds number flows. When viscosity is included, the results are difficult to compare because the Reynolds number for the calculations is much less than that in the experiments.

The calculations in this paper show that heat release stabilizes the flowfield of an H_2-N_2 diffusion flame. When viscosity is added to the calculation for a reacting jet, the flow becomes almost laminar. In future work, we will calculate mean quantities in order to compare to experimental data.

Acknowledgments

This work was sponsored by the Naval Research Laboratory through the Office of Naval Research. The authors would like to thank Professor Norman Chigier from Carnegie-Mellon University for introducing us to the many interesting problems in diffusion flames. In addition, we thank Dr. W.M. Roquemore from the Air Force Wright-Patterson Aeronautics Laboratory for his support and suggestions.

References

[1] Yule, A.J., Chigier, N.A., Ralph, S., Boulderstone, R., and Ventura, J., "Combustion-Transition Interaction in a Jet Flame," *AIAA Journal*, Vol. 19, 1981, 752–760.

[2] Chen, L.D., Seaba, J.P., Roquemore, W.M., and Goss, L.P., "Buoyant Diffusion Flames," *Proceedings of Twenty-Second Symposium (International) on Combustion*, 1988, pp. 677-68,.

[3] Chamberlain, D.S., and Rose, A., "The Flicker of Luminous Flames," *Proceedings of First Symposium on Combustion*, The Combustion Institute, Pittsburgh, PA, 1928, 27–35.

[4] Laskey, K.J., Ellzey, J.L., and Oran, E.S., "A Numerical Study of an Unsteady Diffusion Flame," AIAA Paper 89-0572, 1989.

[5] Mahalingam, S., Cantwell, B., and Ferziger, J., "Effects of Heat Release on the Structure and Stability of a Coflowing, Chemically Reacting Jet," AIAA Paper 89-0661, 1989.

[6] Hermanson, J.C., Mungal, M.G., and Dimotakis, P.E., "Heat Release Effects on Shear Layer Growth and Entrainment," AIAA Paper 85-0142, 1985.

[7] Laskey, K. J., *Numerical Study of Diffusion Jet Flames*, Ph.D. thesis, Department of Mechanical Engineering, Carnegie-Mellon University, Pittsburgh, 1988.

[8] Patnaik, G., Boris, J.P., Guirguis, R.H., and Oran, E.S., "A Barely Implicit Correction for Flux-Corrected Transport," *Journal of Computational Physics*, 71, 1987, pp. 1–20.

[9] Casulli, V., and Greenspan, D., "Pressure Method for the Numerical Simulation of Transient, Compressible Fluid Flows," *International Journal of Numerical Methods in Fluids*, Vol. 4, 1984, p. 1001.

[10] Patnaik, G., Kailasanath, K., Laskey, K.J., and Oran, E.S., "Detailed Numerical Simulations of Cellular Flames," *Twenty-Second Symposium (International) on Combustion*, pp. 1527–1526, The Combustion Institute, Pittsburgh, PA, 1988.

[11] Kee, R.J., Dixon-Lewis, G., Warnatz, J., Coltrin, M.E., and Miller, J.A., *A Fortran Computer Code Package for the Evaluation of Gas-Phase Multicomponent Transport Properties*, Sandia National Laboratory, SAND 86-8246, 1986.

[12] Jones, W.W., and Boris, J.P.,"An Algorithm for Multispecies Diffusion Fluxes," *Computers and Chemistry*, Vol. 5, pp. 139–146, 1981.

[13] Kailasanath, K., Oran, E.S., and Boris, J.P., *A One-Dimensional Time-Dependent Model for Flame Initiation, Propagation, and Quenching*, Naval Research Laboratory, NRL Memo. Rept. 4910, 1982.

[14] Wilke, C.R., "A Viscosity Equation for Gas Mixtures," *Journal of Chemical Physics* Vol. 18, 1950, pp. 578–579.

[15] Burke, S.P., and Schumann, T.E.W., *Industrial Engineering Chemistry*, Vol. 20, 1928, pp. 998–1004.

[16] Oran, E.S., Boris, J.P., *Numerical Simulation of Reactive Flow*, Elsevier, New York, 1987.

[17] Patnaik, G., Laskey, K.J., Kailasanath, K., Oran, E.S., and Brun, T.A., *FLIC–A Detailed, Two-Dimensional Flame Model*, Naval Research Laboratory, NRL Memo. Rept. 6555, 1989.

Dynamics of Turbulent Diffusion Flame Surface

H. Yamashita,* G. Kushida,† and T. Takeno‡
Nagoya University, Nagoya, Japan

Abstract

The time-dependent behavior of two-dimensional diffusion flame developed in a shear layer of a fuel jet into a coflowing airstream was studied in terms of the flame surface model of infinite chemical reaction rate with unity Lewis number. The time-dependent compressible Navier-Stokes equation with the equation for coupling function was solved numerically by using finite difference method. The temperature dependence of viscosity and diffusion coefficient was taken into account so as to study effects of the increases of these coefficients on the flame behavior. The numerical calculation was performed for the case when CH_4 is injected into a coflowing airstream with injection Reynolds number of 2,000. The flame was found to remain laminar when the ratio of airstream velocity to fuel jet velocity is 0.5, whereas it becomes turbulent in the downstream when the ratio is 0.1. It was found that the radial distributions of density and transport coefficients in the upstream laminar region play dominant roles in the flow instability, and hence, in the downstream transition from laminar to turbulent flame. The density distribution has somewhat complicated effects due to coexistence of two distinct instabilities, while the increase in viscosity in the flame has a stabilizing effect. But the increase in diffusion coefficient has a stronger effect since it shifts the flame surface outside the shear layer. The response of the flame to a low frequency oscillating flow was studied.

---

*Associate Professor, Department of Mechanical Engineering.
†Research Associate, Department of Mechanical Engineering.
‡Professor, Department of Mechanical Engineering.

The calculated time-dependent flow and flame behavior reveals that eddies are generated along the flame surface near the tip to entrain the outer air into the jet. This causes some part along the surface to be pinched off to throw out a flame island in the airstream. This flame break off is caused as a result of instantaneous shortage of fuel and has nothing to do with finite rate chemistry. In this sense it may be called "physical extinction" in contrast to the usual "chemical extinction." This extinction and the subsequent process of further fragmentation of the flame islands suggest the importance of large scale eddy motion in the time-dependent flame behavior.

Introduction

The laminar flamelet concept is becoming very popular in the study of turbulent diffusion flame, as well as in the study of turbulent premixed flame. In this concept chemical reactions are so fast that they occur in asymptotically thin layers - called flamelets - embedded within the turbulent flowfield.[1] The local instantaneous structure of the flamelet is assumed to be the same as that of a quasisteady laminar flame. In the limit of fast chemistry the reaction zone becomes concentrated within an infinitely thin surface. This is the flame surface model, and in the past the model has successfully been applied in many studies of laminar diffusion flames, especially in the study of gasdynamic structure of the flame.[2] The time scales of molecular transport processes are much larger than those of chemical reactions. However, the application of the model to the structure study of turbulent diffusion flame must be controversial. The time scales of turbulent transport processes, in general, are not necessarily larger than those of reactions.

The turbulent diffusion flame is usually formed in the shear layer of a fuel jet into a coflowing oxidizer stream. Recent experimental investigations have shown that large scale coherent structures dominate the mixing process, not only in nonreactive shear layers but also in reactive shear layers and flame jets.[3-5] The large scale motion of the luminous flame zone is familiar now. In such a flame the time scale of dominant turbulent transport must be larger than that of reactions, and there is a possibility that the main features of flame dynamics can be described in terms of the flame surface model, without being involved in the complicated interaction process between reaction and turbulence.[6] Furthermore, a two-dimensional fuel jet flame can be studied in terms of the two-dimensional flow calculation, since the large scale coherent structures remain two-dimensional. The small scale three-dimensional turbulence may modify the evolution process of these structures,[7] but it will not alter their essential characteristics. Another suggestion based on experimental observations was that the increase in kinematic viscosity in the flame must play some roles in the flow stability of the flame jet, and hence, in the transition from laminar to turbulent flame. This effect, therefore, must be taken into consideration in the study of flame surface dynamics.[6]

The objective of the present study is to study the dynamics of turbulent diffusion flame developed in a two-dimensional fuel jet, in terms of the flame surface model of infinite reaction rate. In addition, Lewis number will be assumed unity to make full use of the Shvab-Zeldovich formulation of the coupling function.[8] Then the time-dependent compressible Navier-Stokes equations, with the equation for the coupling function, will be solved numerically by using a finite difference method to predict the time-dependent flame surface behavior. The temperature dependence of viscosity and diffusion coefficient will be taken into account so as to study the effects of the increase in these coefficients on the flame surface behavior.

Model and Assumptions

The model adopted in the present study is shown schematically in Fig. 1. The flow is time-dependent and plane two-dimensional. Axial and radial coordinates are represented by x and y, and the origin is taken at the center of injector exit plane. Axial and radial velocities are represented by u and v, respectively, and z is the coupling function defined later. The fuel is injected through an injector of width d with the fully developed velocity distribution and uniform temperature T_o into a coflowing airstream of velocity u_∞ and temperature T_∞. $Y_{i,o}$ is mass fraction of species i present in the fuel, while $Y_{i,\infty}$ is that in the airstream. The thickness of injector plates separating air and fuel is zero, and the boundary layer is developed naturally on the air side as it flows along the upper and lower plates. At the injector exit, the infinite reaction rate produces the flame as soon as the fuel contacts the air, and the flame surface develops immediately from the downstream edges of the separating plates.

In the study, we shall adopt the following assumptions:

(1) The mixture undergoes an overall one-step irreversible reaction described by

$$\nu_F F+\nu_O O+\nu_I I \rightarrow \nu_P P+\nu_I I \tag{1}$$

where ν_i is stoichiometric coefficient of species i, while F, O, I, and P represent fuel, oxygen, inert gas, and product, respectively.

(2) The reaction rate is infinitely fast and the reaction is concentrated in an infinitesimally thin surface.

(3) The mixture behaves like an ideal gas.

(4) The Soret and Dufour effects, as well as the pressure diffusion, can be neglected. Concentration diffusion coefficients D of each species are equal.

(5) Specific heat at constant pressure c_p of the mixture is constant.

(6) Lewis number given by $Le = \lambda/(\rho c_p D)$ is equal to unity, where λ and ρ are thermal conductivity and density of the mixture, respectively.

(7) Viscosity coefficient μ and diffusion coefficient D depend on temperature T in the following manner[9]:

$$\mu \propto T^{0.7}, \qquad D \propto T^{1.7} \tag{2}$$

(8) In the energy equation, work done by pressure and viscous dissipation can be neglected. $\partial p/\partial t$ is neglected as well.

Now, the conservation equations for mass and momentum are given in the following forms:

$$\frac{\partial \rho}{\partial t}+\frac{\partial}{\partial x}(\rho u)+\frac{\partial}{\partial y}(\rho v)=0 \tag{3}$$

$$\rho\left(\frac{\partial u}{\partial t}+u\frac{\partial u}{\partial x}+v\frac{\partial u}{\partial y}\right)=-\frac{\partial p}{\partial x}+\frac{4}{3}\frac{\partial}{\partial x}\left(\mu\frac{\partial u}{\partial x}\right)-\frac{2}{3}\frac{\partial}{\partial x}\left(\mu\frac{\partial v}{\partial y}\right)+\frac{\partial}{\partial y}\left[\mu\left(\frac{\partial v}{\partial x}+\frac{\partial u}{\partial y}\right)\right] \tag{4}$$

$$\rho\left(\frac{\partial v}{\partial t}+u\frac{\partial v}{\partial x}+v\frac{\partial v}{\partial y}\right)=-\frac{\partial p}{\partial y}+\frac{4}{3}\frac{\partial}{\partial y}\left(\mu\frac{\partial v}{\partial y}\right)-\frac{2}{3}\frac{\partial}{\partial y}\left(\mu\frac{\partial u}{\partial x}\right)+\frac{\partial}{\partial x}\left[\mu\left(\frac{\partial u}{\partial y}+\frac{\partial v}{\partial x}\right)\right] \tag{5}$$

Density is related to pressure, temperature, and mass fractions through the equation of state.

$$p=\rho RT\sum_{i=1}^{4}(Y_i/m_i) \tag{6}$$

where m_i is molecular weight of species i and R is the universal gas constant. Now the coupling function z is introduced by the following equation:

$$z=\frac{Y_I-Y_{I,\infty}}{Y_{I,0}-Y_{I,\infty}} \tag{7}$$

Then the energy conservation equation and the mass conservation equations to each species reduce to the following single equation for z:

$$\rho\left(\frac{\partial z}{\partial t}+u\frac{\partial z}{\partial x}+v\frac{\partial z}{\partial y}\right)=\frac{\partial}{\partial x}\left(\rho D\frac{\partial z}{\partial x}\right)+\frac{\partial}{\partial y}\left(\rho D\frac{\partial z}{\partial y}\right) \tag{8}$$

The flame surface is located at the position where z becomes equal to z^* given by the equation

$$z^*=\frac{\frac{Y_{O,\infty}}{j}}{1+\frac{Y_{O,\infty}}{j}} \tag{9}$$

where $j = m_O\nu_O/m_F\nu_F$ is grams of oxygen required to burn 1 g of fuel to effect complete combustion. The surface divides the whole flow region into two regions: fuel side and air side. Then the mass fractions and temperature in the respective region can be given in the following way[8]:

Fuel Side ($1 \geq z \geq z^*$)

$$Y_O=0 \tag{10}$$

$$Y_F=z-\frac{1}{j}Y_{O,\infty}(1-z) \tag{11}$$

$$Y_P=\frac{m_P\nu_P}{m_O\nu_O}Y_{O,\infty}(1-z) \tag{12}$$

$$Y_I=Y_{I,\infty}(1-z) \tag{13}$$

$$T=T_\infty+(T_O-T_\infty)z+\frac{q_O}{m_F\nu_F c_p}\frac{1}{j}Y_{O,\infty}(1-z) \tag{14}$$

Air Side ($z^* \geq z \geq 0$)

$$Y_F=0 \tag{15}$$

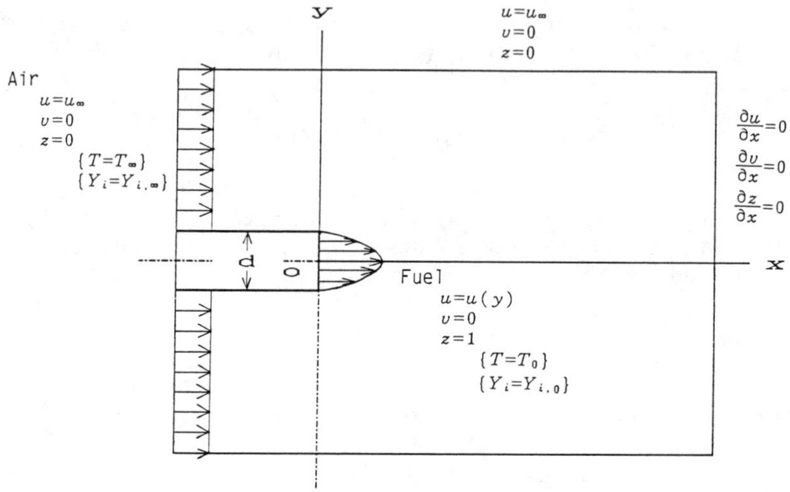

Fig. 1 Two-dimensional jet and boundary conditions.

$$Y_O = Y_{O,\infty} - (j + Y_{O,\infty}) z \tag{16}$$

$$Y_P = \frac{m_P \nu_P}{m_F \nu_F} z \tag{17}$$

$$Y_I = Y_{I,\infty} (1 - z) \tag{18}$$

$$T = T_\infty + (T_O - T_\infty) z + \frac{q_O}{m_F \nu_F c_p} z \tag{19}$$

In these equations $q_O/m_F\nu_F$ is the heat of reaction per unit mass of fuel. Equations (14) and (19) give the flame surface temperature T^* as

$$T^* = T_\infty + (T_O - T_\infty) \frac{\dfrac{Y_{O,\infty}}{j}}{1 + \dfrac{Y_{O,\infty}}{j}} + \frac{q_O}{m_F \nu_F c_p} \frac{\dfrac{Y_{O,\infty}}{j}}{1 + \dfrac{Y_{O,\infty}}{j}} \tag{20}$$

The boundary conditions for u, v, and z are described in Fig. 1. The initial conditions are that at time 0 the fuel is injected suddenly. We are interested, however, in the flame behavior in the later stage when the time has elapsed long enough that effects of the initial conditions have disappeared completely.

The above equations can conveniently be expressed in terms of nondimensional quantities, in which the reference length is d. The reference velocity u_O is the fuel injection velocity averaged over y. Then the reference time is given by d/u_O, while the values of the injected fuel are used as reference for other dependent variables, including viscosity and diffusion coefficient. Then the resulting nondimensional parameters are Reynolds number and diffusion Reynolds number defined by

$$Re=\rho_o u_o d/\mu_o, \quad R_D=u_o d/D_o=(\mu_o/\rho_o D_o)Re \tag{21}$$

where suffix $_o$ denotes the value for injected fuel.

Numerical Calculation

The numerical calculation was performed by using the finite-difference approximation. The technique adopted was SIMPLE method developed by Patankar,[10] and the third order upwind difference scheme (QUICK Scheme) was used for convective terms. The time advance was made by using the fully-implicit method with the interval of 0.5 in the nondimensional time. The computational domain is shown in Fig. 2. The numbers of grid are 121 and 101, respectively, in x and y directions. Two types of the grid sizing were

Grid	X1/d	X2/d	Y1/d=Y2/d
A	29	408	92
B	320	1600	92

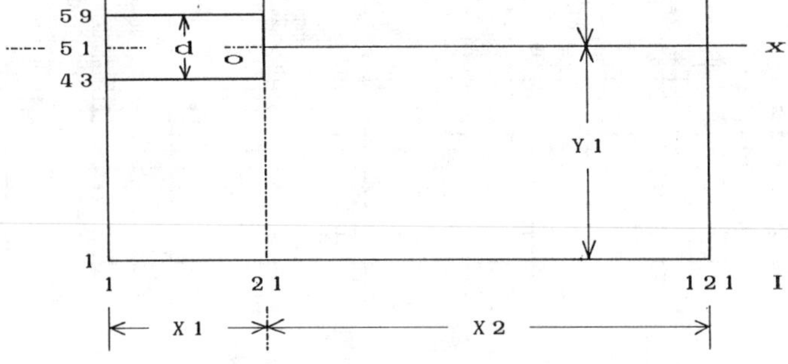

Fig. 2 Computational domain and grid sizing.

used and they are shown in a table in the figure. Type A was used for the study of flow stability in the immediate downstream of the injector, while type B for the study of behavior of the whole flame. The mesh distribution of type A in the vicinity of injector exit is shown in Fig. 3.

The fuel adopted was CH_4, and air was assumed to be composed of 79% N_2 and 21% O_2 by volume to give j = 4.000, $Y_{I,\infty} = 0.7670$, $Y_{O,\infty} = 0.2330$, and $z^* = 0.05504$. The pressure was the atmospheric pressure. The air and fuel temperatures were the room temperature, that is $T_o = T_\infty = 300$ K. The flame surface temperature was assumed to be equal to the adiabatic flame temperature $T^* = 2225.8$ K of the stoichiometric mixture calculated by the chemical equilibrium calculation. The resulting nondimensional flame surface temperature is $\theta^* = T^*/T_o = 7.419$. In most calculations the airstream velocity was put equal to a half of the average fuel injection velocity, or $u_\infty/u_o = 0.5$. The calculation was performed for the fixed Reynolds number of Re = 2,000. The corresponding diffusion Reynolds number is $R_D = 1,428$, where the values of $\mu_o/\rho_o = 0.1571$ cm^2/s and $D_o = 0.2200$ cm^2/s are used.

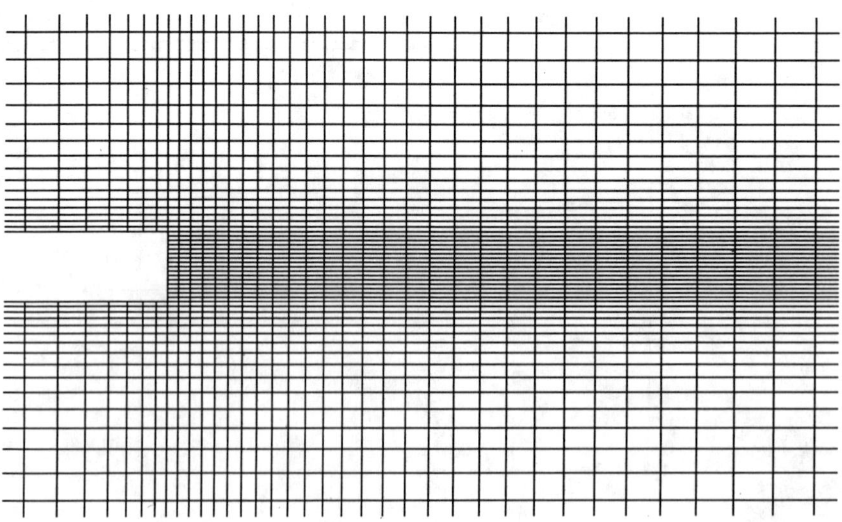

Type A

Fig. 3 Mesh distribution in the vicinity of injector exit.

Results of Calculation

1. Flow Stability

In order to study the effects of variable density and transport properties on the transition from laminar to turbulent flame, the flow stability in the immediate downstream of the injector exit was studied numerically for the fixed Reynolds number of 2,000 and velocity ratio of $u_\infty/u_0 = 0.5$. Table 1 shows the six cases studied for comparison. Case A is the air jet injecting air at room temperature into the coflowing airstream. There are no concentration, density, and temperature distributions in the jet, and the transport coefficients remain constant through the whole flowfield, giving unity for the nondimensional quantities. Case B is the frozen jet, in which CH_4 at room temperature is injected without reaction, and there are small density distributions due to the mixing of CH_4 into the air, but the transport coefficients remain constant. Case C is the flame jet of the present study. Density changes due to the mixing of fuel, air, and product, and due to the temperature increase in the flame. The viscosity, as well as the diffusion coefficient, increases due to the temperature increase in accordance with Eq. (2). Case D is the hypothetical flame jet, in which the density and diffusion coefficient change in the same way as case C, but the viscosity is assumed to remain constant. Case E is opposite to case D, and now the diffusion coefficient is assumed to remain constant. Finally, in case F both the viscosity and the diffusion coefficient are assumed to remain constant. The numerical method adopted was as follows. In the numerical

Table 1 Effects of variable density and transport coefficients on flow instability

Case	ρ	μ	D	Stability
A (Air jet)	1.0	1.0	1.0	Stable
B (Frozen jet)	variable	1.0	1.0	Stable
C (Flame jet)	variable	variable	variable	Stable
D (Flame jet)	variable	1.0	variable	Stable
E (Flame jet)	variable	variable	1.0	Unstable
F (Flame jet)	variable	1.0	1.0	Unstable

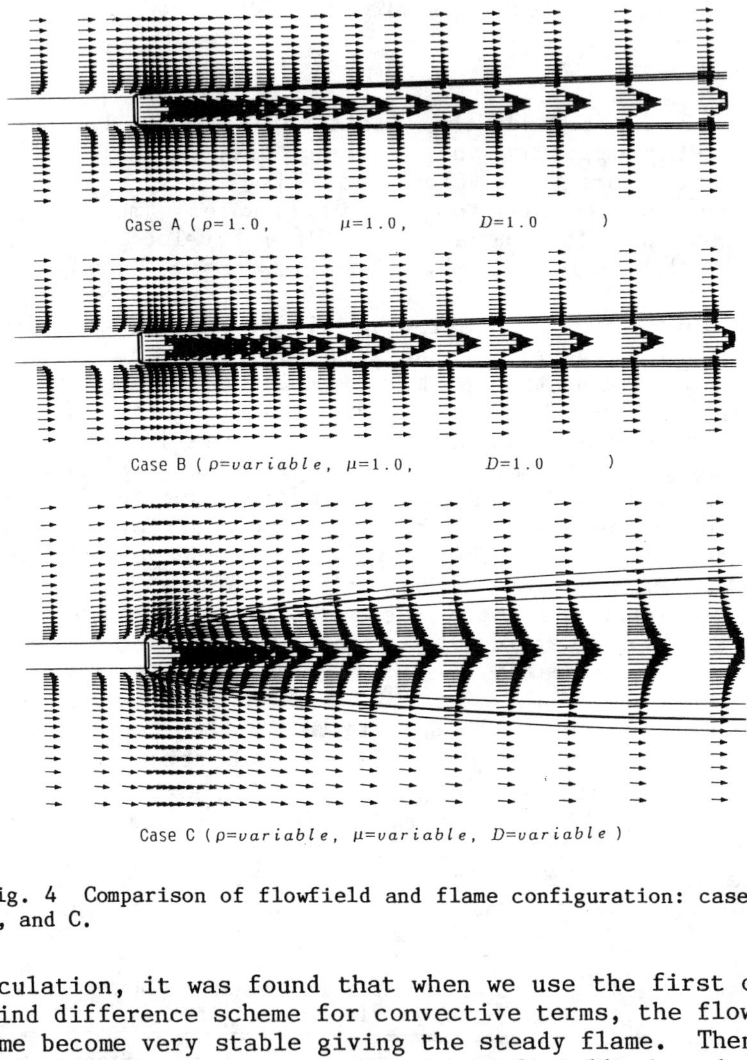

Fig. 4 Comparison of flowfield and flame configuration: cases A, B, and C.

calculation, it was found that when we use the first order upwind difference scheme for convective terms, the flow and flame become very stable giving the steady flame. Then the steady laminar flow was produced artificially by adopting the first order scheme, and then the third order scheme was introduced to study the subsequent transient behavior by advancing the time step. No artificial disturbances were introduced, and if the flow is unstable it became turbulent spontaneously after certain time steps.

In Figs. 4 and 5 the calculated instantaneous flowfields for the respective cases are compared. The corresponding radial distributions of Y_i, ρ, θ, D, v, and u at x = 33.6 are shown in Figs. 6-11. The three lines in Figs. 4 and 5 represent the flame surface and the surfaces 0.5 z^* and 2.0 z^*. Cases A through D are very stable, and,

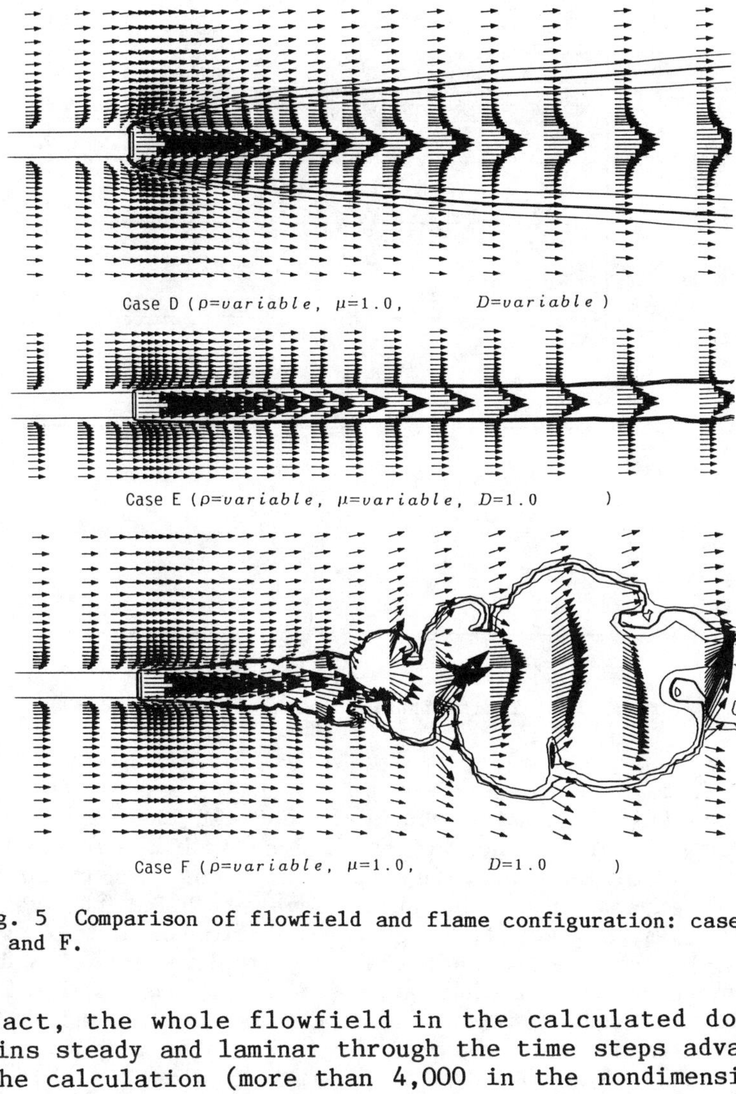

Fig. 5 Comparison of flowfield and flame configuration: cases D, E, and F.

in fact, the whole flowfield in the calculated domain remains steady and laminar through the time steps advanced in the calculation (more than 4,000 in the nondimensional time). The obtained flowfield is almost the same for A and B. The boundary layers are developed along the injector separating plates, and wakes are formed at the downstream end of the plates. The wakes still remain at this axial position for both cases, as is seen in the radial distribution of u shown in Figs. 6 and 7. The only difference is the density distribution due to mixing of CH_4 with air shown in Fig. 7, but it seems that this affects little the flow. In case C, on the other hand, the jet is greatly expanded in radial direction since the radial diffusion of CH_4 is accelerated due to the increase of D in

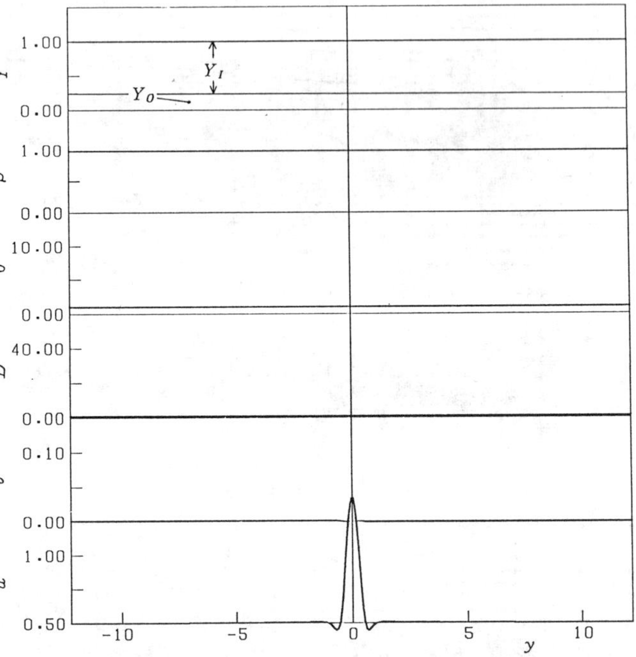

Fig. 6 Radial distributions at x = 33.6 for case A.

the flame, which reaches up to 30 times at the flame surface (Fig. 8). It should be noticed that the position of large density change is shifted outside the shear layer. Also, μ is increased (up to four times) to produce the radial expansion of the shear layer to make the velocity gradient more gradual, and to eliminate the initial wakes. It is interesting to note that the flow acceleration due to the heat release is not so apparent in the flowfield. In case D the radial diffusion of CH_4 is accelerated as well, and the flame shape shown in Fig. 5 is very similar to that of case C. The position of density gradient is shifted as well. However, μ remains small and the shear layer is not so expanded as in case C. Some traces of the initial wakes can be seen in the radial distribution of u in Fig. 9. The velocity gradient remains steep, and we cannot expect the damping of the flow instability by the increase in μ. Therefore, the flow is more unstable as compared to case C, although this is not apparent in the obtained results.

The flow and flame became unsteady in case E. As is seen in Fig. 5, some fluctuations appear at the downstream end. In view of the fact that no fluctuations appear in case D, we notice that the increase in viscosity alone is not so effective to make the flow stable. The increase in diffusion coefficient is much more effective. Figure 10 reveals that the width of shear layer is as narrow as those of cases A and B to make the velocity gradient steeper than that of case D. The most unstable one is case F, and the fluctuations appear at the immediate downstream of the injector exit already. The comparison with case E suggests that the increase in viscosity, of course, has some stabilizing effects. Furthermore, the comparison with case B indicates that the radial density distribution produced by the flame surface makes the flow very unstable, and without the stabilizing effects of the increases in viscosity and diffusion coefficient, the flow becomes turbulent most easily.

2. Instantaneous Flowfield and Flame Configuration

The time-dependent behavior of the whole flame was studied by using type B grid sizing. The results show that the flow and flame in the expanded domain are always unsteady, in contrast to the fact that they remain steady when type A sizing is used (case C shown in Fig. 4.) The flame is slowly oscillating back and forth at the tip. Figure 12 shows an example of the calculated instantaneous flowfield and flame configuration at an arbitrary instant. In Fig. 12 the radial distance is expanded eight times so as to show the details in the flame jet. The top figure shows the distribution of local velocity vector and the flame surface configuration. There are three curves to show the surfaces: the bold curve represents the flame surface where $z = z^*$, whereas, the other lighter curves represent the surfaces $z = 0.5\ z^*$ and $2.0\ z^*$. A rectangle in the immediate downstream of injector indicates the flowfield region shown in Figs. 4 and 5, explained before. The middle figure shows the contours of coupling function as compared to the flame surface described by the bold curves. The contour maximum and minimum are 1 and 0.1, respectively, with the interval of 0.1. It should be noticed that the flame surface is located at the position where z takes the very small value of 0.05504. The bottom figure shows the contours of nondimensional temperature. The maximum is 7, whereas the minimum is 2 and the interval is 1.

As is seen in the figure, the oscillating flame and flow remain almost symmetric in the upper and the lower parts of the axis, which suggests that the observed fluctuations in the flow and flame are not due to turbulence. In the downstream of the flame tip, there are

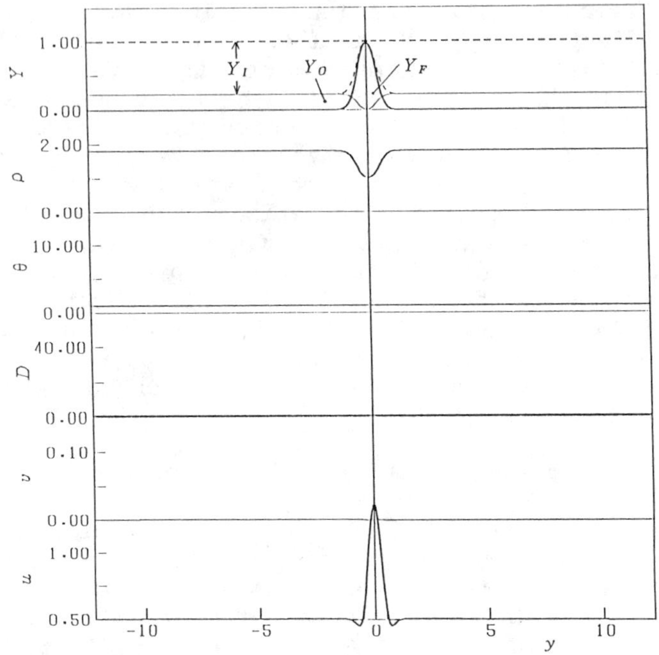

Fig. 7 Radial distributions at x = 33.6 for case B.

some small islands of high temperature gas being convected downstream. These islands are produced as a result of the flame fluctuation at the tip in the previous instants. Figure 13 shows the corresponding stereograms of the coupling function, temperature, and pressure. The radial distance is expanded further to show the details in the flame jet. It is interesting to note that there exists a longitudinal pressure variation of large wavelength, in addition to the small scale pressure fluctuations corresponding to the temperature fluctuations. It is this pressure variation which causes the flame oscillation.

3. Physical Extinction

Figures 14 and 15 illustrate some examples of the time-dependent flame surface behavior in eight consecutive sequences with the nondimensional time interval of 64. The behavior of flame surface and contours of coupling function are shown in Fig. 14, whereas, that of the corresponding nondimensional temperature contours is shown in Fig. 15. As mentioned before, the flame is slowly oscillating back and forth at the tip, in accordance with

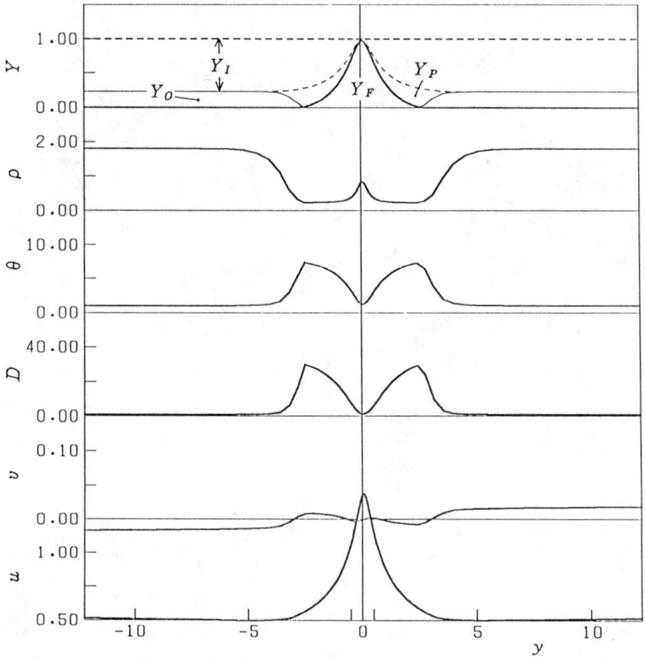

Fig. 8 Radial distributions at x = 33.6 for case C.

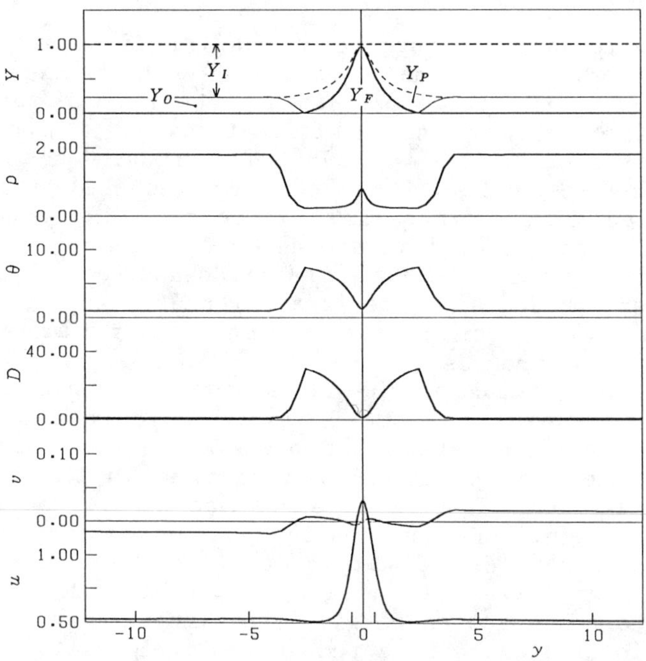

Fig. 9 Radial distributions at x = 33.6 for case D.

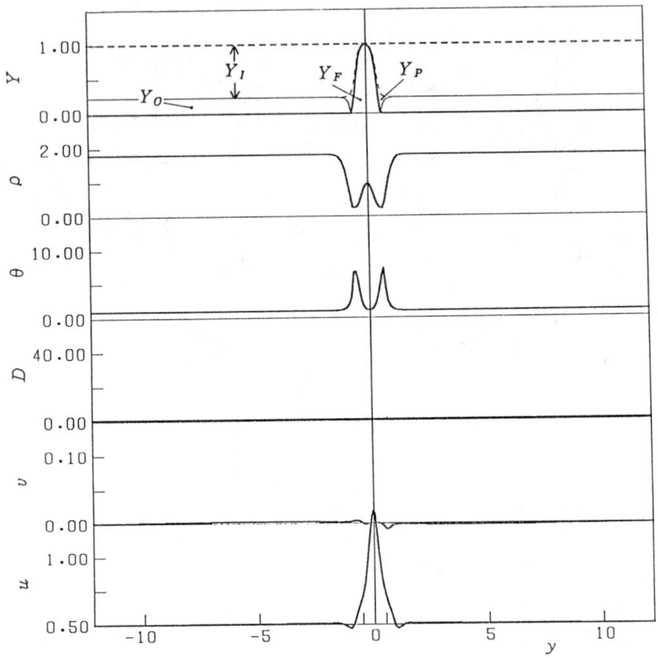

Fig. 10 Radial distributions at x = 33.6 for case E.

the longitudinal pressure wave, and the figures just show the phase when the flame is extending downstream. It can be seen that as the flame extends downstream some part along the surface becomes pinched off to throw out a flame island in the airstream (at t = 416). The produced flame island continues to burn as it is convected downstream and is fragmented further into small pieces. This can be seen clearly in the first five sequences of Fig. 14. The first sequence shows two islands in the downstream of the tip, which were produced by the flame break off in the previous instants. The subsequent sequences reveal this fragmentation process.

The extinction of the flame surface at the tip is a very interesting phenomenon since we have assumed the infinite reaction rate. The break off has nothing to do with finite rate chemistry. Then what is the cause for this extinction? Figure 16 shows the details of the flow and flame configuration leading to the extinction. The instantaneous local velocity vector, relative to the outer uniform air flow, and the flame surface with 0.5 z^* and 2.0 z^* contours are shown with the time interval of 16 for the region indicated by a rectangle in Fig. 14. It can be

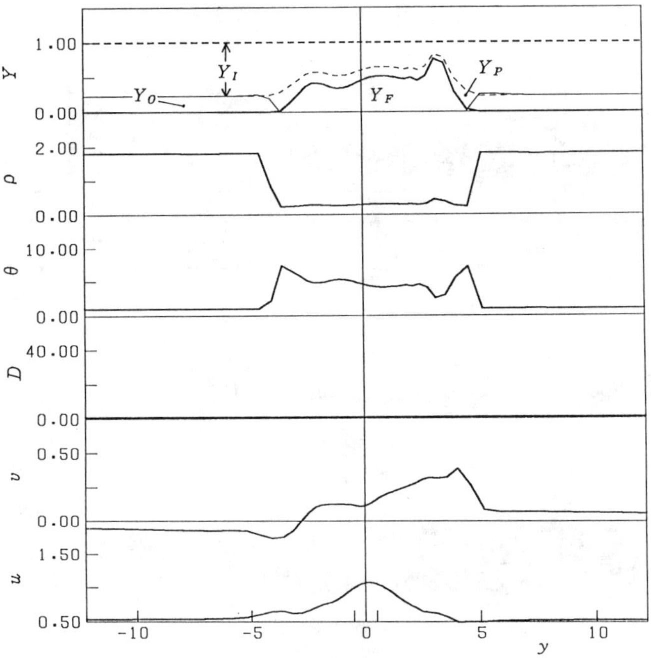

Fig. 11 Radial distributions at x = 33.6 for case F.

seen clearly how symmetric eddies are generated, as the flame extends, along the surface near the tip to entrain the outer air into the jet. The effect of this convective air flow is stronger than that of molecular diffusion, and if the entrainment is rapid enough there appears along the surface a point where no fuels are present momentarily. Then the flame has to be pinched off at that position to produce a flame island of bullet shape pointing upstream. We may say that this kind of break off can happen only through the nonsteady flow motion. It is not due to finite rate chemistry as is the case of usual extinction, but due to shortage of fuel. In this sense, we may call the phenomenon "physical extinction" in contrast to the usual "chemical extinction."

Similarly, Figs. 17 and 18 reveal the details of the further fragmentation process of the produced flame islands, shown in the first four sequences of Fig. 14. As is seen in the figures, symmetric eddies are developed along the islands as well, and the outer air is introduced into the islands through the eddies. Then the original bullet type shape of the island has to change into the characteristic shape like a fork with two prongs, because

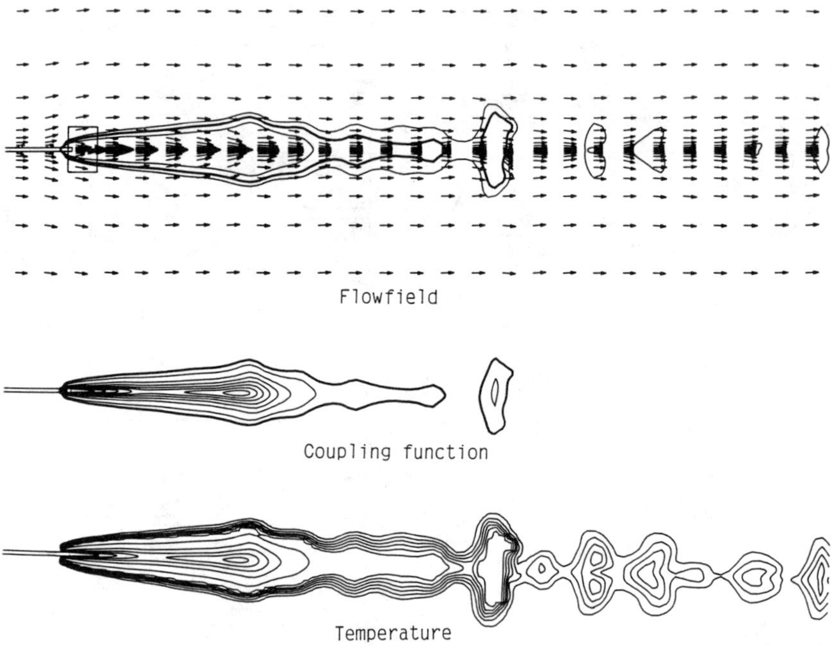

Fig. 12 Instantaneous flowfield and flame configuration.

of the air entrainment at the back end. The further entrainment causes the break off of one of the prongs at its root to separate a small flame island, and the break off of the remaining prong follows resulting in three small flame islands. In this way, the fragmentation of the islands occurs through the physical extinction process here again. The further fragmentation will not occur any more, since now the size of the fragmented islands becomes smaller than that of the eddies. The islands move as a whole in the eddy motion, decreasing their size through the combustion governed by the molecular diffusion process until the fuel is consumed completely.

Discussions

1. Flow Stability

The flow instability and the resulting transition from laminar to turbulent flame in the fuel jet diffusion flame have been the subject of many investigators and yet our present understandings are still incomplete.[11,12] The

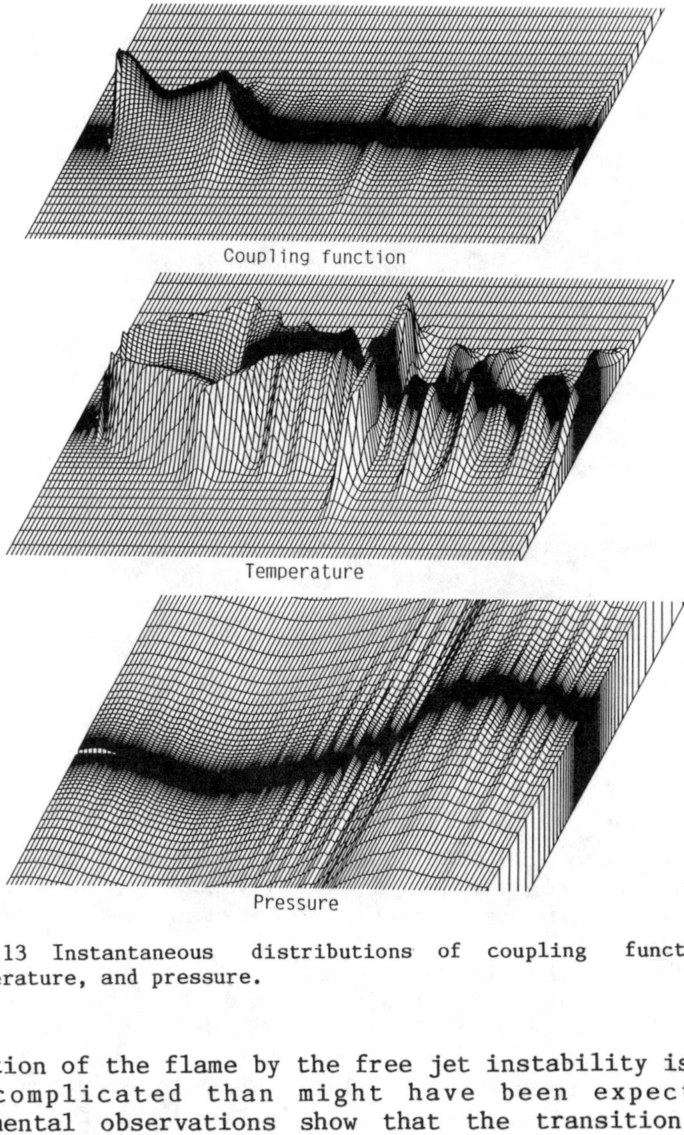

Fig. 13 Instantaneous distributions of coupling function, temperature, and pressure.

transition of the flame by the free jet instability is much more complicated than might have been expected.[5] Experimental observations show that the transition from laminar to turbulent flow is delayed remarkably by the presence of the diffusion flame, and this has mainly been attributed to the increase in the viscosity.[5,6,11,12] However, the present study has shown that the increase in diffusion coefficient is much more important, since it shifts the flame surface outside the jet shear layer. Although the finding is based on the result of a specific case and its degree of generality is not clear at this stage, the implication seems plausible and important

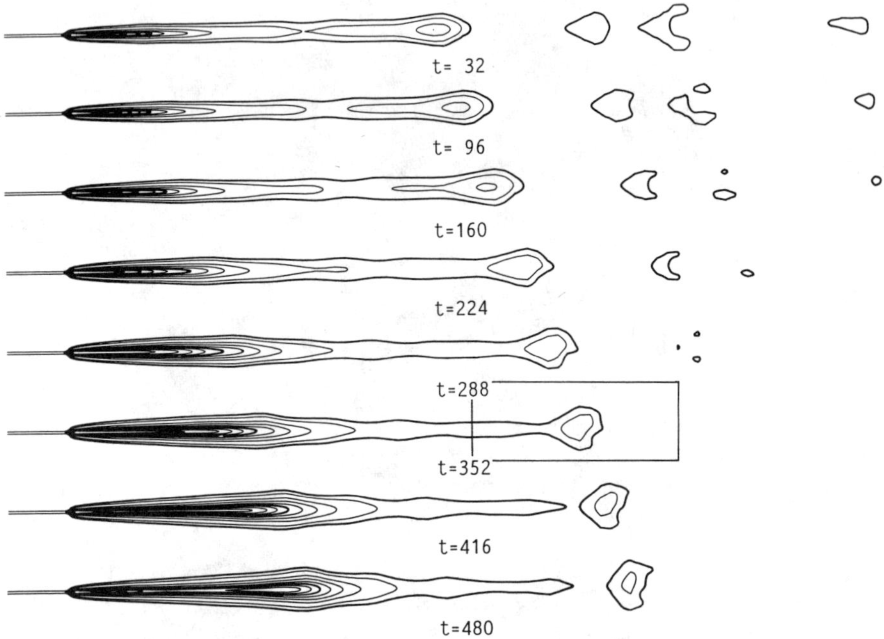

Fig. 14 Time-dependent flame behavior; flame surface and contours of coupling function.

universally. The effect will be especially important in most hydrocarbon flames, in which values of z^* are rather small and the flame surface is located at the outer edge of shear layer.

On the other hand, the effect of the variable density is not clear yet. The results shown in Figs. 4-11 could not elucidate this effect, since the first four cases of Table 1 remain laminar and we could not distinguish their respective degrees of stability. This is probably because the given velocity ratio u_∞/u_0 is too large for the fixed Reynolds number to cause the transition. In order to confirm this, the calculations were made for the case when the velocity ratio is as low as 0.1 for the same Reynolds number. The result is shown in Fig. 19, where the instantaneous flowfield is compared for the frozen jet (case B) and the flame jet (case C). Now, the both flows become unstable to cause the transition in the calculated domain. The transition point appears clearly in the frozen jet, as well as in the flame jet. The resulting induction

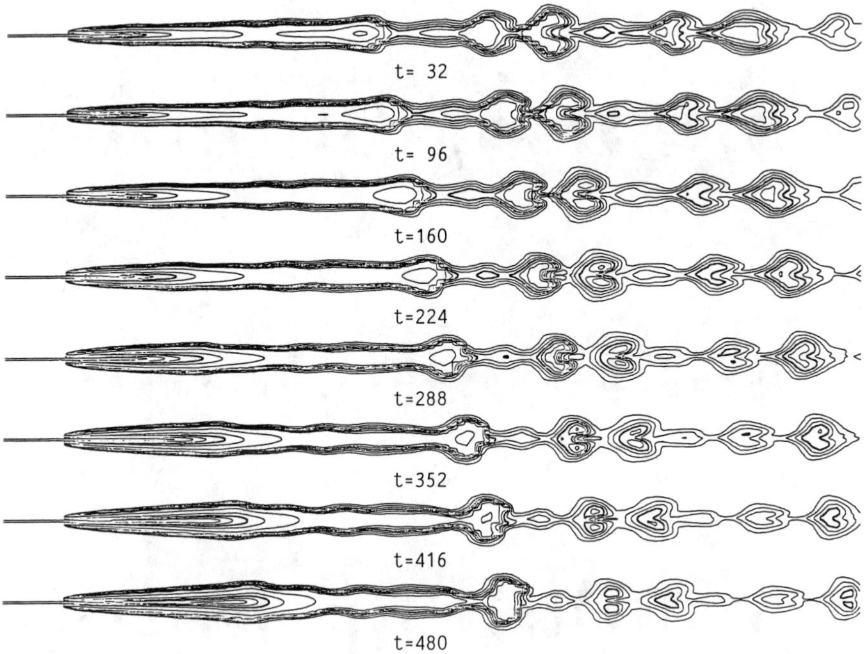

Fig. 15 Time-dependent flame behavior; contours of temperature.

length from the injector exit to the transition point becomes shorter for the frozen jet than for the flame jet, suggesting now the flame jet is more stable than the frozen jet.

The above results may seem somewhat contradictory to the previous results shown in Figs. 4 and 5. The comparison of case B and case F suggests that the radial density distribution produced by the flame surface makes the flow unstable. However, we should remember that there are no damping effects due to the increases in viscosity and diffusion coefficient in the latter case. The sole effect of a density distribution on the two-dimensional shear flow instability has been well studied theoretically[13] and experimentally.[14] The flow becomes unstable when the density gradient is opposite to the velocity gradient, whereas, the flow tends to be stable when both gradients are the same direction. That is, the flow is unstable when light gas is flowing faster than heavy gas. In the case of flame jet, however, the situation is somewhat complicated. The presence of the flame surface produces the peculiar density distribution as shown in Fig. 8. The density gradient is negative in the core jet, whereas, it

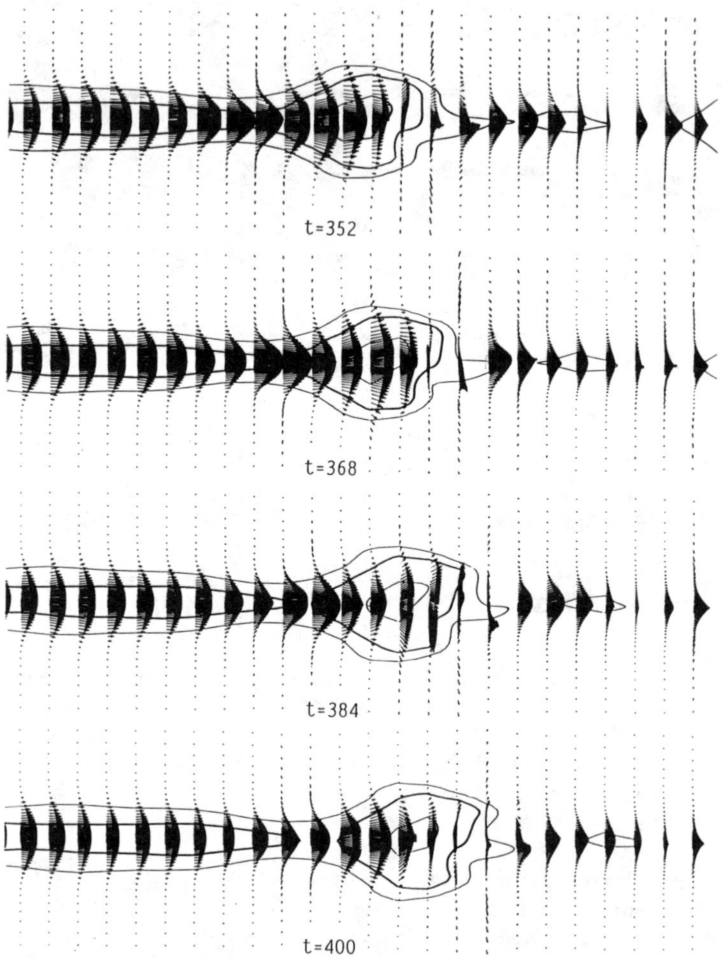

Fig. 16 Physical extinction process due to shortage of fuel.

is positive in the outer layer. Then the flow becomes relatively stable in the inner core jet, whereas, it is more unstable in the outer shear layer. The stability of the whole flame is determined by superposition of these two distinct flow instabilities. The comparison of case B with case F suggests that the superposition makes the flow very unstable, if there are no damping effects due to the increases in viscosity and diffusion coefficient. It is this latter effects which makes the flame jet more stable than the frozen jet. Anyway, all these things will

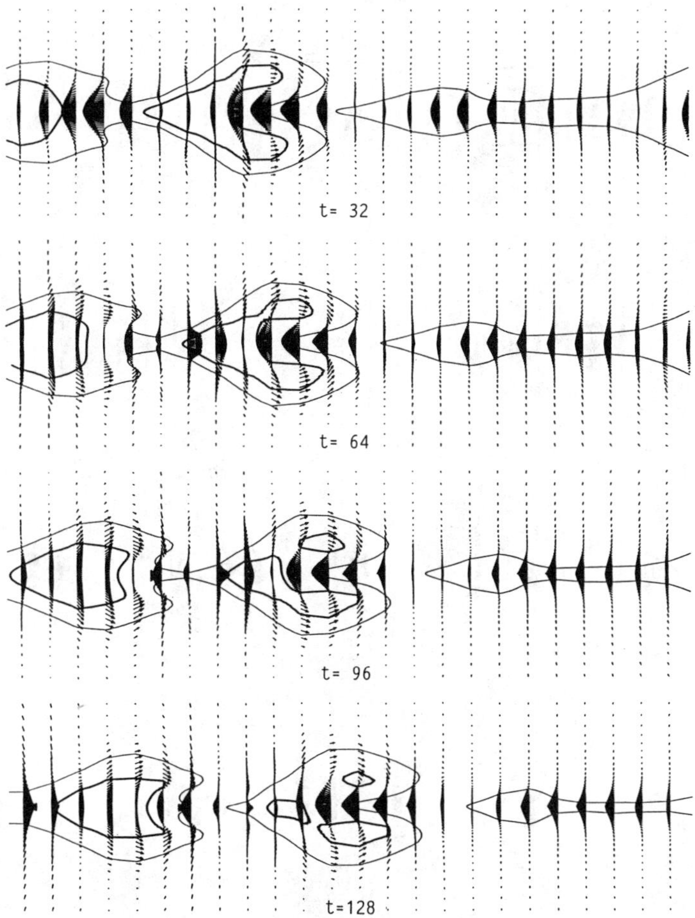

Fig. 17 Fragmentation process of flame islands.

contribute to make the flame transition so complicated. A complete survey of these effects is currently underway.

2. Physical Extinction

One limitation of the present calculation is the space resolution. The finite mesh size adopted in the calculation gives the minimum scale that can be simulated. This limitation, combined with the limitation of two-dimensional calculation, forces to treat the unsteady flow of rather large scale fluctuation. This is the reason why we adopted the relatively high value of 0.5 for the airstream velocity ratio. The induced flow is the slowly varying large scale

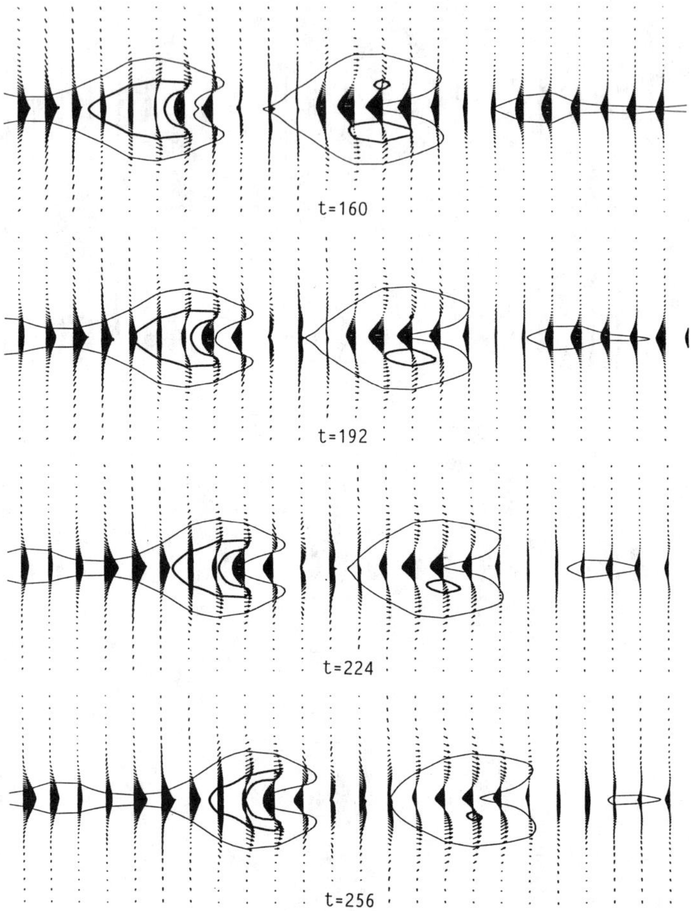

Fig. 18 Fragmentation process of flame islands (continued).

fluctuation. It is not the turbulent flow in the usual sense, but a kind of oscillating flow. We may think that the above mentioned limitations, as well as the flame surface assumption, are not serious in this fluctuating flow, and the calculation can simulate what actually happens. Another problem is what is the origin of this fluctuating flow. The reason is not clear at this stage, but should be related to the interaction between the adopted boundary conditions and the numerical technique, or SIMPLE method. Although this is the problem to be made clear, we think that the response of the flame surface to the induced fluctuating flow can be simulated correctly in this calculation.

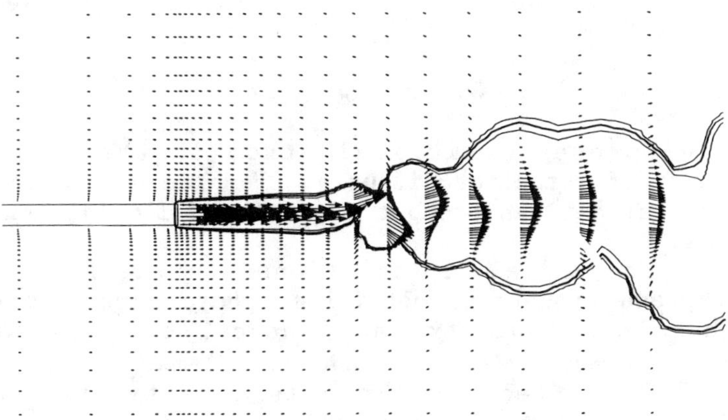

Case B (ρ=*variable*, μ=1.0, D=1.0)

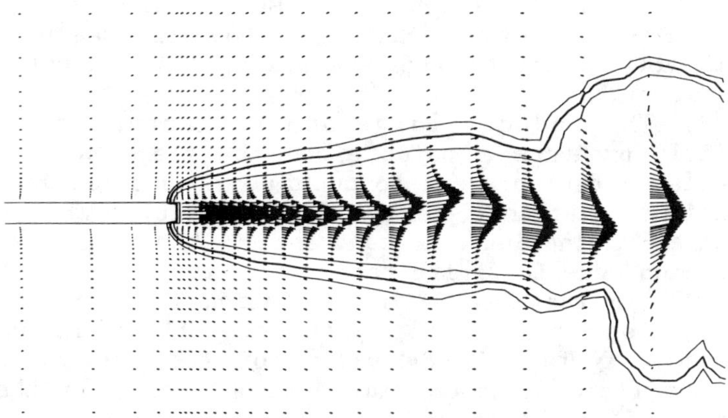

Case C (ρ=*variable*, μ=*variable*, D=*variable*)

Fig. 19 Comparison of flow stability of frozen jet and flame jet.

We think that the observed physical extinction is very likely to occur in actual situations. One example is a singing diffusion flame,[15] which shows a periodic flame separation. Since the actual chemical reactions always proceed at finite rates, no extinction processes can be free from their effects. Usually the chemical extinction should occur just before the physical extinction, since the temperature in the region is reduced rapidly due to the air entrainment. However, it is important to note that in nonsteady flow motions the flame can be extinguished even when the reaction rate is infinitely fast. In this sense, we think that the concept of "physical extinction" is important. Anyway, we believe that the present study has revealed the importance of fluid dynamics in the time-dependent flame behavior.

Concluding Remarks

The present numerical study on the gasdynamic structure of turbulent diffusion flame developed in a two-dimensional jet has brought about the following concluding remarks.

1) The radial distributions in the flame jet of density and transport coefficients have important effects on the flow stability, and, hence, on the resulting transition from laminar to turbulent flame.

2) The effect of variable density is somewhat complicated because of coexistence of two distinct flow instabilities. The increase in viscosity due to the high temperature in flame jet has a stabilizing effect. The increase in diffusion coefficient, however, has a stronger stabilizing effect through the shift of flame outside the shear layer.

3) The induced large scale fluctuation in the flowfield produces a point along the flame surface where there is no fuel present to continue combustion leading to an extinction at that point. This extinction may be called "physical extinction" in contrast to the usual "chemical extinction" due to finite rate chemistry.

4) In the fluctuating diffusion flame, large scale eddy motions at the tip produce a flame island, and subsequent further fragmentation of the island. In this way, the fluid dynamics play dominant roles in the time-dependent flame behavior.

Acknowledgments

The present study was supported in part by a Grant-in-Aid from the Japanese Ministry of Education (No. 01627507). The numerical calculation was performed by using the FACOM VP-200 of the Computation Center of Nagoya University. The authors wish to express their sincere thanks to the staff of the Center.

References

[1]Peters, N., "Laminar Flamelet Concepts in Turbulent Combustion," Proceedings of the Twenty-First Symposium (International) on Combustion, The Combustion Institute, Pittsburgh, 1988, pp.1231-1250.

[2]Takeno, T., "A Gasdynamic Analysis of a Counter-Flow Diffusion Flame," Combus. Sci. Technol., Vol.5, 1972, pp.99-106.

[3]Takeno, T., and Kotani, Y., "A Study on the Structure of Turbulent Jet Diffusion Flames," Combus. Sci. Technol., Vol.10, 1975, pp.45-57.

[4]Takeno, T., and Kotani, Y., "An Experimental Study on the Stability of Jet Diffusion Flame," Acta Astronautica, Vol.2, 1975, pp.999-1008.

[5]Takeno, T., and Kotani, Y., "Transition and Structure of Turbulent Jet Diffusion Flames," Progress in Astronautics and Aeronautics, Vol.58, AIAA, 1978, pp.19-35.

[6]Takeno, T., and Kotani, Y., "Comments on Structure of Turbulent Jet Diffusion Flame," Acta Astronautica, Vol.6, 1979, pp.1009-1010.

[7]Fedorchenko, A. T., "Action of Small-Scale Turbulence on the Development of Coherent Structures in a Mixing Layer," Soviet Physics Dokl., Vol.33, 1988, pp.717-719.

[8]Williams, F. A., Combustion Theory, 2nd ed., Benjamin/Cummings, Menlo Park, CA, 1985, p.73.

[9]Strehlow, R. A., Combustion Fundamentals, McGraw-Hill, New York, 1984, p.67.

[10]Patankar, S. V., Numerical Heat Transfer and Fluid Flow, McGraw-Hill, New York, 1980, p.1.

[11]Coats, C. M., and Zhao, H., "Transition and Stability of Turbulent Jet Diffusion Flames," Proceedings of the Twenty-Second Symposium (International) on Combustion, The Combustion Institute, 1989, pp.685-692.

[12]Savas, O., and Gollahalli, S. R., "Flow Structure in Near-Nozzle Region of Gas Jet Flames," AIAA J., Vol.24, 1986, pp.1137-1140.

[13]Maslowe, S. A., and Kelly, R. E., "Inviscid Instability of an Unbounded Heterogeneous Shear Layer," J. Fluid Mech., Vol.48, 1971, pp.405-415.

[14]Davey, R. F., and Roshko, A., "The Effect of a Density Difference on Shear-Layer Instability," J. Fluid Mech., Vol.53, 1972, pp.523-543.

[15]Gaydon, A. G., and Wolfhard, H. G., Flames, Their Structure, Radiation and Temperature, 4th ed., Chapman and Hall, London, 1979, p.183.

Structure of the Reaction Zone in a Reacting Mixing Layer

A. F. Ghoniem,* O. M. Knio,† and G. Heidarinejad†
Massachusetts Institute of Technology, Cambridge, Massachusetts

Abstract

The dynamics, mixing, and chemical reaction between two streams of fuel and oxidizer flowing at two different velocities are computed using the vortex/transport element methods. Two- and three-dimensional simulations are obtained for the nonreacting mixing layer to assess the accuracy of the numerics and to determine the range of validity of the two-dimensional approximation. Solutions for the reacting mixing layer are obtained for different Damkohler numbers and for fast chemistry and finite-rate kinetics. Results indicate that at low Damkohler numbers, the reaction occurs within the large eddies in a distributed reaction zone mode. At high Damkohler numbers, the reaction occurs at the outer edges of the eddies in a thick reaction sheet mode. In both cases, the products are concentrated within the large eddies where the vorticity concentration is highest. The results emphasize the role of the entrainment currents associated with the formation of the large eddies on the burning rate in the mixing layer.

I. Introduction

In this paper, we report the results of a numerical simulation of the dynamics, mixing, and combustion in a mixing layer developing between two streams of fuel and oxidizer flowing at different velocities. The goals of this study are to determine 1) how the flowfield that develops during the rollup of the shear layer contributes to combustion between the two streams, and 2) how the reaction zone is affected by this flowfield. For this purpose, we

---

* Associate Professor, Department of Mechanical Engineering.
† Graduate Research Assistant.

vary some of the chemical parameters, most notably the Damkohler number, and study the structure of the reaction zone.

The shear layer analyzed herein possesses some of the important components that characterize a turbulent field — e.g., high concentration of vorticity, which evolves in time and space through a series of instabilities producing a highly unsteady flow with a range of length scales. Moreover, while the vorticity is evolving these dynamical changes, it induces strong local entrainment currents and a strong unsteady strain field, both of which are expected to have an important influence on the burning process. The shear layer thus provides a more complete model to study turbulent-combustion interactions than an isolated vortex[1] or a stagnation point flow.[2]

To limit the complexity of the interactions and to isolate the various effects, we analyze an incompressible flow model in which density variation with energy release is neglected. A compressible flow model is introduced in a companion paper in this volume. The numerical schemes we have developed for this flow are grid-free Lagrangian field methods that limit numerical diffusion, maximize numerical resolution around zones of strong gradients, and utilize computational elements that adapt to the evolving flow. The vortex/transport element method is most suited to study these high Reynolds and Peclect number problems in which convective currents determine the rate of mixing. Utilizing its high resolution, one can investigate the important question of how the flowfield determines the geometry and location of the reaction zone.

A compact formulation of the problem is presented in Sec. II. The vortex/transport element method is summarized in Sec. III with emphasis on the properties of the scheme that make it most suitable to study turbulence-combustion interactions. In Sec. IV, we review the results for a nonreacting layer in two and three dimensions, and analyze the results for the reacting mixing layer in two dimensions. In Sec. V, we summarize the conclusions.

II. Formulation

The model we use to study vorticity-combustion interactions in shear flow is based on the following assumptions: 1) the Mach number is small; 2) the heat release is low so that the temperature rise does not affect the thermodynamic or transport properties of the fluid, or the dynamics of the flow; 3) both reactants and products behave as perfect gases with equal molecular weight and equal densities; and 4) the chemical reaction is singlestep and irreversible. It is described by $F + O \longrightarrow P$ (for more details, see Refs. 3 and 4). Under these conditions, the governing equations are

$$\frac{d\omega}{dt} = \frac{1}{Re} \nabla^2 \omega \tag{1}$$

$$\frac{ds_j}{dt} = \frac{1}{Pe_j} \nabla^2 s_j + Q_j W_j \tag{2}$$

where $\omega = \nabla \times u$ is vorticity, $u = (u,v)$ is velocity, $\nabla . u = 0$ for the incompressible flow considered here, t is time, d/dt is the material or Lagrangian derivative, Re is the Reynolds number, and ∇^2 is the Laplacian operator. The model requires the transport of four scalars: temperature T, fuel concentration c_F, oxidizer concentration c_O, and product concentration. For the temperature, $s_1 = T$, Pe_1 is the Peclet number, Q_1 is the enthalpy of reaction, and $W_1 = -W$ is the reaction rate. For the fuel concentration: $s_2 = c_F$, Pe_2 is the mass diffusion Peclet number, $Q_2 = -1$, and W_2 is the reaction rate. For the oxidizer concentration, $s_3 = c_O$, with the remaining parameters being the same as for the fuel concentration. For the product concentration: $s_4 = c_P$, $Q_4 = 1$, and $W_4 = W$. The reaction rate $W = A_f\, c_F\, c_O \exp(-T_a/T)$, where A_f is the frequency factor of the chemical reaction and T_a the normalized activation energy.

The variables are nondimensionalized with respect to the appropriate combination of the velocity scale, $\Delta U = U1 - U2$, representing the velocity difference between the high- and low-speed streams, the length scale, Λ, representing a typical eddy size in the domain, the freestream temperature T_∞, and freestream reactant's concentration, $c_{F\infty}$. We assume the stoichiometry to be 1.

III. Numerical Scheme

Solutions are obtained using the vortex/transport element method. The method is a grid-free Lagrangian field scheme in which the gradients of the primary variables are discretized accurately using finite-area elements with overlapping cores. For the velocity, we use the curl, i.e., the vorticity, since it possesses the convenient property that the vorticity remains constant along the particle trajectory. For scalars, we use their gradients. These computational elements are then transported along the particle trajectories. In the case of a two-dimensional incompressible flow, the vorticity remains constant along the particle path, whereas the scalar gradient changes its magnitude according to the strain field and its direction according to the vorticity. Source terms in the conservation equations can also change the value of the scalar and its gradient locally. The strain field distorts the shape of the material elements employed to transport the vorticity and the scalar gradients. This is simulated in the method by changing the cores of the elements as they are strained. Diffusion is simulated by expanding the cores according to the solution of the diffusion equation. At the end of the time step, the velocity and the scalar concentrations are computed by integrating over the contribution of the field elements. The method has been described in detail in Refs. 3 and 6. Here we present the main equations used in the algorithm.

For the two-dimensional incompressible vortex method,

$$\omega(\mathbf{x},t) = \sum_{i=1}^{N} \omega_i \, h^2 \, f_\delta(\mathbf{x}-\chi_i(\mathbf{X}_i,t)) \tag{3}$$

where ω_i is the vorticity at the center of an element, N the total number of vortex elements, h the average distance between the centers of neighboring elements in two principal directions, $h^2 = h_x h_y$, δ the core radius of a vortex element, and $f_\delta = 1/\delta^2 \, f(r/\delta)$ the core function describing the distribution of vorticity associated with an element. Equation (3) is equivalent to expanding a function $\omega(\mathbf{x},t)$ in terms of N kernel functions, f_δ, located at $\mathbf{X}_i$ and with weights $\omega_i h^2$. The velocity field of a distribution of vortex elements is given by the desingularized Biot-Savart law as follows:

$$\mathbf{u}(\mathbf{x},t) = \sum_{i=1}^{N} \omega_i \, h^2 \, \mathbf{K}_\delta(\mathbf{x}-\chi_i(\mathbf{X}_i,t)) \tag{4}$$

where

$$\frac{d\chi_i}{dt} = \mathbf{u}(\chi_i(\mathbf{X}_i,t),t) \,, \tag{5}$$

$\mathbf{K}_\delta(\mathbf{x}) = \mathbf{K}(\mathbf{x}) \, \kappa(r/\delta)$, $\mathbf{K}(\mathbf{x}) = (-y,x)/r^2$, $r=|\mathbf{x}|$, and $\kappa(r) = 2\pi \int_0^r f(r')r' \, dr'$.

For the transport of scalars, the gradient $\mathbf{g} = \nabla s$ is expressed as

$$\mathbf{g}(\mathbf{x},t) = \sum_{i=1}^{N} \mathbf{g}_i(t) \, h^2 \, f_\delta(\mathbf{x} - \chi_i(\mathbf{X}_i,t)) \tag{6}$$

where

$$\mathbf{g}_i(t) = \frac{\delta s_i \; \delta l_i(t)}{h^2} \, \mathbf{n}_i(t) \tag{7}$$

where δs_i, δl_i, and n_i are the jump in the scalar concentration across the element, the material length associated with the computational element, and the unit vector normal to δl_i, respectively. Equation (6) is based on the expansion of g in terms of the core function f_δ. Given the location and strength of the transport elements, scalar concentration is computed by direct integration over the fields of the transport elements

$$s(\mathbf{x},t) = \sum_{i=1}^{N} g_\delta(t)\, h^2 \cdot \nabla G_\delta(\mathbf{x}-\chi_i(\mathbf{X}_i,t)) \tag{8}$$

where, in two dimensions, $\nabla G_\delta(\mathbf{x}) = (x,y)/2\pi r^2\ \kappa(r/\delta)$. Note that this formulation is fully compatible with the vortex method, since all of the information required to compute the scalar transport is already a part of the vortex computations, including all of the expressions for Green's functions. For extended derivations and discussions, see Refs. 7 and 8. The species transport equation, Eq. (2), can be solved by using the transport element method in fractional steps: convection, diffusion, and reaction. The transport of a scalar by convection is implemented by moving the element along the particle trajectory and adjusting the gradient according to Eq. (7). The effect of molecular diffusion is simulated by expanding the cores of the elements according to $\delta^2(t+\Delta t) = \delta^2(t) + 2\,\alpha\,\Delta t$, where Δt is the time step and α the molecular diffusivity. The last step is implemented by changing the strength of the transport elements according to the following expression[5]:

$$\frac{d}{dt}\,\delta s_i = \sum_{j=1}^{k+1} \frac{dW}{ds_j}\,\delta s_j \tag{9}$$

where k is the total number of species. The compressible form of both algorithms is summarized in an accompanying paper in this volume.[9] For three-dimensional extensions, see Refs. 10 and 11.

IV. Solutions

Two-Dimensional Shear Layer

We start by reviewing results obtained for the dynamics and mixing in a 2D, spatially developing, nonreacting shear layer. The velocity ratio across the layer is 2, and the initial vorticity distribution across the layer downstream of the splitter plate is described by a Gaussian that corresponds to an error function velocity distribution. Figure 1 shows the vortex/transport elements and their instantaneous velocity with respect to the mean velocity of the two streams for an unforced mixing layer. The familiar rollup and pairing of these structures, which follow the growth of numerically induced perturbations, are exhibited clearly in these plots. The entrainment of vortex fluid as well as irrotational fluid from both streams into the large-scale structures during these processes is also clear. With the formation of the coherent vorticity structures, zones of strong strain, the braids, are formed between the large eddies.

Figure 2 shows a comparison between the mean streamwise velocity and its mean fluctuations and the corresponding experimental measurements of both quantities[12] at five different cross sections downstream of the splitter plate. The numerical results were obtained by averaging over 500 time steps to achieve statistical stationarity. The deviation between the numerical and experimental results is due to a small difference between the growth rate of the layer in both cases. We note that the growth rate of the experimental layer was different from the almost universally accepted value for this flow. The results also show that the unsteadiness induced by the growth of the natural instabilities of the flow produces "turbulent" fluctuations on the order of 30% of the mean flow velocity.

Figure 3 depicts a comparison between the mean scalar concentration and its fluctuations and the corresponding experimental measurements of both quantities[12] for different values of the Peclet number. The initial profile of scalar concentration at the splitter plate is close to a Heaviside function. It is interesting to note that the mean

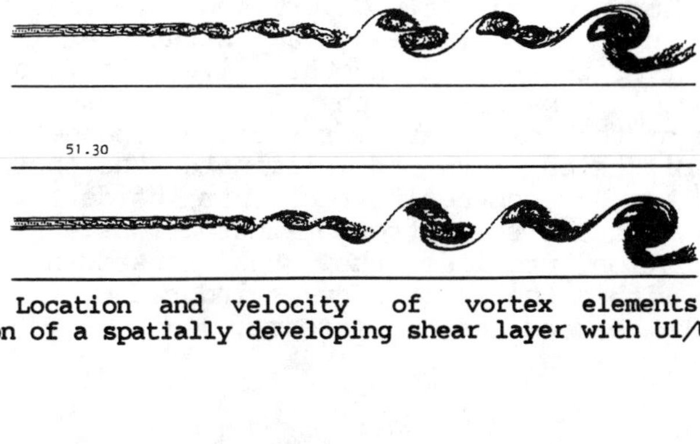

Fig. 1 Location and velocity of vortex elements used in simulation of a spatially developing shear layer with U1/U2 = 2.0.

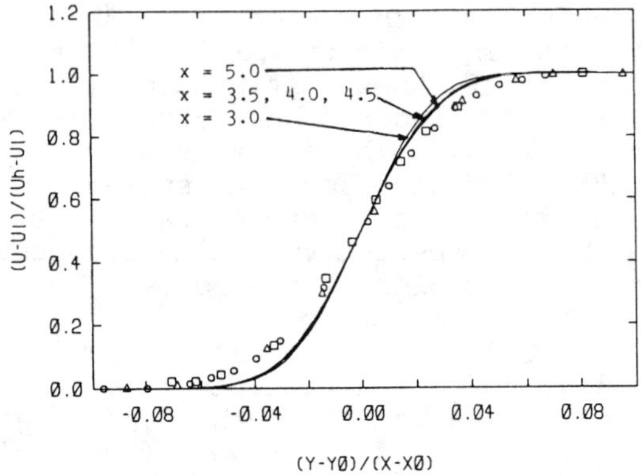

(a)

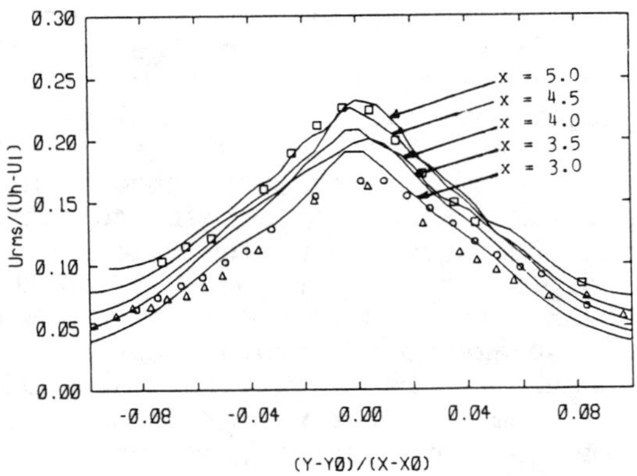

(b)

Fig. 2 a) Time-average streamwise velocity distributions at five sections downstream of splitter plate. Open symbols are results of experimental measurements.[12] b) Time-average streamwise velocity fluctuations at five cross sections downstream of the plate.

concentration, which is determined by the entrainment currents into the large-scale structures, are almost independent of the Peclet number; i.e., it is dominated by convection. Conversely, the fluctuations are strongly dependent on the molecular diffusion and changes with Peclet number. Diffusion acts to eliminate the sharp gradients generated by the convective currents inside the structures and, thus, reduces the strong fluctuations. We note that the fluctuation profiles show the consequence of mixing asymmetry within the large eddy is induced by the difference of velocity between the two streams.

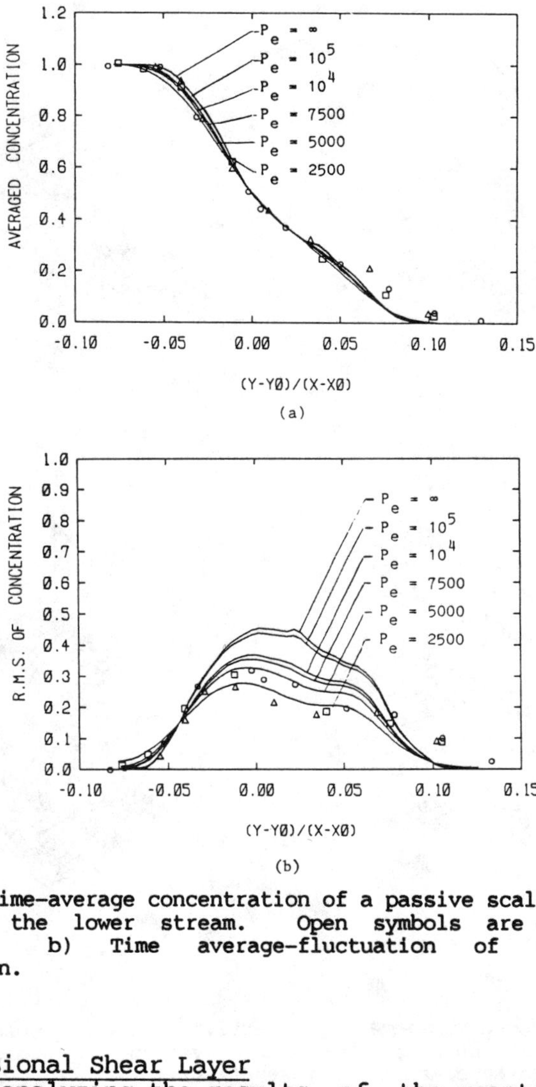

Fig. 3 a) Time-average concentration of a passive scalar initially flowing in the lower stream. Open symbols are experimental results.[12] b) Time average-fluctuation of the passive concentration.

Three-Dimensional Shear Layer

Before analyzing the results of the reacting 2D shear layer, we establish the relevance of 2D solutions to the study of a turbulent shear layer. Figures 4, 5 show results for the application of the 3D vortex/transport element method to a nonreacting, temporally growing shear layer perturbed in the streamwise and spanwise directions.[10,11] Figure 4 shows the evolution of the material layer aligned initially with the midsection of the shear layer, and Fig. 5 exhibits the intersections of all of the material planes within the shear layer with two planes normal to the streamwise direction and two planes normal to the spanwise direction.

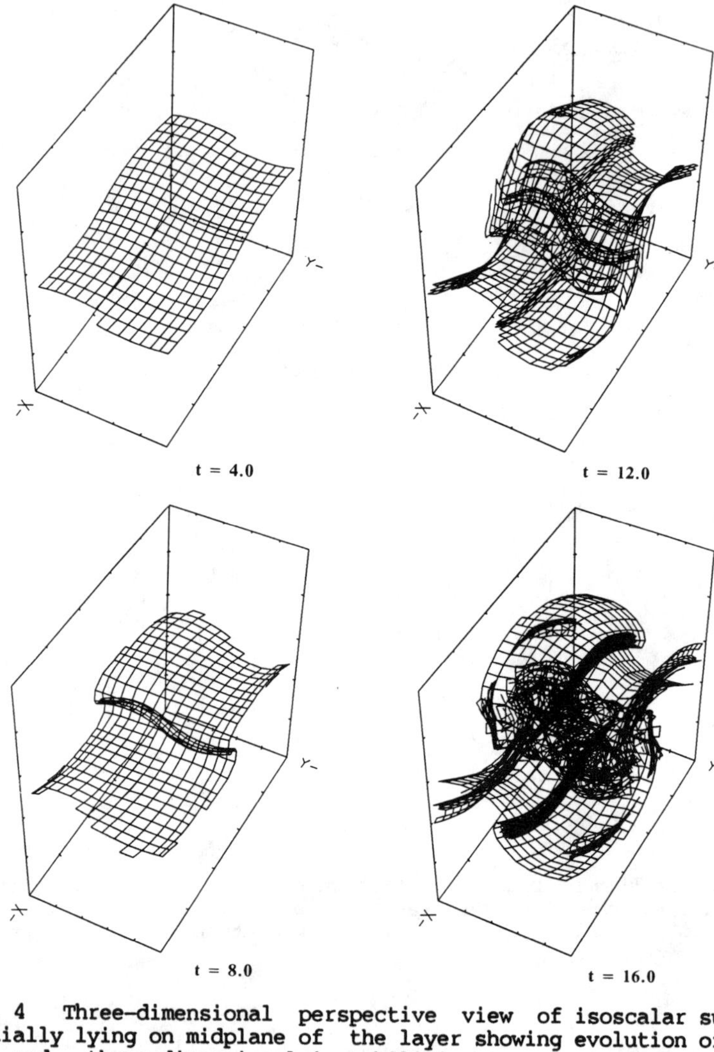

Fig. 4 Three-dimensional perspective view of isoscalar surface initially lying on midplane of the layer showing evolution of both two- and -three dimensional instabilities.

The results show that in the initial stages and until the 2D instability matures into a large-scale coherent structure, the three-dimensional instability is suppressed. Even after this stage, and while the three-dimensional instability is developing into streamwise vortex rods that wrap around the spanwise structure, the coherent structures remain undisturbed except for slight undulation in the spanwise direction, referred to as the translative instability. At later stages, and with the maturation of

(a) y = 3.3, (AA). (c) x = 6.6, (CC).

(b) y = 1.6, (BB). (d) x = 2.0, (DD).

Fig. 5 Intersection of material layers within shear layer with two planes normal to spanwise direction (a,b), and two planes normal to streamwise direction (c,d).

the three-dimensional instability, one still can identify clearly the 2D structures. Thus, 2D solutions can be used to describe a major component of the physics of shear layers.

Reacting Shear Layer

Figure 6 shows the vorticity contours and the products concentration contours within a shear layer with U1/U2 = 2.0. Figure 7 shows the vorticity contours, the oxidizer's contours (top stream), and the product concentration contours within a single eddy in the same reacting shear layer. The flow is incompressible and with uniform density, hence the dynamics are the same as in Fig. 1. To obtain these results, we assumed that $T_a = 0$, $Da = A_f = 10$, and $Q = 1$, while $Re = Pe = 5000$. The Damkohler number $Da = \tau_f/\tau_c$,

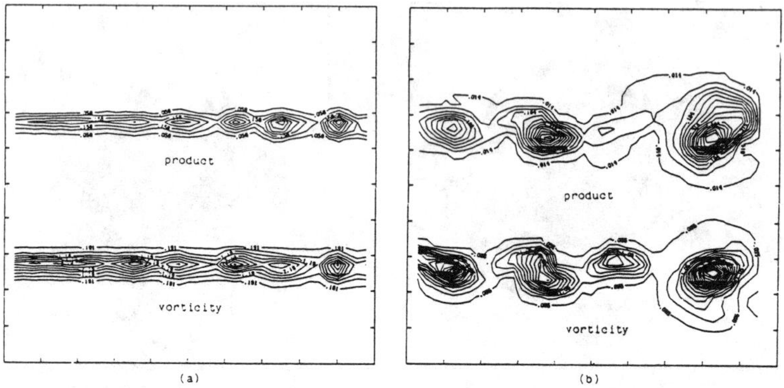

Fig. 6 Vorticity contours and product concentration contours within reacting shear layer.

where $\tau_f = \Lambda/\Delta U$ is the flow time and $\tau_c = 1/[A_f c_{F\infty} \exp(-T_a/T)]$ is the chemical time. The figure reveals a strong similarity between the instantaneous, spatially resolved product concentration, $c_P(x,t)$, and the vorticity concentration, $\omega(x,t)$. Moreover, c_P reaches a maximum at the centers of the eddies where ω is maximum and reaches a minimum at the braids between the eddies where the strain field is maximum and the vorticity is almost zero. Clearly, vorticity plays a prominent role in determining the properties of this combustion field.

To explain the distribution of products within the eddies, we note that the convective currents are strongest at the outer edges of the large eddies leading to strong entrainment of reactants from both streams toward the centers of the eddies. Meanwhile, the same currents also lead to the entrainment of products formed at the outer edges of the eddies into their centers. The flow at the braids resembles that in a stagnation point flow, where each stream bifurcates toward the two adjacent eddies. Thus, products formed within the braids are likely to be convected toward the centers of the large eddies.

To investigate the effect of the Damkohler number on the structure of the reaction zone and the distribution of the product concentration, we compare two computations for Da = 0.2 and 10 in Fig. 8. In both cases, the reaction rate, $W/A_f = c_F c_O$, is also plotted to determine the zone of maximum reaction rate. It is clear that the Damkohler number, although it affects the concentration level of products, does not affect the distribution of the products within the layer; this is determined by the convection field as shown before. However, as depicted in the plots of the reaction rate contours, Da determines the mode of combustion and the topology and location of the reaction zone.

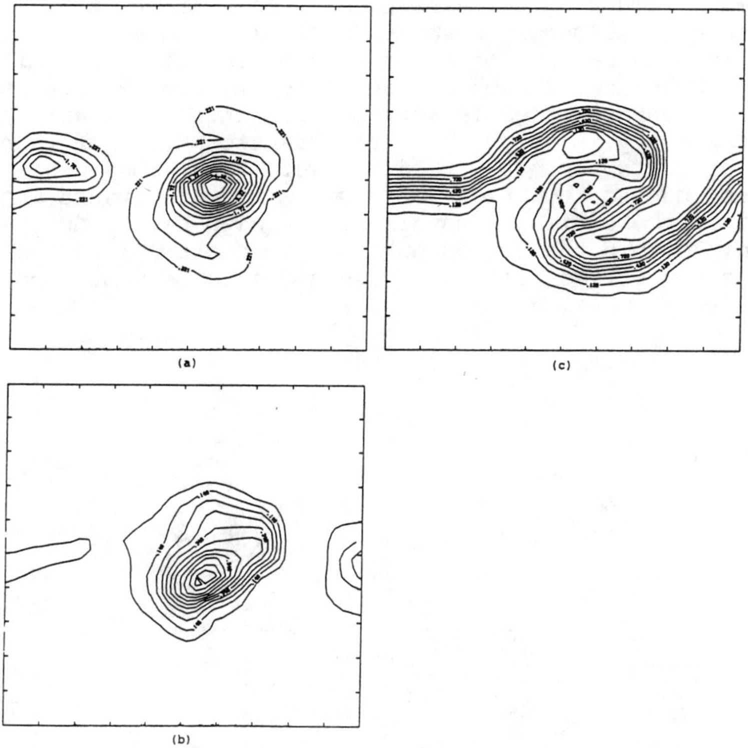

Fig. 7 a) Vorticity contours, b) oxidizer contours, and c) product contours within one eddy in the reacting shear layer in Fig. 6.

In the case of small Da, the reaction takes place in the form of a distributed reaction zone inside the large eddies. In this case, since chemistry is slow, the two reactants are first drawn into the eddy, become thoroughly mixed and then react. This is confirmed by the fact that W is maximum at the centers of the eddies. When Da is high, the reactants are brought in contact at the outer edges of the eddies by the entrainment currents. Diffusion at the outer edges forms a mixture that reacts almost instantaneously since chemistry is fast. The products of combustion that form within the reaction zones around the eddy are then convected toward the centers of the eddies.

Thus, although the convective field establishes product concentration, it is the Damkohler number that determines the mode of combustion. It is interesting to note that, when the Damkohler number is high, the reaction takes place in the form of thick reaction sheets aligned with the strain field; when the Damkohler number is small, the reaction takes place within the eddies where the strain is weak. Thus, the effect of the strain field on the reaction is strongest when the Damkohler number is high.

In both cases, while the rate of reaction is finite due to the presence of both reactants within the braids, the product concentration is very small. This is because at low Da, chemistry is slow and the transformation of reactants into products occurs at a slower rate than that which corresponds to the flow of the mixed reactants into the cores of the eddies. Conversely, at high Da, products start to form within the braids but are then drawn into the cores by the convective field. It should also be mentioned that,

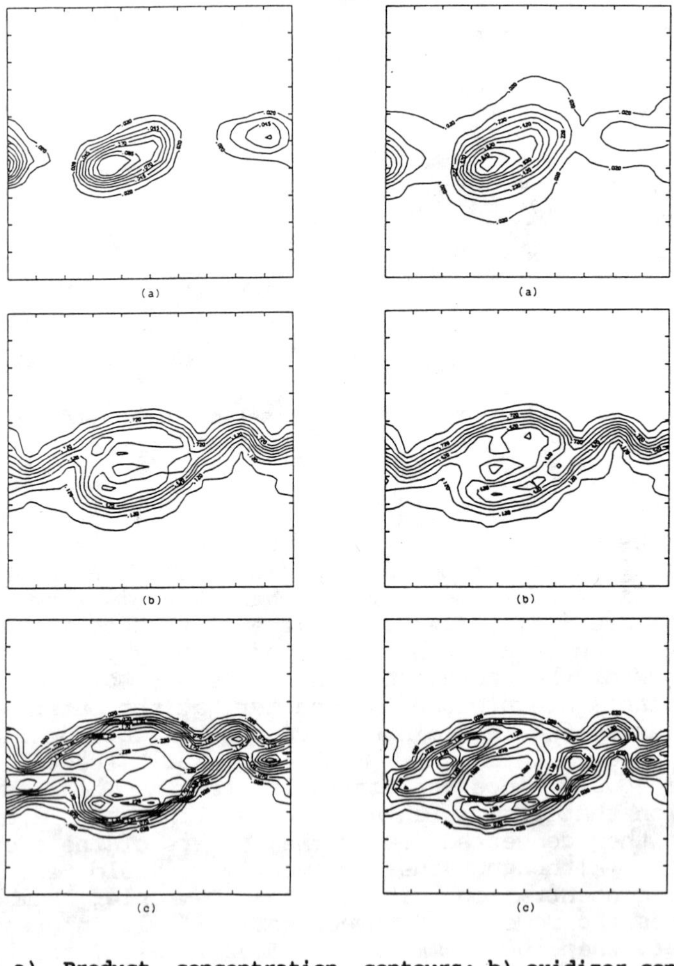

Fig. 8 a) Product concentration contours; b) oxidizer contours; and c) reaction rate contours for Da = 0.2 (left) and Da = 10 (right).

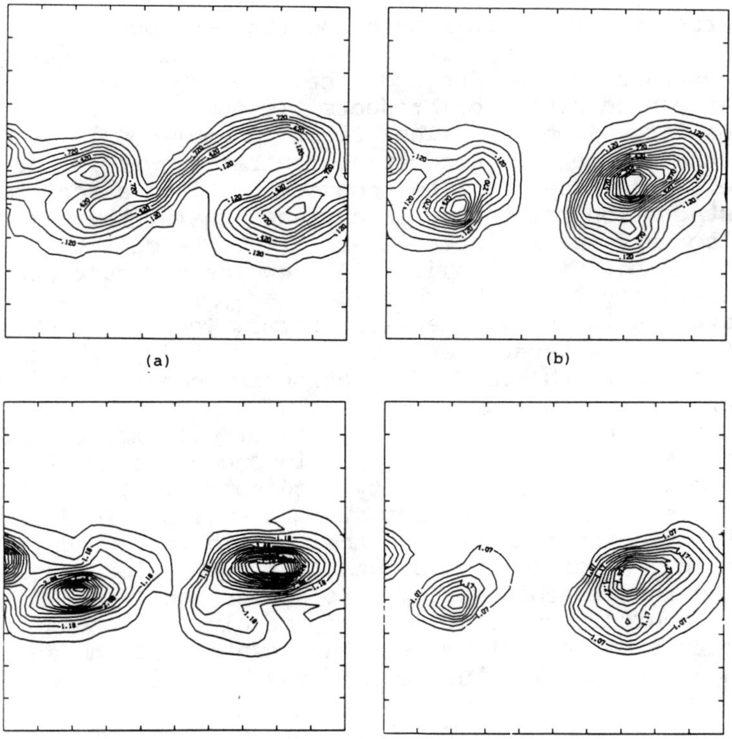

(a) (b) (c) (d)

Fig. 9 a) Oxidizer contours, b) products concentration contours, c) vorticity contours, and d) temperature contours in shear layer with temperature-dependent kinetics.

since the Peclect number is high, mixing within the braids is slow and incomplete. This is because the overlap between both reactants there is not expected to extend over a thick region (by boundary-layer analogy).

It is also interesting to note that, in this case of temperature-independent kinetics in which the reaction is entirely mixing controlled, the intensity of concentration contours is independent of the Damkohler number. In all cases, one observes a uniform distribution of product concentration contours within the eddy. This, as we will show next, is in contrast with the temperature-dependent kinetics case. For more details, see Ref. 13.

Computations of a reacting mixing layer with finite rate kinetics were obtained for A_f = 8000, T_a = 10, and Q = 1. Results are shown in Fig. 9 in terms of oxidizer concentration (top stream reactant), product concentration, vorticity, and temperature contours. The most important differences between this case and the previous case are that 1) the braid zone within which product concentration is zero extends further beyond the previous case, and 2) a higher concentration of products is observed at the centers of the large eddies. The first phenomenon is due to the fact that since the product concentration within the braids is very small due to the convective field associated with the stagnation point flow established there, the temperature is also very small [note that $T(x,t) = c_P(x,t)$ since the Lewis number is unity]. A low temperature zone will substantially lower the reaction rate, which depends on the temperature in an exponential form. The second phenomenon represents the reverse effect. The center of the eddy is a hot zone where the rate of reaction exponentiates due to its temperature dependence.

Computations using temperature-dependent kinetics confirm the conclusion that local concentration of products is proportional to vorticity concentration, and that the latter is the most important dynamic variable in determining the interactions between the field and the chemical reaction. We performed computations for a wide range of Re, Da, and the stoichiometric ratio across the layer.[13] All of the results confirmed the conclusion that product concentration is maximum at the center of the eddies and reached very small values within the braids.

V. Conclusions

Numerical simulation for the reacting shear layer using the vortex/transport element method shows that the structure of the reaction zone is determined by the flowfield and the Damkohler number, while the local product concentration is always proportional to the local vorticity. At high Da, the reaction zone is a thick reaction zone aligned with the outer edges of the eddies. Products are produced within this zone and are then entrained into the core of the eddy. At low Da, reactants from the freestreams first are drawn into the eddy core and then are reacted. In both cases, the similarity between the local product concentration and the vorticity distribution suggests the following relationship as a model for turbulent combustion: $c_P(x,t) = F(Re,Da,\psi)\,\omega(x,t)$, where ψ is the ratio of reactants across the layer.

The present study suggests that a proper flow model of turbulent combustion could be the reacting shear layer since it combines two important elements: an unsteady, spatially evolving vortex and an unsteady strain field. In an accompanying paper in this volume, we present results for a compressible flow model of a similar flow in which the effects of combustion on turbulence are analyzed in detail.

Acknowledgments

This work was supported by the Air Force Office of Scientific Research Grant 84-0356, the National Science Foundation Grant CBT-8709465 and the Department of Energy Grant DE-FG04-87AL44875. Computer support was provided by the John von Neumann National Computer Center.

References

[1]Marble, F. E., "Growth of a Diffusion Flame in the Field of a Vortex," Recent Advances in the Aerospace Sciences, edited by C. Casci, Plenum, 1985, p. 395.

[2]Peters, N., "Laminar Diffusion Flamelets in Non-premixed Turbulent combustion," Progress in Energy and Combustion Science, Vol. 10, 1984, pp. 319-339.

[3]Ghoniem, A. F., Heidarinejad, G., and Krishnan, A., "Turbulence-Combustion Interactions in a Reacting Shear Layer," Lecture Notes in Engineering, 40, Proceedings of France-U.S.A. Joint Workshop on Turbulent Reactive Flows, Springer-Verlag, July 1987, pp. 637-671.

[4]Ghoniem, A. F., "Computational Methods in Turbulent Reacting Flows," in Reacting Flows: Combustion and Chemical Reactors, Part I, edited by G. S. S. Ludford, Lecture Notes in Applied Mathematics, Vol. 24, AMS, RI, 1986.

[5]Ghoniem, A. F., Heidarinejad, G., and Krishnan, A., "Numerical Simulation of a Thermally-Stratified Shear Layer Using the Vortex Element Method," Journal of Computational Physics, Vol. 79, 1988, pp. 135-166.

[6]Krishnan, A. and Ghoniem, A. F., "Origin and Manifestation of Flow-Combustion Interaction in a Premixed Shear Layer," Proceedings of 22nd Symposium (International) on Combustion, The Combustion Institute, Pittsburgh, PA, 1988, pp. 665-675.

[7]Heidarinejad, G., "Vortex Simulation of the Reacting Shear Layer," Ph.D. Thesis, Department of Mechanical Engineering, Massachusetts Institute of Technology, Cambridge, MA, 1989.

[8]Krishnan, A., Numerical Study of Vorticity-Combustion Interactions in Shear Flow, Ph.D. Thesis, Department of Mechanical Engineering, M.I.T., 1989.

[9]Ghoniem, A. F., and Krishnan, A. "Vorticity-Combustion Interactions in Turbulent Reacting Jet," International Colloquium on Dynamics of Explosion and Reactive Systems, 1989, to appear in same volume.

[10]Knio, O. M. and Ghoniem, A. F., "Numerical Study of a Three-Dimensional Vortex Method," J. Comput. Phys, 86, 1990, pp. 75-106.

[11]Knio, O. M. and Ghoniem, A. F., "Three Dimensional Vortex Simulation of Passively Stratified Shear Flow," Journal of Computational Physics, to appear.

[12]Masutani, S. M. and Bowman, C. T., "The Structure of a Chemically Reacting Plane Mixing Layer," Journal of Fluid Mechanics, Vol. 172, 1986, pp. 93-126.

[13]Heidarinejad, G. and Ghoniem, A. F., "Vortex Simulation of the Reacting Shear Layer; Effect of Reynolds Number and Damkohler Number," AIAA Paper 89-0573, 1989.

Flow-Combustion Interactions in a Turbulent Jet

Ahmed F. Ghoniem* and Anantha Krishnan†
Massachusetts Institute of Technology, Cambridge, Massachusetts

Abstract

The interactions between the flowfield and the combustion process in a reacting jet are analyzed using numerical simulations. Computational results are obtained for the compressible flow model of low-Mach-number combustion using the vortex/transport element method. We investigate how entrainment and a strong strain field, which enhance the mixing and reduce the local reaction zone, respectively, can alter the rate of burning, and how combustion exothermicity, leading to local volumetric expansion and baroclinic vorticity generation, can affect the flowfield. We show that the strain field can lead to local extinction of the reaction and to reduction of the total rate of burning. However, mixing enhancement by strong entrainment can overcompensate for this reduction. Volumetric expansion suppresses the growth of the instability and reduces the rate of formation of coherent vorticity structures. However, the total rate of burning in the reacting layer is always higher than the corresponding laminar flame. Results confirm the dependence of the reaction zone structure on the Damkohler number, which was found in the incompressible flow analysis of a similar reacting shear flow problem.

I. Introduction

The analysis, understanding, and modeling of turbulence-combustion interactions in heterogeneous reacting flow systems are fundamental to the efficient and clean operation of many practical systems. The complexity of the problem arises from the fact that the two processes are

---

* Associate Professor, Department of Mechanical Engineering.
† Scientist, CFD Research Co., Huntsville, AL.

individually complex, both are dynamically unstable, intrinsically unsteady, and occur on scales much smaller than the integral scales of the overall flow. The problem is compounded by the continuous, time-dependent, local and global feedback between the two processes. Numerical simulation of their governing equations is used here to help gain some insight into this problem.

Simple models, which restrict the interactions between the two processes to one mode, have been analyzed previously. We mention the interactions between an isolated steady vortex and a diffusion flame[1] and the interaction between a steady strain field and a diffusion flame.[2] In a turbulent field, however, one expects to encounter both modes simultaneously. Moreover, the two modes can evolve in time in a way that cannot be represented using simple models. In this paper, we use an unsteady jet model to represent both modes in a dynamically consistent way and to study their effect on the reaction zone in a reacting jet environment.

The effect of heat release on turbulent flow, which sometimes is referred to as flame-generated turbulence, is the other side of this coin. In the low-Mach-number limit, it has been shown that this effect is represented by the volumetric expansion and the vorticity generation in areas of finite density gradients.[3] This low-Mach-number approximation is applied herein to resolve some of the flow-combustion interactions.

To analyze these interactions, we choose a model problem and perform detailed numerical simulations over a range of its governing parameters. In a companion paper in this volume,[4] we applied the same strategy to a simple model: an incompressible reacting shear layer in which temperature variations were not allowed to affect the dynamics of the flow-field. Herein we use a compressible flow model in which the changes in the temperature field affect the density through the equation of state. The changes of the density field then are allowed to affect the velocity field through the continuity equation and the vorticity field through the baroclinic term in the vorticity transport equation.

The paper is organized as follows. In Sec. II, we review the formulation and numerical schemes. Detailed derivation and discussion of both are referenced accordingly. In Sec. III, results of the nonreacting jet and the reacting jet at low and high Damkohler numbers are analyzed in terms of the evolution of the flow-field, the vorticity, the product concentration, and the reaction zone structure. Conclusions regarding the interactions between the flow-field and the combustion process are summarized in Sec. IV.

II. Formulation and Numerical Scheme

The low-Mach-number approximation of the combustion equations, consisting of the Navier-Stokes equations, the species conservation equations, and the energy equation, in unconfined flow is utilized to filter out the pressure waves and render the pressure independent of the density.[2,5] Under these conditions, one can show that the heat release leads to generation of an irrotational velocity field, $\nabla\phi$, which when superimposed on the existing vorticity-induced rotational velocity, u_ω, represents the total velocity in a reacting flow. Employing this velocity decomposition and assuming that 1) the flow is two-dimensional, 2) both reactants and products behave as perfect gases, and 3) the chemical reaction, F + O --> P, can be described by a single-step Arrhenius kinetics, the governing equations can be written as

$$\rho\, T = \text{const} \tag{1}$$

$$\mathbf{u} = \mathbf{u}_\omega + \nabla\phi \tag{2}$$

$$\nabla^2\phi = \frac{1}{T}\frac{dT}{dt} \tag{3}$$

$$\frac{d\omega}{dt} = \frac{1}{\rho^2}\nabla\rho \times \nabla p - \omega(\nabla.\mathbf{u}) + \nu\,\nabla^2\omega \tag{4}$$

$$\frac{ds_j}{dt} = \frac{1}{Pe_j}\nabla^2 s_j + Q_j\, W_j \tag{5}$$

The pressure gradient is obtained from the momentum equation: $\nabla p = -\rho\;\mathbf{a} + \mu\;\nabla^2\mathbf{u}$, where $\mathbf{a}$ is the particle acceleration and μ dynamic viscosity. Both velocity components, $\mathbf{u}_\omega$ and $\nabla\phi$, must satisfy the velocity boundary condition in the direction normal to the boundaries of the domain.

The variables are defined as follows: $\omega k = \nabla \times \mathbf{u}$ is vorticity where k is the unit vector normal to the x-y plane, $\mathbf{u} = (u,v)$ is velocity, t is time, d/dt is the material or Lagrangian derivative, and Re is the Reynolds number. The model requires the transport of four scalars: temperature T, fuel concentration c_F, oxidizer concentration c_O, and product concentration. For the temperature: $s_1 =$

T, Pe_1 is the Peclet number, Q_1 is the enthalpy of reaction, and $W_1 = - W$ is the reaction rate. For the fuel concentration: $s_2 = c_F$, Pe_2 is the mass diffusion Peclet number, $Q_2 = - 1$, and $W_2 = W$ is the reaction rate. For the oxidizer concentration, $s_3 = c_O$, and the rest of the parameters are the same as for the fuel concentration. For the product concentration: $s_4 = c_P$, $Q_4 = 1$, and $W_4 = W$. The reaction rate is $W = (A_f/\rho)^4 c_F c_O \exp(-T_a/T)$, where A_f is the nondimensional frequency factor of the chemical reaction and T_a is the normalized activation energy.

The variables are nondimensionalized with respect to the appropriate combination of a characteristic velocity $U_o = U_j/2$, where U_j is the jet velocity, a characteristic length $L_o = 2\sqrt{2}\sigma$, where σ is the standard deviation of the Gaussian curve used to describe the vorticity distribution in the jet flow, the jet fluid temperature T_∞, and the jet fluid reactant concentration, $c_{F\infty}$. We assume the stoichiometry to be 1.

The vortex/transport element method used to obtain the solution of the incompressible approximation of Eqs. (1–5) is described in Refs. 4 and 6. The extension to compressible flow at low Mach numbers is described in Refs. 5, 7, and 8. Here we briefly present this extension. Using the principle of vortex decomposition,[1] the Green's function solution of Eq. (3) can be written as

$$\nabla\phi(\mathbf{x},t) = \sum_{i=1}^{N} \frac{1}{T_i} \left(\frac{dT}{dt}\right)_i h_i^2(t) \, \nabla G_\delta(\mathbf{x} - \chi_i(\mathbf{X}_i,t)) \tag{6}$$

where

$$\nabla G_\delta = \frac{(x,y)}{2\pi r^2} \left(1 - \exp\left(-\frac{r^2}{\delta^2}\right)\right)$$

Moreover, the change in the vorticity transported by each element as a result of the generation of baroclinic torque is given by

$$\frac{d\Gamma_i}{dt} = - \frac{\nabla\rho_i}{\rho_i} \times \left(\frac{du}{dt}\right)_i h_i^2 \tag{7}$$

The change of the scalar gradient along a particle trajectory is obtained by relating the change in the flow map to the distortion of the isoscalar surfaces:

$$g_i(t) = \frac{\delta s_i \; \delta l_i(t) \; \rho_i(t)}{\rho_i(0) \; h^2} \; n_i(t) \tag{8}$$

The vorticity transport equation is solved in fractional steps: convection by transporting the vortex elements without changing their circulation along particle trajectories, diffusion by expanding the cores of vortex elements, and generation by changing the circulation of the element according to Eq. (7). For more details, see Refs. 5 and 8.

III. Results

Solutions were obtained for the initial stages of development of a two-dimensional planar turbulent jet. Vortex simulations of the flow-field in a spatially developing jet were obtained in Ref. 9. In this work, we concentrate on the case of a temporally developing jet by moving the computational window at the average speed of the jet and the ambient fluid, i.e. at $U_j/2$, and imposing periodic boundary conditions across the domain. In this reference frame, the jet fluid and the ambient fluid initially move in opposite directions at the same speed. In the reacting jet, we also assumed that the activation energy is very small or that the reaction rate is temperature-independent. Extension to finite activation energy kinetics will be published in the future (see Ref. 8). In all cases, the jet is perturbed by a varicose mode, i.e., a sine wave and its image across the centerline, whose amplitude is 0.05 λ, where λ is the wavelength of the perturbation and $\lambda = 6.6L_o$. Symmetry is imposed across the centerline.

For the reacting jet, we use the same geometric parameters to describe the flow. The Peclet number Pe = 1000, the heat release Q = 4, i.e., $T_P/T_R = 4$, and the Lewis number is unity.

Nonreacting Jet

Figure 1 shows the evolution of the jet structure in terms of the location and velocity of all the vortex/transport elements used in the simulation at three time steps. The vorticity rolls up to form two symmetric large-scale structures on both sides of the centerline. The structures generate strong entrainment currents that pull the ambient and jet fluids into the cores of the large eddies. Concomitantly, a strong strain field is established between neighboring eddies that produces very thin material layers separating the two freestreams. The structures move in the direction of the jet fluid at a small velocity induced by the mutual interactions between the two side-by-side structures (this is equivalent to the self-induced velocity of vortex rings[10]). This dynamical asymmetry across each structure leads to asymmetric entrainment of the jet and the ambient fluids into its core. As discussed next, the structure entrains more of the jet fluid than the ambient fluid.

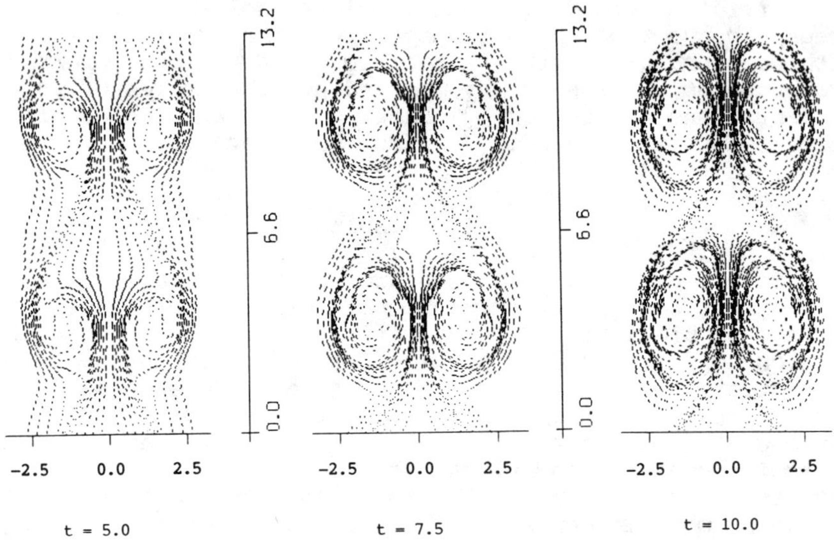

Fig. 1. Vortex/transport element locations and velocity vectors at t = 5.0, 7.5, and 10.0, respectively for the nonreacting jet.

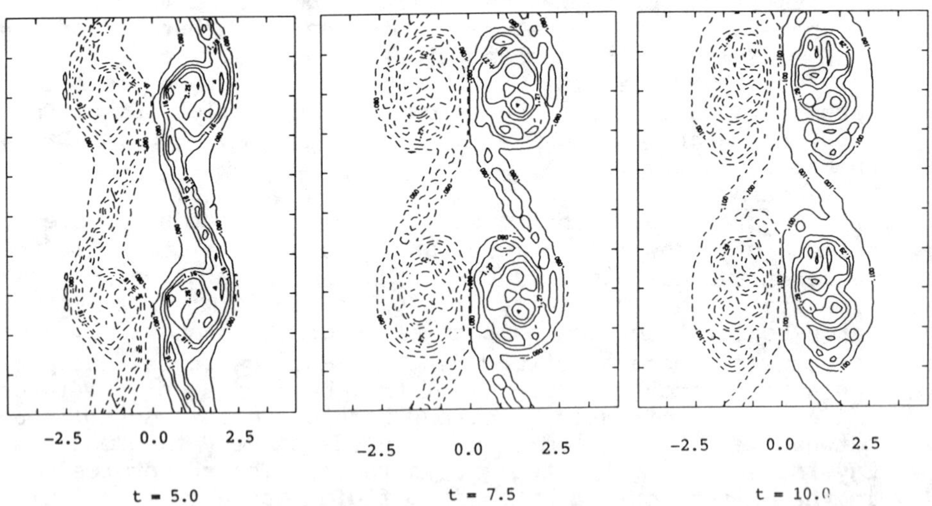

Fig. 2. Vorticity contours at t = 5.0, 7.5, and 10.0, respectively, for the nonreacting jet.

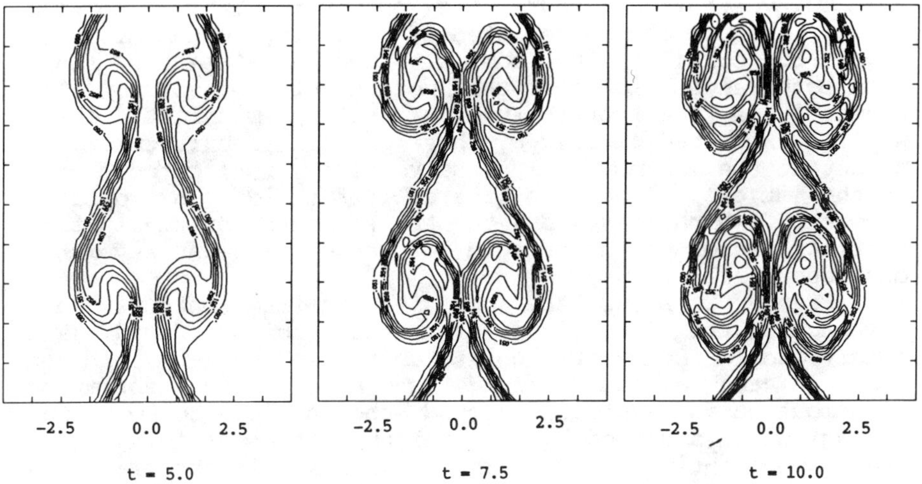

Fig. 3. Scalar concentration contours at t = 5.0, 7.5, and 10.0, respectively, for the nonreacting jet.

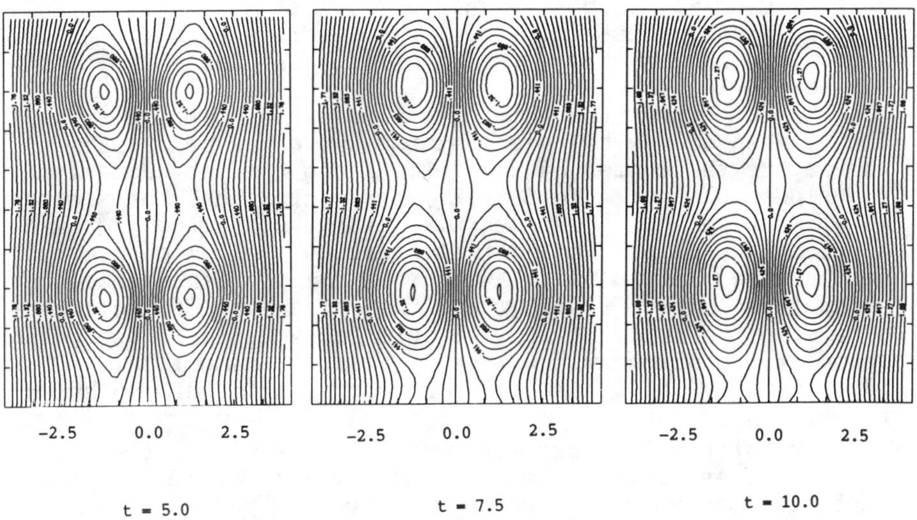

Fig. 4. Streamline contours at t = 5.0, 7.5, and 10.0, respectively, for the nonreacting jet.

The vorticity contours are shown in Fig 2. During rollup, the braids are depleted of vorticity, which migrates toward the cores of the eddies. The entrainment of fluid from both sides follows the motion of vorticity. The asymmetry of the vorticity distribution within the large eddy is exhibited clearly by these plots. In particular, we note that the maximum value of vorticity is located closer to the jet centerline than the ambient fluid. This dynamic asymmetry, which arises due to the asymmetry of the initial vorticty field, is the origin of the entrainment asymmetry discussed earlier.

Figure 3 shows the passive scalar contours that distinguish the jet fluid from the ambient fluid. The passive scalar concentration is unity in the jet fluid and zero in the ambient fluid, whereas the concentration distribution is asymmetric around the $s = 0.5$ contour. The mixing contours formed by the rollup of the vorticity field are shown clearly. The strain field within the braids is responsible for the formation of thin material layers separating areas of widely different values of the scalar concentration, i.e., strong scalar gradients, at the outer edges of the eddies and within the braids. These gradients are exhibited in the form of closely packed scalar contours within the braids and on the outer edges of the eddies. The entrainment asymmetry is also illustrated in this figure: more fluid with scalar values $1.0 < s < 0.5$, i.e., originating from the jet, than with values $0.5 < s < 0.0$, i.e., originating from the ambient, is found inside the eddies.

The concentration contours indicate that the ambient fluid, when driven by this instability, never penetrates all the way to the center of the jet. Along the jet centerline, one finds only jet fluid [it is expected that the sinuous mode, which is responsible for producing growing oscillations along the jet centerline, will change this picture (see Ref. 8)]. Away from the jet centerline, one encounters mixed fluid (in the presence of molecular diffusion), pure jet fluid, or pure ambient fluid, depending on one's location with repect to the centerline. This intermittency of mixing is characteristic of turbulent flows.

The streamlines are shown in Fig. 4. With the growth of the eddies, more closed streamlines are formed around the center of the eddy, indicating more overall entrainment into the core of the structure. We also note the formation of a center, and a saddle stagnation points near the center of the structure and the center of the braid, respectively. These points correspond to a vorticity maximum and a strain rate maximum. The mean material length, calculated as an average over the lengths of all the material layers, is plotted as a function of time in curve 1 of Fig. 5.

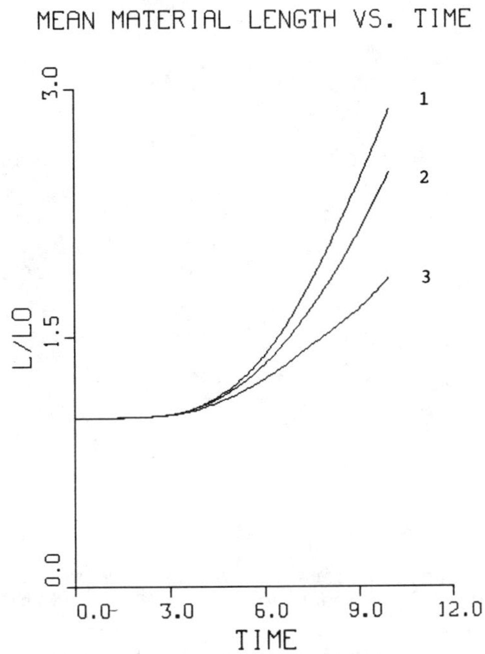

Fig. 5. Plot of the mean material length vs time. Curve 1 is the nonreacting jet, curve 2 the reacting jet at $A_f = 0.0625$, and curve 3 the reacting jet at $A_f = 0.5$.

Reacting Jet at Low Damkohler Number

Results for a reacting jet at $A_f = 0.0625$ are shown in Fig. 6 in terms of the location and velocity of the vortex elements. The early stages of development strongly resemble those found in the nonreacting jet. However, at the later stages, the eddies occupy more area in both directions as a result of volumetric expansion. We note that most of the expansion occurs at the cores of the eddies, indicating that the reaction takes place at the centers. Meanwhile, the braids and the outer edges remain unchanged from their nonreacting layer form.

Fig. 7 shows the vorticity contours. Because of volumetric expansion associated with heat release, the local vorticity concentration is lower than in the nonreacting case. The vorticity is still maximum at the cores of the eddies and reaches a minimum within the braids. The effect of heat release in this case is most evident at the later stages.

Fig. 8 shows the product concentration contours. From the early stages, the product concentration is maximum at the centers of the eddies and is very small within the

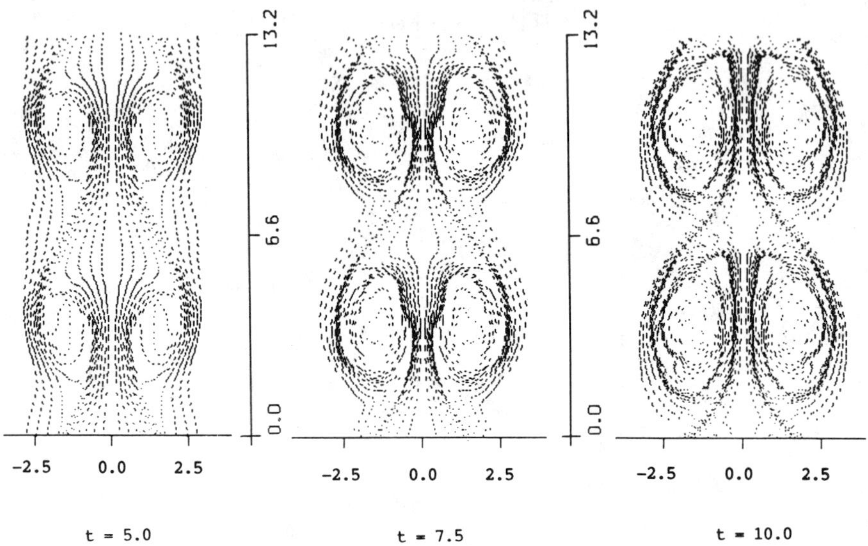

Fig. 6. Vortex/transport element locations and velocity vectors at t = 5.0, 7.5, and 10.0, respectively, for $A_f = 0.0625$.

braids. This is consistent with the results in Fig. 6, which indicate that most of the reaction occurs within the cores of the eddies. As time progresses, the products reach zero within the braids. This is despite the fact that within the braids, both reactants meet along a large interface. To explain the very low concentration of products within the braids, we recall the fact that the flowfield there can be described by a stagnation point flow at high Reynolds numbers. The stagnation point coincides with the saddle point shown in the streamlines of Fig. 4. Thus, while the two streams of the fuel and oxidizer meet along the interface, diffusion between these streams -- and, hence, mixing -- is very slow. The rate of product formation within the braids is, therefore, very small. Moreover, the products formed there are entrained into the cores of the eddies by the same flowfield.

Comparison of the vorticity distribution and the product concentration suggests the following relationship: $c_p(x,t) = F(Re,Da)\ \omega(x,t)$, where F is a function of the Reynolds number and the Damkohler number. A similar relationship was derived for the case of a reacting mixing layer at low heat release and for temperature-dependent kinetics.[4,11] In that case, the functional form included ψ, the reactants ratio across the layer. It is expected that such an extension will also apply to the case of a reacting jet. As will be shown later, the origin of this relationship lies in the fact that this flow is dominated by the convective field driven by the vorticity.

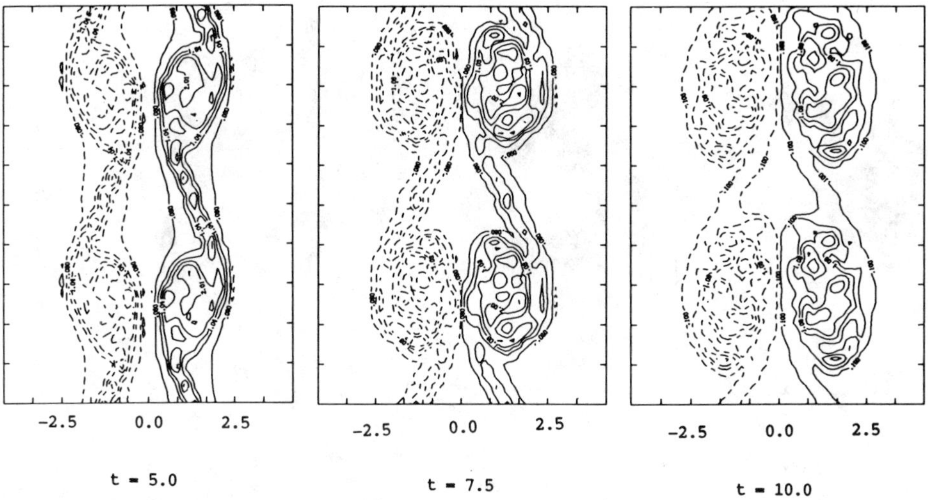

Fig. 7. Vorticity contours at t = 5.0, 7.5, and 10.0, respectively, for $A_f = 0.0625$.

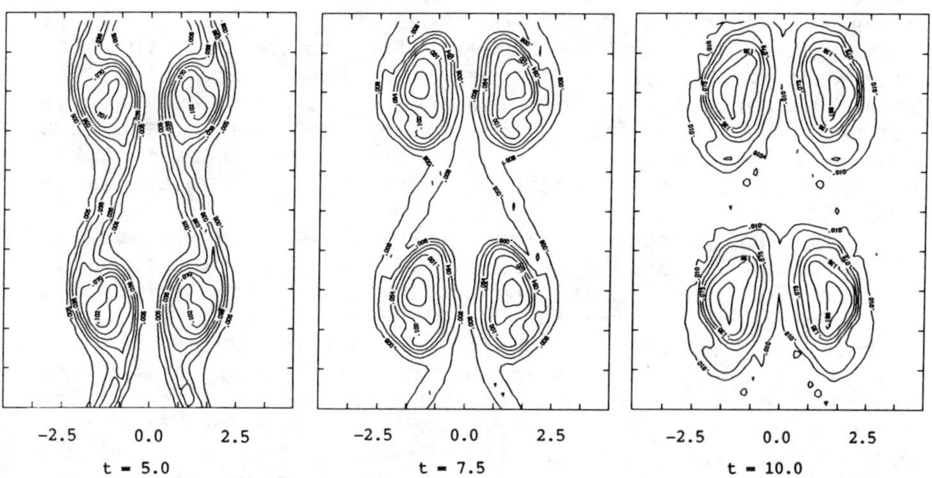

Fig. 8. Product concentration contours at t = 5.0, 7.5, and 10.0, respectively, for $A_f = 0.0625$.

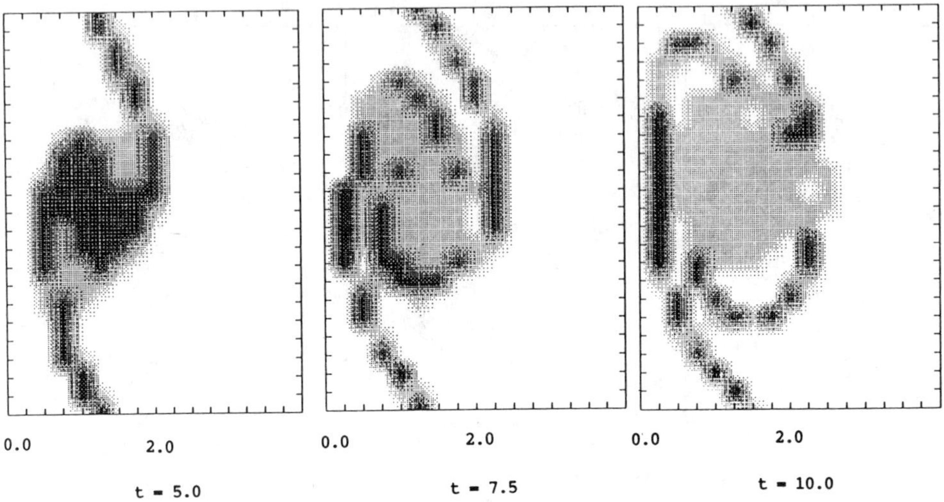

Fig. 9. Reaction rate contours at t = 5.0, 7.5, and 10.0, respectively, for $A_f = 0.0625$.

To analyze the effect of the flowfield on combustion, we plot the reaction rate contours in Fig. 9. Here we divide the full range of W into three sections and use a different shade of gray to identify each section. The reaction rate is highest within the eddies and decreases monotonically outward toward the freestreams. Especially at the early stages, combustion occurs in a distributed reaction zone mode where mixing between the two streams proceeds before the onset of the chemical reaction. We recall that low Damkohler number refers to the case in which chemistry evolves at a slower rate than convection. At later stages, and after the cores of the eddies have reacted, the reaction zone moves outward toward the boundaries of the eddies. At this stage, the reaction takes the form of a thick reaction sheet surrounding the large eddies.

The total products formed in the field, $M = \int \rho\, c_p\, dx\, dy$, is shown in curve 2 of Fig. 10. We also plot the same quantity for the corresponding laminar flame (curve 1), i.e., for a calculation in which we do not allow the layer to roll up. At the early stages, and while the layer is going through strong deformation, the total amount of products is less than that formed in a laminar flame within the same period. This is due both to the formation of a very strong strain field that reduces the area of the reaction zone and to the fast elimination of reaction surfaces within the core of the eddy. The average burning rate during early rollup is actually less than the laminar burning rate due to the strong strain field. During this stage, the entrainment currents into the eddy core are strong and the reacting core is swelling to ingest this flow. At later stages and with the weakening of the instability and the mixing of the ingested reactants within the eddy core, the rate of reaction increases within the eddy core and far exceeds the laminar rate. It is interesting to note that, at the very late stages, the rate of burning within the shear layer approaches that of the laminar layer.

The effect of heat release on the development of the flow, most notably the rollup of the vorticity layer and the associated increase in the length of the material lines, is shown in curve 2 of Fig. 5. The average length of the material layers is reduced as a result of the heat release. As mentioned earlier, heat release reduces the local concentration of vorticity by expanding the material area, i.e., through the term $\omega(\nabla.u)$ in the vorticity transport equation. It is well established that the rate of growth of instability of a thick shear layer is smaller than that of a thin shear layer.[6] The reduction in the average length of the material layer is small since the Damkohler number is low and the amount of heat release is small. It is also interesting to note that the effect of heat release begins to appear after the initial stages of rollup. This is because combustion is governed by mixing in this case.

Reacting Jet at High Damkohler Number

Now we discuss the results obtained for the same flow field but for $A_f = 0.5$. Figure 11 shows the vortex elements and their velocity vectors. From the early stages, the effect of heat release on the flowfield is strong. Compared with the low Da case, the thickness of the shear layer is growing faster and the vorticity rollup is weaker. Since the Damkohler number is high, the rate of heat release is fast with respect to the flow and volumetric expansion increases the area occupied by vorticity. At $t = 10$, it appears that the heat release has hindered the folding of the layer that was observed in the nonreacting case. It is also clear that although some of the reaction is occurring at the center of the eddy, where some of the expansion is observed, most of the reaction takes place at the outer edges of the eddies. This is evident from the fact that the distance between the layers of the elements at the outer edges is expanding as time progresses. Thus, in this case, the reaction zone is essentially around the edges of the eddies. This will be confirmed subsequently.

The vorticity distribution is shown in Fig. 12. It reveals that the strength of the instability in this case is reduced due to the early expansion of the vorticity layer as

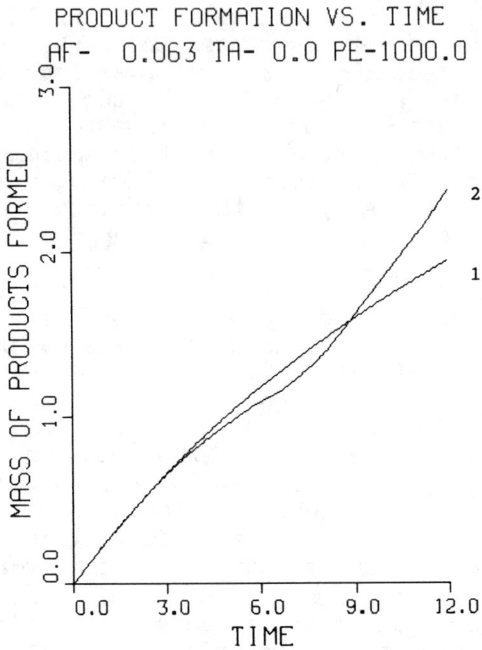

Fig. 10. Plot of mass of products formed vs time for A_f = 0.0625. Curve 1 is the laminar flame, curve 2 is the reacting jet.

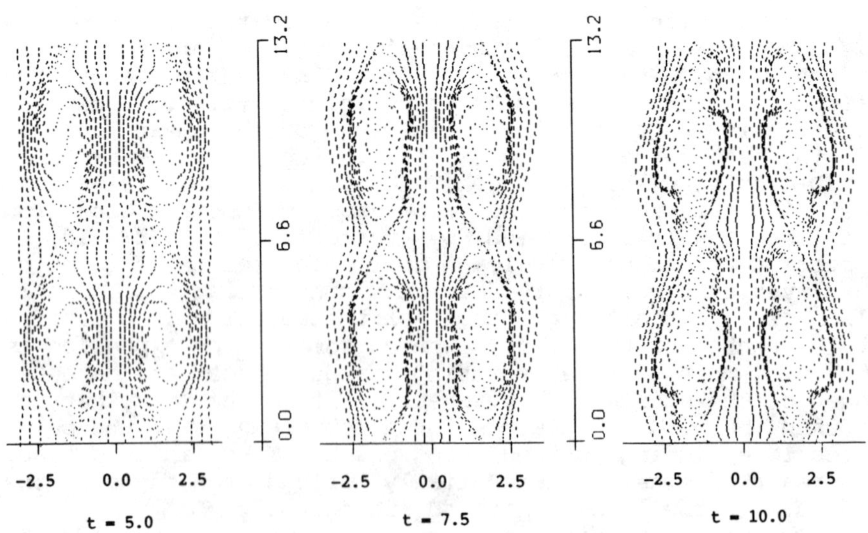

Fig. 11. Vortex/transport element locations and velocity vectors at t = 5.0, 7.5, and 10.0, respectively for A_f = 0.5.

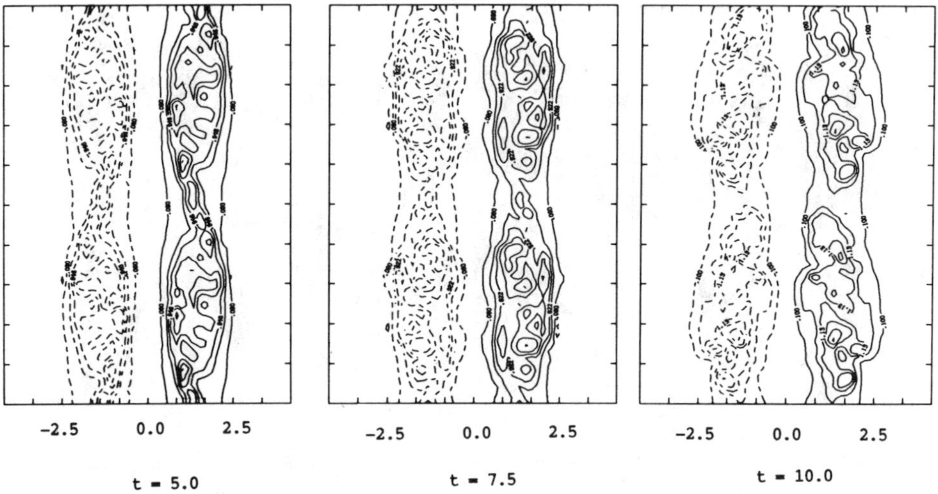

Fig. 12. Vorticity contours at t= 5.0, 7.5, and 10.0, respectively, for A_f = 0.5.

a result of heat release. This effect starts from the early stages and lasts far into the nonlinear range. Although heat release does not completely inhibit the growth of the instability or the rollup of the vorticity layer, it greatly reduces both processes. The layer remains flat and the vorticity stays distributed within the original zone of shear. It should be mentioned that, within the reaction zone, baroclinic vorticity generation plays an important role in determining the local concentration of vorticity. The distinction between the role of volumetric expansion and that of baroclinic vorticity generation will be addressed in a future contribution.

Figure 13 shows the product concentration contours. As in the low Da case, products reach a maximum at the centers of the eddies and zero within the braids. Since the Damkohler number is higher, the level of product concentration is also higher, as indicated by the closely packed contours. However, as exhibited earlier in the distribution of vortex elements, the eddies are weaker due to the volumetric expansion. As observed previously, a strong correlation can be found between the local product concentration and the vorticity.

The effect of the flowfield on the combustion process is best explained by analyzing the reaction rate contours shown in Fig. 14. The figure indicates that from the onset of the rollup process, the reaction zone is confined to a thick reaction sheet that surrounds the large eddies. It is important to mention here that the reaction rate falls

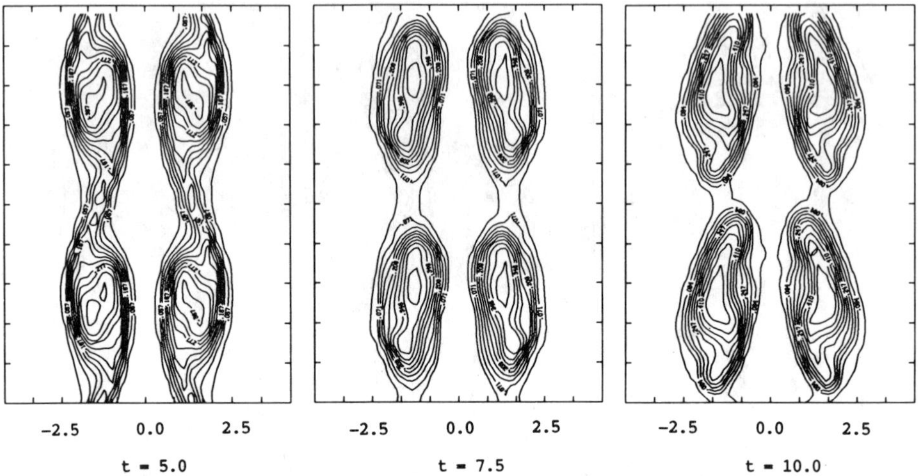

Fig. 13. Product concentration contours at t= 5.0, 7.5, and 10.0, respectively, for $A_f = 0.5$.

smoothly, although rapidly, from the its maximum to zero within the reacting eddy. In the figures, we show areas with different shades that correspond to values of W that differ by one third of each other. Contrary to the low Da case, the combustion process is determined by the chemical time scales.

The total mass of products in the field is shown in Fig. 15 for the reacting shear layer (curve 2) and the corresponding laminar flame (curve 1). The early thinning of the shear layer, associated with the onset of the rollup process, reduces the rate of product formation. At the end of this phase, the reaction is almost extinguished, as evident by the fact that the rate of increase of M is almost zero. However, as the entrained reactants mix and react, the rate of burning increases beyond the value of the laminar flame and the total products exceed those produced by the laminar flame. It is interesting to note that the rate of burning approaches that of the laminar flame at the very late stages when the flame leaves the eddy cores. The figure shows that both the strain field and the entrainment field are important to the rate of burning. Although the former imparts a negative effect, the latter proves to be strong enough to compensate for these negative effects.

Curve 3 of Fig. 5 shows the average length of the material layer. As a result of the effect of volumetric expansion, which was discussed earlier, the average length of the material layer is substantially smaller than in the previous case of low Da combustion. The heat release

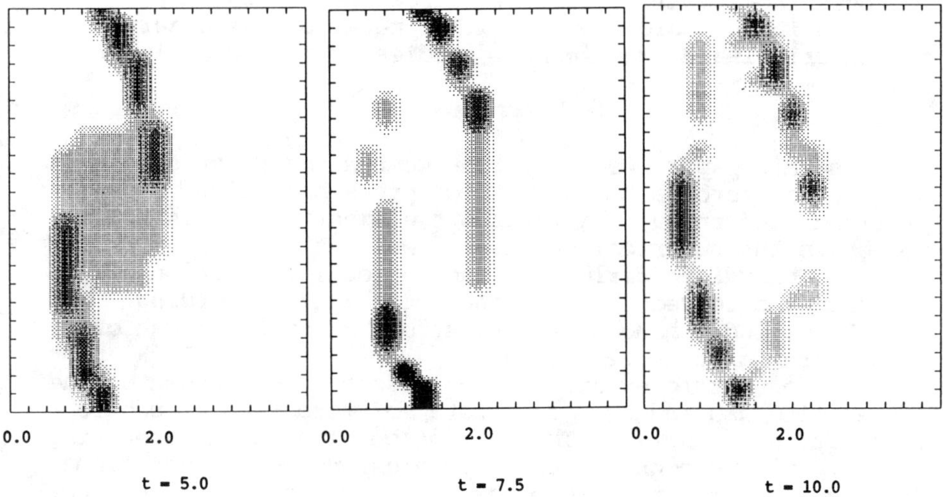

Fig. 14. Reaction rate contours at t= 5.0, 7.5, and 10.0, respectively, for $A_f = 0.5$.

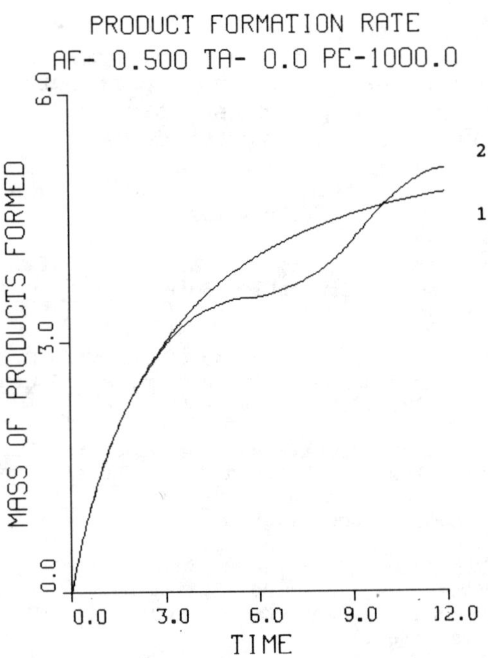

Fig. 15. Plot of mass of products formed vs time for $A_f = 0.5$. Curve 1 is the laminar flame, curve 2 is the reacting jet.

improves the stability of this flow and reduces the rollup observed in the cold flow. Similar behavior was observed in the numerical and experimental studies of the shear layer.[12]

V. Conclusions

In this paper, we study flow-combustion interactions in a reacting turbulent jet. A compressible flow model is utilized, and the vortex/transport element method is applied to obtain the numerical solution. We conclude that

1) The strain field can lead to local extinction of the reaction and to reduction in the total rate of burning.

2) Mixing enhancement by strong entrainment can overcompensate for this reduction.

3) Volumetric expansion suppresses the growth of the instability and reduces the rate of formation of coherent vorticity structures.[4] The rate of burning in the reacting layer is always higher than the corresponding laminar flame. Results confirm the dependence of the reaction zone structure on the Damkohler number, which was found in the incompressible flow analysis of a similar reacting shear flow problem.

Acknowledgments

This work was supported by the Air Force Office of Scientific Research Grant 84-0356, the National Science Foundation Grant CBT-8709465, and Department of Energy Grant DE-FG04-87AL44875. Computer support was provided by the John von Neumann National Computer Center.

References

[1]Karagozian, A. R. and Marble, F. E., "Study of a Diffusion Flame in a Stretched Vortex," Combustion Science and Technology, Vol. 45, 1986. 45, pp. 65-84.

[2]Carrier, G. F., Fendell, F. E. and Marble, F. E., "The Effect of Strain Rate on Diffusion Flames," SIAM Journal of Applied Mathematics, Vol. 28, 1975, pp. 463-500.

[3]Ghoniem, A. F., "Computational Methods in Turbulent Reacting Flows," Reacting Flows: Combustion and Chemcial Reactors, Part I, edited by G. S. S. Ludford in Lecture Notes in Applied Mathematics, Vol. 24, AMS, Rhode Island, 1986.

[4]Ghoniem, A. F., Knio, O.M., and Heidarinejad, G., "Structure of the Reaction Zone in a Reacting Mixing Layer", paper presented at 12th International Colloquium on Dynamics of Explosions and Reactive Systems, 1989.

[5]Ghoniem, A. F., Heidarinejad, G. and Krishnan, A., "Turbulence-Combustion Interactions in a Reacting Shear Layer," Turbulent

Reactive Flows, edited by R. Borghi and S. N. B. Murthy, in Lecture Notes in Engineering, Vol. 40, Springer-Verlag, 1988, pp. 638-671.

[6]Ghoniem, A. F., Heidarinejad, G. and Krishnan, A., "Numerical Simulation of a Thermally-Stratified Shear Layer Using the Vortex Element Method," Journal of Computer Physics, Vol. 79, 1988, pp. 135-166.

[7]Krishnan, A. and Ghoniem, A. F. "Origin and Manifestation of Flow-Combustion Interaction in a Premixed Shear Layer," Proceedings of the 22nd Symposium (International) on Combustion, The Combustion Institute, Pittsburgh, PA, 1988, pp. 665-675.

[8]Krishnan, A., "Numerical Study of Vorticity-Combustion Interactions in Shear Flow," Ph.D. Thesis, Dept. of Mechanical Engineering, Massachusetts Inst. of Technology, Cambridge, MA, 1989.

[9]Ghoniem, A. F., Chen, D. Y. and Oppenhiem, A. K., "Formation and Inflammation of a Turbulent Jet," AIAA Paper #84-0572, 1984.

[10]Knio, O. M. and Ghoniem, A. F., "Numerical Study of a Three-Dimensional Vortex Method," Journal of Computer Physics, 1989 (to be published) (see also AIAA Paper 89-0574).

[11]Heidarinajad, G. and Ghoniem, A. F., "Vortex Simulation of the Reacting Shear Layer; Effects of Reynolds and Damkohler Number," AIAA Paper 89-0573 1989.

[12]Riley, J. J. and McMurtry, P. A., "The Use of Direct Numerical Simulations in the Study of Turbulent Chemically Reacting Flows," Turbulent Reactive Flows, edited by R. Borghi and S. N. B. Murthy, in Lecture Notes in Engineering, Vol. 40, Springer-Verlag, 1988, pp. 486-514.

Chapter IV. Turbulent Flames

Investigation on the Laminar and Turbulent Burning Velocities of Premixed Lean and Rich Flames of CH_4-H_2-Air Mixtures

Y. Liu,* B. Lenze,† and W. Leuckel†
Universität Karlsruhe, Karlsruhe, Germany

Nomenclature

a	=	[m^2/s]	temperature conductivity
C_{fuel}	=	[%]	concentration of fuel gas
C_{st}	=	[%]	stoichiometric concentration of fuel gas
D	=	[mm]	diameter of nozzle
L	=	[mm]	length of Bunsen flame
L_t	=	[mm]	turbulence length scale
R	=	[mm]	radius of tube burner
r	=	[mm]	radial coordinate
Re	=	[-]	Reynolds number
S_L	=	[m/s]	laminar burning velocity
S_T	=	[m/s]	turbulent burning velocity
u	=	[m/s]	axial velocity
u'	=	[m/s]	fluctuation velocity of u
$\underline{u'}_0$	=	[m/s]	u' in apprroach flow
$\bar{u}$	=	[m/s]	mean axial velocity
u_m	=	[m/s]	mean velocity of the laminar tube flow
v	=	[m/s]	radial velocity
x	=	[mm]	axial coordinate
Φ_{total}	=	[-]	overall equivalence ratio

Abstract

In the present paper experimental investigations of both the influence of turbulence on the premixed combustion and the affection of the upstream turbulence conditions by the flame front are presented.

---

*Research Associate, Dept. of Chem. Eng.
†Prof. Dr.-Ing., Dept. of Chem. Eng.

The application of CH_4-H_2-air combustion mixtures enabled the laminar burning velocity S_L to be varied between 0.17 and 1.47 m/s for lean and between 0.25 and 1.50 m/s for rich flames. The overall equivalence ratios Φ_{total} of the combustion mixtures were between 0.6 and 1.3. By use of a stagnation flame system and turbulence generating grids, turbulence intensities u'_0 up to 2.9 m/s for lean and 2.8 m/s for rich flames could be achieved in the approach flow. The resulting turbulent burning velocities S_T, determined by using a LDV technique, ranged from 0.6 up to 17.5 m/s and 0.3 to 18.4m/s for the lean and rich flames, respectively.

The required laminar burning velocities S_L of CH_4-H_2-air mixtures were determined at different CH_4:H_2 ratios and various total equivalence ratios by using the tube method, which provided data of a quality comparable to that of the previously published S_L-values, as far as the latters are available.

The results of the investigated turbulent stagnation flames show a strong affection of the upstream turbulence condition by the presence of the combustion front - a feedback effect of the flame characteristics into the approach flow under both lean and rich flame conditions. Also in the now-realized, more extended u'_0 and S_L range the results of the lean flames comply well to the correlation

$$\frac{S_T}{S_L}=1+5.3\cdot S_L^{0.5}\cdot\frac{u'_0}{S_L}$$

which we found in our previous investigation. In the case of rich flames, the turbulence is found to have considerably less effect on the premixed combustion, and it seems that the influence of turbulence decreases with increasing equivalence ratio Φ_{total} for $\Phi_{total}>1$.

Introduction

Since the introduction of the stagnation flame configuration,[1] several investigators have used this burner system for fundamental studies of premixed turbulent flames.[2,3] In comparison with conical Bunsen- and V-shaped flames the stagnation flame configuration shows significant advantages because of its flat flame front and the lack of flow recirculation in the stabilization zone. Since the flame front gets orientated normal to the approach flow, the turbulent burning velocity S_T can be clearly defined, by analogy to the laminar burning velocity, as the mean flow velocity entering the flame front.

In a recent work Liu and Lenze[2] investigated the influence of turbulence on premixed stagnation flames at different laminar burning velocities (S_L) experimentally. By utilizing CH_4-H_2 mixtures as fuel gases they were able to vary S_L from 0.18 m/s to 0.72 m/s systematically. For this S_L-range, and for turbulence intensities u'_0 up to 1.2 m/s, they found the following correlation for lean flames:

$$\frac{S_T}{S_L} = 1 + 5.3 \cdot S_L^{0.5} \cdot \frac{u'_0}{S_L} \qquad (1)$$

Another experimental result of this work was the fact that the turbulence effect on rich flamestheis considerably weaker than that on lean flames.

The present experiments are aimed at investigating the influence of turbulence on the flame characteristics in a more extensive parameter (S_L and u'_0) range, proving the validity of Eq.(1) for lean flames and getting more insights into the decline process of the turbulence effect for rich flames. S_L was varied between 0.18 m/s and 1.5 m/s by using CH_4-H_2 mixtures. The overall equivalence ratio Φ_{total} ranged from 0.6 to 1.3. The turbulence fluctuation velocity u'_0 in the approach flow was set at values between 0.2 m/s and 2.9 m/s by use of turbulence generating grids. Special attention was also given to the "flame-generated turbulence" in the stagnation flame system.

Since experimental S_L-data were not available in the above indicated, extensive S_L and Φ_{total} range, it was necessary, first, to determine the laminar burning velocities of CH_4-H_2-mixtures at different CH_4 to H_2 ratios and various equivalence ratios.

Determination of the Laminar Burning Velocities S_L of CH_4-H_2-air Mixtures at Different Equivalence Ratios

To determine S_L at various Φ-values, the tube burner method was chosen because of its simplicity. Since the used tube burners (diameter D = 4, 7, 12 mm) have a length of more than 100 times that of the tube diameter D, one can assume that the velocity at the burner exit adopts a parabolical profile

$$u_{(r)} = 2 \cdot u_m \cdot [1 - (\frac{r}{R})^2] \qquad (2)$$

with U_m, R, and r as the mean velocity, tube radius, and radial coordinate, respectively. Then the length L of the laminar Bunsen-flame (500<Re<2000) will become

$$L=\frac{4}{3}\cdot R\cdot\sqrt{\left(\frac{u_m}{S_L}\right)^2-1} \tag{3}$$

and S_L can be deduced from Eq. (3) to be[4]

$$S_L=\frac{u_m}{\sqrt{1+\left(\frac{3}{4}\cdot\frac{L}{R}\right)^2}} \tag{4}$$

By adjusting the gas flow rate, the Bunsen-flames were maintained to have a reasonable length, since too small Bunsen-flames would cause large relative error when measuring the flame length, and at too large gas flow rate the flow could become a turbulent one. The flame length L has been measured optically at different CH_4:H_2 ratios and various overall equivalence ratios (Φ_{total}). Typical flame length was between 10 and 30 mm. S_L values, calculated from L according to Eq. (4), are shown in Fig. 1.

A comparison of our results with those by Günther and Janisch,[5] who determined their S_L values of CH_4-H_2-air mixtures at the stoichiometric gas concentrations (S_{Lst}) and the gas concentration with maximum S_L (S_{Lmax}) by observing particle trajectories in Bunsen-flames, shows a relative deviation ranging from -4.5% to +6% concerning S_{Lst}, and from -5.5% to +5.9% concerning S_{Lmax}. The average relative deviations (wherein each negative value of deviaton is converted to a positive one, for example, -4.5 -> +4.5) are 3.8% and 4.2% for S_{Lst} and S_{Lmax}, respectively. Yu et al[6] also measured S_L values by using LDV (Laser Doppler Velocimetry) technique in a laminar stagnation flame at various equivalence ratios but only for low H_2 content. Therefore, comparison concerning S_L at different Φ_{total} values can only be made for the results of pure CH_4 and with the CH_4:H_2 ratios 70:30 and 54.7:45.3. At almost every equivalence ratio our measurements provide a somewhat higher S_L value. However, also here relative deviations are of the order of a few percents only and attain an average of 6.8% and 4% for $\Phi_{total}< 1$ and $\Phi_{total}>1$, respectively. Considering the large discrepancies existing between the experimental S_L

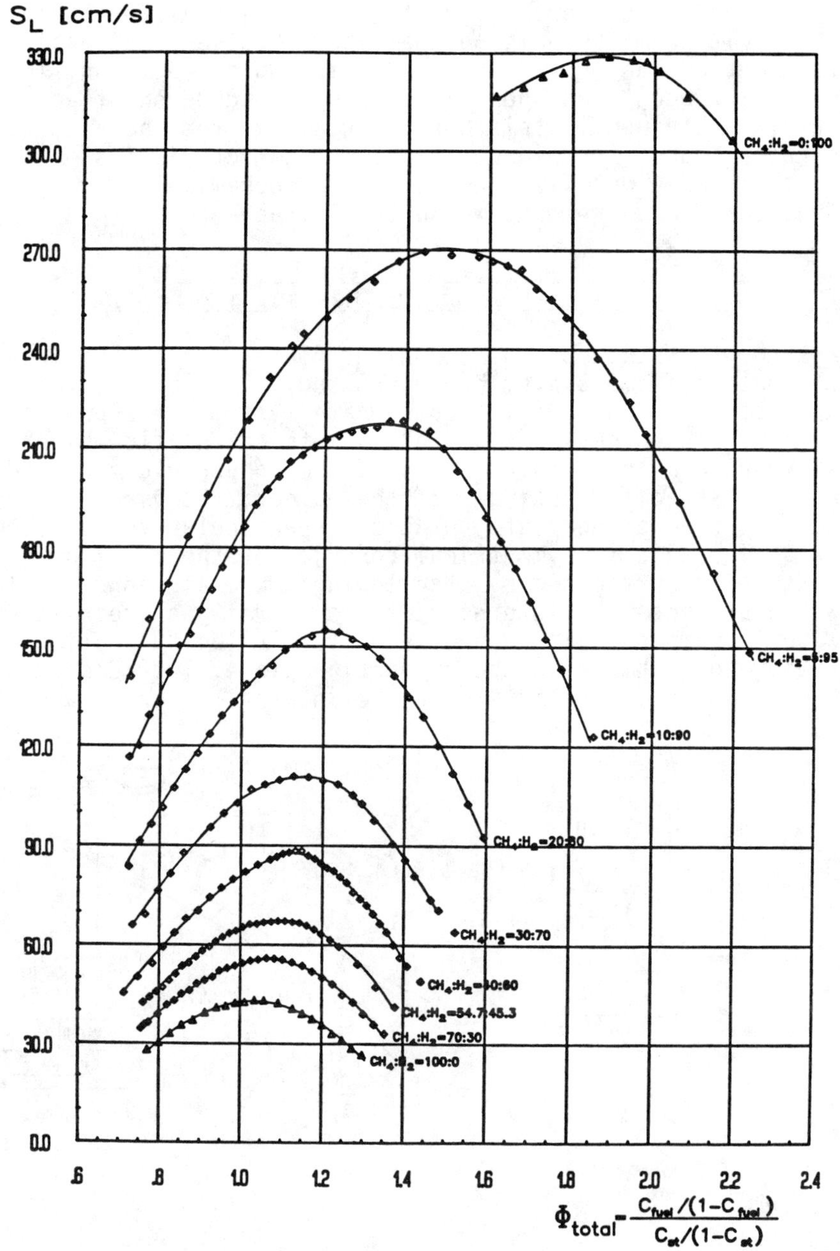

Fig.1 Laminar burning velocities of CH_4-H_2-air mixtures at different CH_4:H_2 ratios and various total equivalence ratios.

data even for pure fuel gases (for example, CH_4 and H_2) determined by different investigators using different methods, it seems surprising that the relative "inexpensive" tube method can provide data of comparable precision. These data have been used when evaluating the turbulent flame results obtained in this work. Certainly, for the accurate correlation of turbulence and flame characteristics further, more detailed and more physicochemically considered measurements would be desirable.

Investigations on the Turbulent Flame Propagation
Experimental Setup and Measuring Methods

For the study of the turbulent flame characteristics, a stagnation flame system has been used, its experimental apparatus and the applied coordinate system are shown in Fig. 2. The burner consists of an internal nozzle which is surrounded by an annular orifice. The concentric core and annular streams, issuing from the burner, are directed towards a stagnation plate, placed perpendicular to the flow axis. The axially symmetrical core jet of the CH_4-H_2-air mixture is surrounded by a coflowing, pure air annular flow which prevents the internal jet from forming a shear layer with the ambient air. Hence, a uniform velocity profile is preserved right up to the stagnation plate. Two burners of D = 40 and 80 mm diameter for the internal nozzles and 124

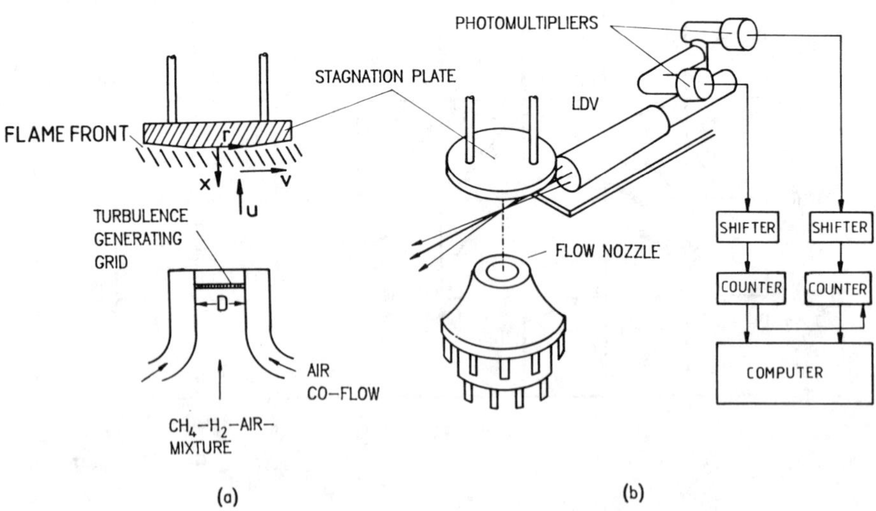

Fig.2 Experimental apparatus with stagnation plate oriented coordinate system. Two burners of D=40 and 80 mm diameter for the internal nozzle are used.

and 160 mm diameter for the annular orifices were used in this work.

Perforated plates, placed at the internal nozzle exit, served to generate the required turbulence intensities. With the combination of different blockage ratios and various perforated plate hole diameters, fluctuation velocities u' up to 2.9 m/s could be achieved in the approach flow. The mean exit velocity and the global stretch rate $(d\bar{u}/dx)$ ranged from 2.5 to 12 m/s $(13300 < Re_D < 36000)$ and from 70 to 440 sec^{-1}, respectively.

The turbulent flames were stabilized by a circular watercooled stagnation plate of 240 mm diameter. Since the reaction zones of most of the flames are located between x = 10 and 50 mm, which means that stand-off distance of the flame front from the plate surface is much greater than the quenching distance, the cooling of the plate can be considered to have no direct effect on the flame stabilization zone. The lower outer section of the plate was chamfered with an angle of 7°, which allowed LDV measurements on jet axis up to a distance of 5 mm from the stagnation point.

To measure the turbulence characteristics by LDV, magnesium oxide particles of between 0.7 and 1.2 μm diameter were produced by a rotating brush-type particle generator and introduced into the core flow. In order to obtain both velocity components ($\bar{u}$,$\bar{v}$; u' and v') as well as the probability density functions (PDF) of flow velocity, a two-component LDV-system, incorporating fre- quency shift and countertype signal processors, was used. The typical data rate in the approach flow was between 2 and 5 kHz and decreases below 1 kHz in the reaction zone. To avoid possible biasing effects, the rate of sampling was always maintained to be at least five times smaller than the data rate. The measuring volume had a length of about 1 mm and a diameter of about 80 μm.

Characteristics of Turbulent Stagnation Flames and Determination of the Turbulent Burning Velocity S_T

In order to illustrate the characteristics of turbulent stagnation flames, examples are presented from a lean (see flame A, Fig.3a) and a rich flame (see flame B, Fig. 3b).

Initial measurements in the isothermal flows again indicate the special flow characteristics favorable for the study of turbulent premixed combustion, as already shown in previously published works[1-3]. The mean axial velocity $\bar{u}$ decreases steadily towards the stagnation plate (see Fig.3a and 3b), while the mean radial component $\bar{v}$ remains nearly zero along the whole length of the flow axis ($\bar{v}$ is not

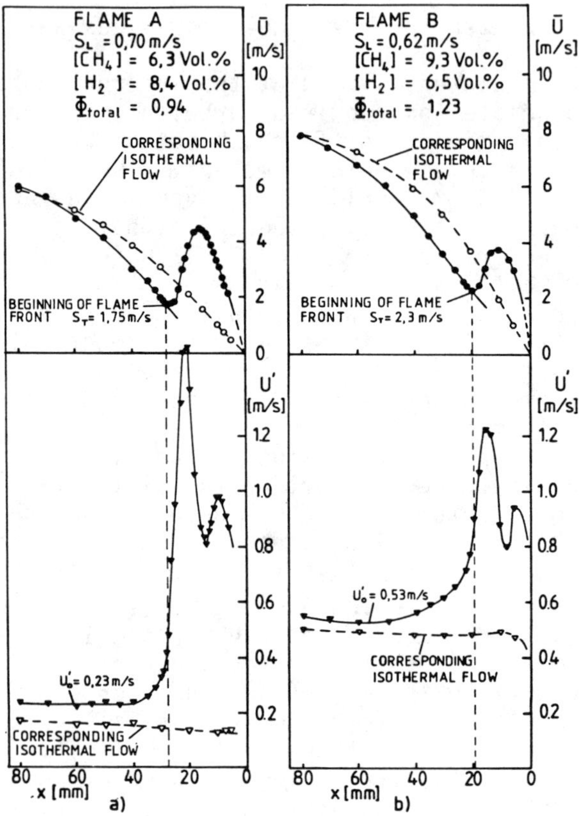

Fig.3 Axial distributions of $\overline{u}$ and u' of a) a lean flame (flame A) and b) a rich flame (flame B) and their corresponding isothermal flows.

shown here, see Ref.7). Another distinct feature is that u' remains nearly constant down to an axial distance of about 60 mm from the stagnation point. Also the measured radial profiles (not shown) of $\overline{u}$ and u' in this region indicate that these remain approximately constant within a radius of 50 mm and 30 mm for nozzles of D = 80 mm and 40 mm, respectively. These flow characteristics, therefore, allow a clearly defined correlation between the flame boundary and the approach flow and an unambiguous determination of the turbulent burning velocity.

In the presence of a flame, $\overline{u}$ decreases with decreasing distance to the stagnation plate down to a particular location where u deviates from its nearly linear decay and increases due to the gas expansion caused by the combustion heat release. This particular location of the first, clearly recognizable deviation has been regarded as the beginning of the flame front as also done in Ref. 1, 2, and 7. Because the flame front is normal to the approaching flow, the turbulent burning velocity S_T is then, in analogy to the definition of the laminar burning velocity, equal to the mean axial velocity $\overline{u}$ at this location. Thus, $\overline{u}$ increases further to a local maximum before dropping back to zero at the stagnation point.

Also in the case of flames, u′ can be seen to remain nearly constant in a certain region in front of the flame (x =80 - 40 mm for flame A and x = 80 - 50 mm for flame B). The fact that u′-values of corresponding isothermal flows in this region are slightly lower may be explained by the affection of the pressure gradient by the flame front. As the flow approaches the flame front, u′ is increased to a considerably higher level already at the above defined upstream edge of the reaction zone. By looking at the PDF′s of u at various locations on the flow axis in the case of the lean flame A (see Fig. 4a), one can distinguish two steps of u′-increase from the constant u'_0 up to the u′-maximum within the flame front, namely a first step upstream and a second step downstream of the edge of the flame front. At the beginning of the flame front, a second peak of higher velocities appears, which indicates the alternating presence of the cold reactants and hot reacted products. This intermittency effect increases steadily towards the flame front center, resulting in the formation of an u′-maximum due to the bimodality of the u-distribution. This fact means that the u′-increase in the second step (i.e. inside the flame front) is mainly attributed to the intermittency effect. Contrary to the second step, no bimodality is visible in the first step. As the flow approaches the flame front, the u-distribution is getting broader while remaining monomodal and of Gaussian shape up to the edge of the reaction zone. This is obviously similar when the PDF′s of the rich flame B are considered. Hence, the result discussed shows that a turbulent flame front with its fluctuation and instability (due to effects like gas expansion etc.) can represent a source of perturbation which propagates up to some distance also in the upstream direction, affecting the upstream turbulence by stimulating u′ to increase.

Downstream of the flame front, the u-distributions become monomodal and Gaussian again as shown in Fig. 4b (flame B, x = 10 and 7 mm). These distributions are not only noticeably broader than the one in the "unaffected" approach flow (with standard deviation u'_0) but also broader than the other one at the beginning of the flame front

(Actually the use of the expression "unaffected" is not correct because also in the region in which u' remains constant the turbulence intensity is increased due to changes of the pressure gradient. The corret expression should be "unaffected by fluctuation and instability of the flame front"). This fact may indicate "flame generated turbulence" downstream of and probably also within the reaction zone. Further experiments with conditioned LDV- measurements are planned in order to get more insights into the combustion-turbulence interaction within the flame front and quantify the relationship.

Comparison of Lean and Rich Flame Results

Experimental results of 28 lean flames and 19 rich flames are listed in Table 1 and 2. In Fig. 5, S_T/S_L ratios as a function of $u'_0/(S_L)^{1/2}$ are presented.

The turbulence intensity u'_0 was chosen because it characterizes the turbulence condition in the still "unaffected" approach flow and, thus, can be regarded as a flame-independent input parameter. It is clearly seen from Fig.5, that, also in the now strongly extended parameter range (u'_o up to 2.9 m/s and S_L up to 1.47 m/s), the lean flame

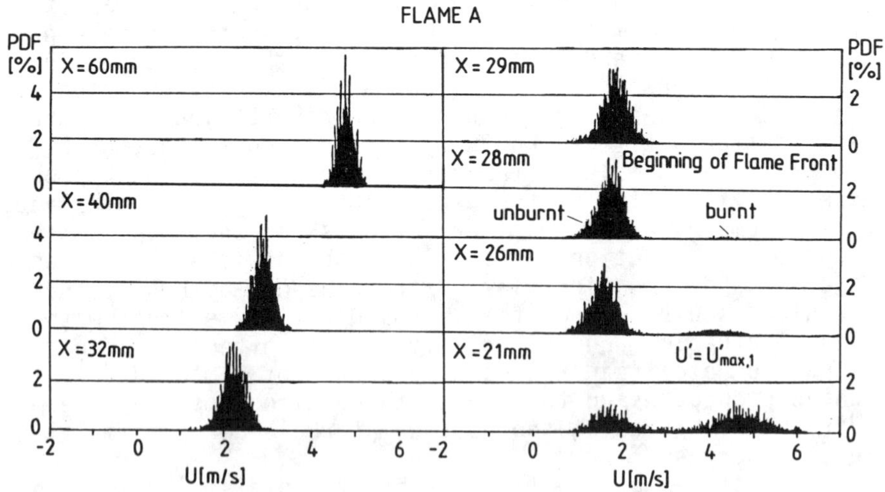

Fig.4a Probability density functions of axial velocity u at various locations on the flow axis upstream of and within the flame front of the lean flame A. The u-distribution is getting broader without the occurence of the intermittency effect until the flow reaches the edge of the flame front (x=28 mm).

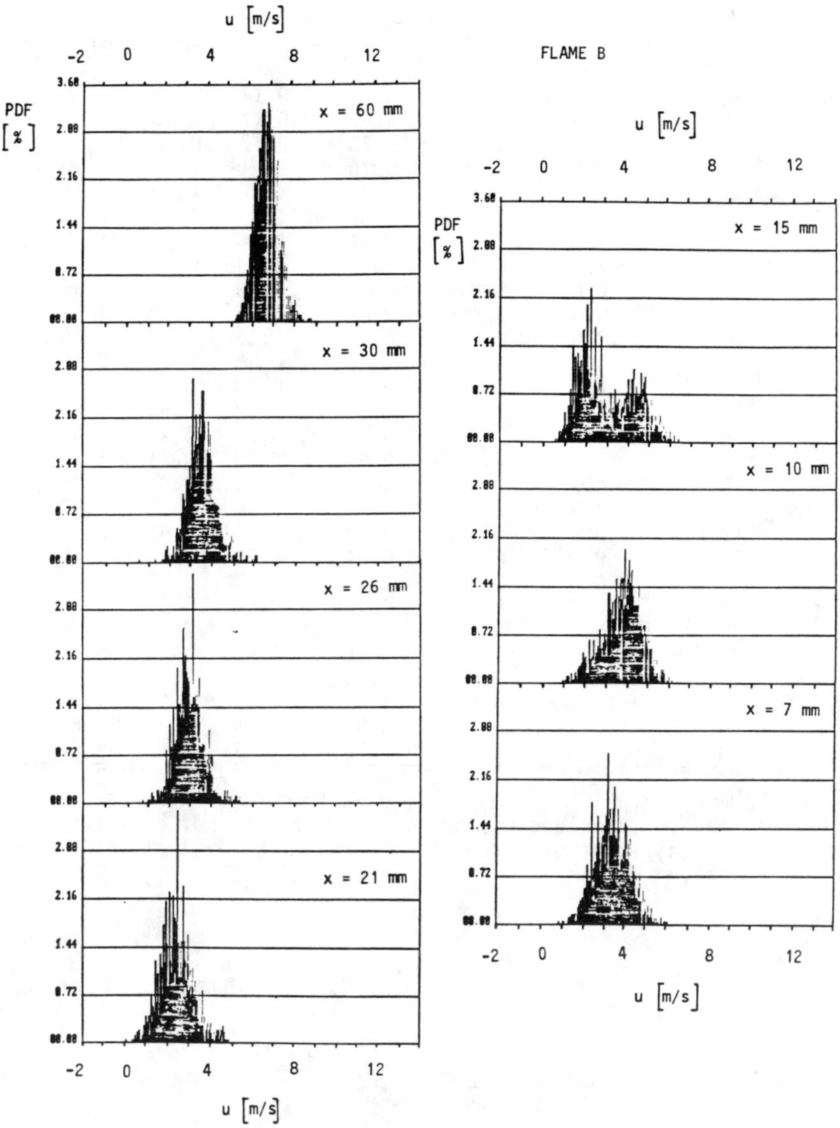

Fig.4b Probability density functions of axial velocity u at various locations on the flow axis upstream of, within, and downstream of the flame front of the rich flame B.

Table 1 Mixture Compositions, Φ_{total}, S_L, u'_0 and S_T of Investigated Lean Flames

Flame No.	$[H_2]$ Vol%	$[CH_4]$ Vol%	Φ_{total}	S_L m/s	u'_0 m/s	S_T m/s
1	0.0	6.5	0.66*	0.17	0.23	0.67
2	0.0	6.5	0.66*	0.17	0.28	0.80
3	0.0	6.5	0.66*	0.17	0.25	0.75
4	0.0	6.8	0.70	0.20	0.18	0.60
5	0.0	6.8	0.70	0.20	0.24	0.75
6	4.9	4.5	0.59	0.21	0.50	1.45
7	4.9	4.7	0.62	0.24	0.40	1.30
8	4.9	4.7	0.62	0.24	0.34	1.20
9	0.0	7.8	0.81	0.30	0.35	1.30
10	0.0	8.4	0.87	0.35	0.58	2.27
11	0.0	8.4	0.87	0.35	1.00	3.50
12	4.9	5.4	0.70	0.36	0.75	2.75
13	4.9	5.9	0.76	0.38	0.50	1.95
14	4.9	7.2	0.91	0.52	0.66	3.08
15	4.9	7.3	0.92	0.53	0.49	2.43
16	4.9	7.3	0.92	0.53	0.21	1.32
17	4.9	7.3	0.93	0.53	0.41	2.12
18	4.9	7.4	0.94	0.54	0.37	2.00
19	8.4	6.3	0.94	0.70	0.45	2.70
20	8.4	6.3	0.94	0.70	0.23	1.78
21	8.4	6.4	0.95	0.71	0.32	2.10
22	10.6	6.4	0.96	0.88	1.32	7.50
23	13.4	4.3	0.89	1.02	2.25	13.40
24	14.1	3.9	0.86	1.05	1.42	8.60
25	17.0	2.5	0.80	1.22	1.80	12.30
26	15.6	4.5	1.00	1.24	2.90	17.50
27	17.1	3.9	0.97	1.24	1.22	8.80
28	19.8	1.5	0.78	1.47	1.80	12.60

* For flames with $\Phi_{total} < 0.7$ the S_L data from Ref.7 were used. As shown in chap. 2, the use of the data of Ref.7 is not expected to cause any significant differences concerning the evaluation of the turbulent flame data.

results conform very well to the relationship

$$\frac{S_T}{S_L} = 1 + 5.3 \cdot S_L^{0.5} \cdot \frac{u'_0}{S_L}$$

which we found in Ref.2 for smaller u'_0 and S_L values (u'_0 < 1.2 m/s and S_L < 0.72 m/s). The validity of Eq. (5), for different fuel types and mixture compositions, means that those are irrelevant concerning the turbulence effect, as long as S_L is maintained constant.

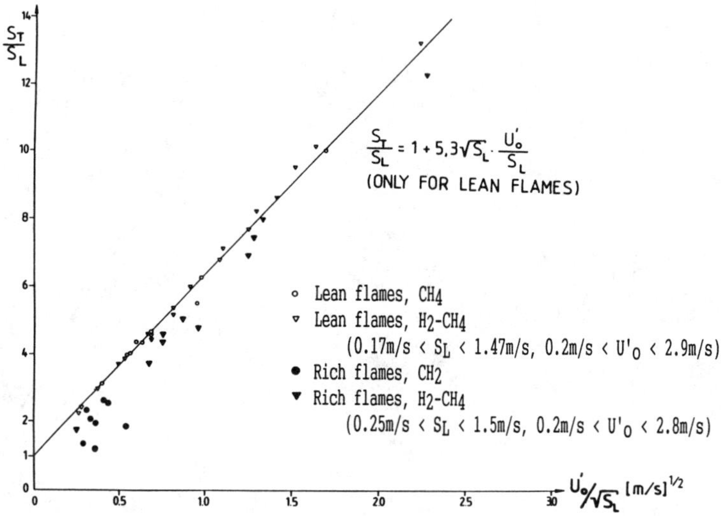

Fig.5 Correlation of the turbulent burning velocity S_T and the flameindependent turbulence intensity u_o' with S_L as parameter.

Certainly, the term $S_L^{1/2}$ in Eq.(5) is not dimensionless, i.e., the constant 5.3 has to be given the dimension $[s/\text{-}m]^{1/2}$. To make S_L dimensionally correct, another suitable property of the dimension [m/s] is needed which we are still looking for. One possibility may be the use of the turbulence length scale L_t which can be introduced, for instance, in the form of (a/L_t) with a as temperature conductivity (dimension $[m^2/s]$). However, this requires systematical and precise measurements of L_t both in the approach flow and in the flame reaction zone.

Contrary to the lean flame case, Eq.(5) obviously can not be applied to the rich flame results. All of the S_T/S_L ratios of rich flames are located below the linear line according to Eq.(5). This means that, at the same S_L value, turbulence has a smaller acceleration effect on rich flames compared to lean flames. In order to quantify this decline, the relative increase of burning velocity $(S_T - S_L)/S_L$ is chosen as a property describing the turbulence effect. In Fig. 6, $(S_T - S_L)/S_L$, normalized by the term $5.3 \cdot S_L^{1/2} \cdot u'_0/S_L$ (The constant 5.3 has the dimension of $(s/m)^{1/2}$), is plotted against the total equivalence ratio Φ. As to be expected, the lean flame results are close to 1 for all the equivalence ratios. In the region $\Phi_{total} > 1$, however, the ratio is seen to decrease with increasing Φ_{total}. Hence, the richer the combustion mixture is, the weaker the turbulence

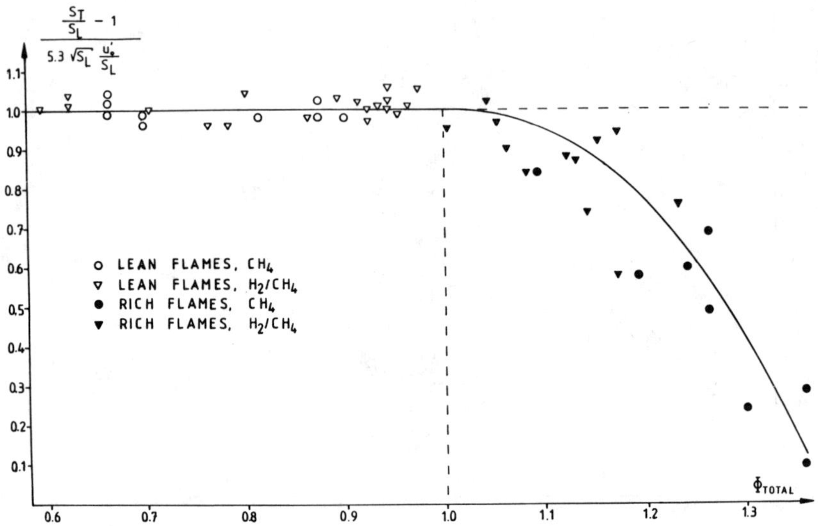

Fig.6 Correlation of the normalized turbulence effect and the total equivalence ratio Φ_{total} for lean and rich flames.

Table 2 Mixture Compositions, Φ_{total}, S_L, u'_0 and S_T of Investigated Rich Flames

Flame No.	$[H_2]$ Vol%	$[CH_4]$ Vol%	Φ_{total}	S_L m/s	u'_0 m/s	S_T m/s
29	0.0	12.5	1.36	0.25	0.18	0.30
30	0.0	12.0	1.30	0.30	0.30	0.55
31	0.0	11.7	1.26	0.34	0.25	0.86
32	0.0	11.7	1.26	0.34	0.21	0.65
33	0.0	11.5	1.24	0.36	0.20	0.74
34	0.0	11.1	1.19	0.40	0.26	0.90
35	0.0	10.3	1.09	0.42	0.20	1.00
36	6.5	9.3	1.23	0.62	0.53	2.30
37	8.5	8.1	1.17	0.66	0.20	1.15
38	6.8	7.7	1.05	0.67	1.10	5.30
39	10.2	7.2	1.12	0.86	0.70	3.90
40	12.0	5.9	1.04	0.94	0.64	4.30
41	11.3	6.5	1.08	1.01	0.75	4.40
42	14.1	6.1	1.14	1.11	1.02	5.30
43	16.0	5.6	1.17	1.24	1.43	9.20
44	15.9	5.3	1.12	1.27	0.98	6.40
45	17.0	4.5	1.06	1.40	1.47	9.70
46	18.3	4.7	1.15	1.51	2.82	18.40
47	0.0	12.0	1.30	0.30	0.16	0.41

influence on its combustion wave is. When looking for a possible explanation for this phenomenon, one should certainly consider the fact that, contrary to lean flames with complete combustion, the burning of rich mixtures is more incomplete for increasing equivalence ratio.

Conclusions

In the present investigation we first determined the laminar burning velocity S_L of CH_4-H_2-air mixtures at different CH_4:H_2 ratios (from pure CH_4 to pure H_2) for various equivalence ratios. Results show that the relatively simple and inexpensive tube method can provide S_L data of satisfactory quality, and may be used for immediate S_L measurements in order to get a first information of laminar burning behaviors.

Systematic measurements of a large number of turbulent stagnation flames indicate a strong affection of the upstream turbulence by the flame front both in the lean and rich flame case. Although the parameter range was extended to u'_o = 2.9 m/s and S_L=1.4 m/s, results of the lean flames still conform well to the simple relationship found previously

$$\frac{S_T}{S_L}=1+5.3\cdot S_L^{0.5}\cdot\frac{u'_0}{S_L}$$

In the case of rich flames, the turbulence effect is shown to conciderably decrease with increasing equivalence ratio Φ_{total}. The results suggest that, for rich flames, the richer the combustion mixture is, the smaller the influence of turbulence on combustion is.

Acknowledgment

These investigations were conducted within the German cooperative flame research program **TECFLAM** (Project No. 2.2). The authors wish to express their gratitude to the Ministry of Research and Technology of the Federal Government of Germany/FRG for their financial support received.

References

[1]Cho,P., Law,C.K., Hertzberg,J.R., and Cheng,R.K., "Structure and Propagation of Turbulent Premixed Flames Stabilized in a Stagnation Flow", Proceedings of the Twenty-first Symposium (Int.) on Combustion, pp. 1493-1499, 1986. The Combustion Institute.

[2]Liu, Y., and Lenze,B., "The Influence of Turbulence on the Burning Velocity of Premixed Flames at Different Laminar Burning Velocities of CH_4-H_2 Mixtures. Proceedungs of the Twenty-second Symposium (Int.) on Combustion, pp. 747-754, 1988. The Combustion Institute.

[3]Cho,P., Law,C.K., Cheng,R.K., and Shepherd,I.G., "Velocity and Scalar Fields of Turbulent Premixed Flame in a Stagnation Flow", Proceedings of the Twenty-second Symposium (Int.) on Combustion, pp. 739-745, 1988. The Combustion Institute.

[4]Günther,R., Verbrennung und Feuerung, Springer Verlag, 1984, p.96.

[5]Günther,R., and Janisch,G., "Meßwerte der Flammengeschwindigkeit von Gasen und Gasgemischen". Chemieingenieurtechnik 43:975-978-(1971).

[6]Yu,G., Law,C.K., and Wu,C.K., "Laminar Flame Speeds of Hydrocarbon + Air Mixtures with Hydrogen Addition". Combustion and Flame 63:339-347(1986).

[7]Liu,Y., Lenze,B., and Leuckel,W., "Investigation on the Combustion-Turbulence Interaction in Premixed Stagnation Flames of H_2-CH_4 Mixtures". Proceedings of the Seventh Symposium on Turbulent Shear Flows,
Stanford, 1989.

Geometry Effects on Premixed Turbulent Propagating Flames

S. S. Ibrahim,* R. S. Cant,* and K. N. C. Bray*
University of Cambridge, Cambridge, England, United Kingdom

Abstract

Geometry effects on premixed propagating flames are presented for a stirred bomb of fixed volume. Three geometries have been studied, incorporating rectangular, cylindrical, and spherical symmetry. A one-dimensional finite difference computation has been carried out using a modified version of the Bray-Moss-Libby model for premixed turbulent combustion. The progress of the chemical reaction is predicted by solving a balance equation for the reaction progress variable, in which the mean turbulent reaction rate is modelled using a laminar flamelet library formulation. Analytic correlations are used to represent the variation of the laminar flame speed and integrated reaction rate with pressure, unburned temperature, and turbulent strain rate. The Pressure Implicit Split Operator semi-implicit predictor-corrector algorithm, a non-iterative method, is used to solve the time-dependent model equations. Predicted results for two different sets of initial conditions giving thin and thick turbulent flames are presented in terms of the reaction progress variable and the overall burning velocity. The calculations lead to the conclusion that large effects on flame propagation are apparent due to the geometry alone.

I. Introduction

The geometry and operating conditions of the combustion chamber in many practical devices (e.g., spark ignition engines and furnaces) are important because of their effects on the flame speed and heat release distribution.

---

*Department of Engineering.

Cant and Bray[1,2] have reported results from one-dimensional calculations for premixed flames in a constant volume bomb with a rectangular geometry. Two turbulent burning rate models have been used: the mixing-limited eddy breakup (EBU) model attributable to Mason and Spalding[3] and a strained laminar flamelet model employing analytical expressions for the turbulent burning rate[1]. Cant and Bray concluded that the EBU model which requires a single specific value for its numerical constant may lead to considerable error. Cant and Bray also found that modelling the turbulent scalar fluxes using countergradient transport had only a small effect in the confined vessel, which is also the case under study here.

In the present work, three different codes, namely, rectangular, cylindrical and spherical, are developed to study the effects of geometry on the turbulent flame propagation inside closed vessels. A reduced version of the numerical model developed by Cant and Bray for premixed combustion is used. A brief description of the theoretical analysis is presented in the following sections. Further details of the model used can be found in Cant and Bray[1] and Cant[5]. The results obtained from the model for the different geometries are presented and discussed. Conclusions from this study are summarized in the last section.

II. The Model

The numerical model attributable to Cand and Bray[1,2] for premixed flames is used with different geometries. In this model, conservation equations for mean mass, momentum, and energy are solved numerically along with equations for the turbulent Reynolds stress r and the turbulence kinetic energy dissipation rate ε.

An additional transport equation is also solved for the reaction progress variable c, defined as a normalized product species mass fraction. This equation has a source term representing the turbulent reaction rate. A model of the turbulent reaction rate has been proposed by Cant and Bray[2] based on the assumption that burning occurs in strained laminar flames. The ensemble mean reaction rate at location $\mathbf{x}$ and time t is written

$$\overline{w}(x, t) = w_y(x, t)\nu_y(x, t) \tag{1}$$

where ν_y is the average number of crossing per unit distance in a direction y along which $\overline{c}$ is constant. It is given[6] by

$$\nu_y(x,t) = g\bar{c}(1-\bar{c})/\hat{L}_y \tag{2}$$

where g is a constant having a value of 1.5 and $\hat{L}_y$ is the integral length scale of flame wrinkling and is taken to be equal to the turbulence length scale.

The chemical reaction per flamelet crossing w_y appearing in Eq. (1) may be written

$$w_y = \frac{\rho_R S_L^o}{|\bar{\sigma}_y|} \int_o^{\infty} \frac{w(\varepsilon)}{w(0)} P(\varepsilon) d(\varepsilon) \tag{3}$$

where ρ_R is the density of the reactant mixture, S_L^o is the unstrained laminar burning velocity, $|\bar{\sigma}_y|$ is the mean magnitude of the direction cosine of the flamelet interface in direction y, and $P(\varepsilon)$ is assumed[7] as a log-normal distribution of viscous dissipation rates.

The simplified approach used by Cant and Bray[2] on modelling S_L^o and $w(\varepsilon)/w(0)$ yields

$$S_L^o = S_{Lo}^o \left[\frac{T_R}{T_o}\right]^{\alpha} \left[\frac{\bar{P}}{P_o}\right]^{\beta} \tag{4}$$

and

$$\frac{w(\varepsilon)}{w(0)} = 1 - \varepsilon/\varepsilon_q ; \varepsilon \le \varepsilon_q$$

$$= 0; \ \varepsilon > \varepsilon_q \tag{5}$$

where ε_q is the value of ε above which flamelet reaction is considered negligible. S_{Lo}^o, T_o and P_o are appropriate reference values and α and β are constants.

Substituting Eqs. (4) and (5) in Eq. (3) gives after integration

$$w_y = \rho_R S_L^o I_o(\hat{\varepsilon}/\varepsilon_q, \sigma) \tag{6}$$

where $I_o(\hat{\varepsilon}/\varepsilon_q, \sigma)$ is a known algebraic function[2] in which σ is the standard deviation of the log-normal distribution of ε.

An expression for the mean turbulent burning rate $\bar{w}$ is finally obtained by assembling Eqs. (6) and (2):

$$\bar{w} = \frac{g\bar{\rho} S_{Lo}^o \left(\frac{T_R}{T_o}\right)^{\alpha} \left(\frac{P}{P_o}\right)^{\beta} I_o\left(\frac{\varepsilon}{\varepsilon_q}, \sigma\right) (1+\bar{\tau})\, \tilde{c}\, (1-\tilde{c})}{|\sigma_y|\, \hat{L}_y\, (1+\bar{\tau}\tilde{c})} \tag{7}$$

which represents the source term used for the $\tilde{c}$ transport equation.

The unburned gas temperature T_R is calculated explicitly assuming isentropic compression:

$$T_R = T_{Ro} \left(\bar{P}/P_o \right)^{\frac{\gamma_R - 1}{\gamma_R}} \tag{8}$$

The ignition process is represented by means of an artificial source term $\bar{w}_{ign}$ which replaces $\bar{w}$ in the ignition region:

$$\bar{w}_{ign} = \bar{\rho} \frac{1}{t_{ign}} \tag{9}$$

where t_{ign} (= 1×10^{-3} s in the present study) is the ignition time interval over which the reaction progress variable rises approximately linearly. Model parameters are listed in Table 1.

III. The Numerical Solution

For one-dimensional, axisymmetric flow, the density-weighted conservation equations can be written in the following generalized variable form:

$$\frac{\partial}{\partial t} (\bar{\rho}\, \phi) + \frac{1}{r^m} \frac{\partial}{\partial r} \left(r^m \bar{\rho}\, \tilde{v}\, \phi \right) = S_\phi \tag{10}$$

where m = 0,1,2 for the rectangular, cylindrical, and spherical coordinate systems, respectively. Here ϕ stands for any of the variables $\tilde{v}$, $\tilde{c}$, $\tilde{\varepsilon}$, $\tilde{e}$, and $\tilde{r}_{22}$, and S_ϕ represents their corresponding source/sink terms. (Details of these terms together with the model constants used can be found in Ref. 2).

The governing equations are reduced to finite difference equations by integrating the governing differential equations over a finite control volume. The set of finite

Table 1 Model constants and initial conditions

P_o	=	1.01×10^5 Nm^{-2}	α	=	2.13
T_o	=	297.0 K	β	=	-0.17
S^o_{Lo}	=	0.319 ms^{-1}	γR	=	1.4
ε_q	=	5000.0 s^{-1}			

difference equations thus derived is solved numerically using the non-iterative PISO scheme developed by Issa[8] that is suitable for time-varying problems. The implementation used in the present work is described by Cant.[5]

The initial values of r and ε are found to be important in determining the behavior of the propagating flame. In our study, we will show the sensitivity of the results to changes in their values. Two initial turbulence levels were chosen to examine these effects:

1) Case I: $r_{in} = 0.1\ m^2 s^{-2}$, $\varepsilon_{in} = 20\ m^2 s^{-3}$, corresponding to an initial turbulence intensity $u'_{in} = 0.316\ ms^{-1}$ and integral length scale $\ell = 0.47 \times 10^{-3}$ m.
2) Case II: $r_{in} = 1.5\ m^2 s^{-2}$, $\varepsilon_{in} = 4000\ m^2 s^{-3}$, corresponding to $u'_{in} = 1.22\ ms^{-1}$ and $\ell = 0.47 \times 10^{-3}$ m.

The initial turbulence length scale has been calculated as follows:

$$\ell_{in} = C_\mu^{0.75}\ k_{in}^{1.5}/\varepsilon$$

where C_μ is constant, having a value of 0.09, and k_{in} is the turbulence kinetic energy $(k = 3r/2)$.

The calculations for the three geometries were performed for closed vessels of fixed length or radius, L = 0.025 m. The ends of the one-dimensional geometry are closed, and an ignition point is chosen at the left-hand end. Uniform finite difference grid arrangements of 25 nodes have been used for all cases. The time step has a value of $\delta t = 3.3 \times 10^{-6}$ s for all calculations presented here. The boundary conditions are those of no flow or flux through the walls.

IV. Results and Discussion

Calculated results from case I are shown in Fig. 1 for the reaction progress variable $\tilde{c}$. A thick flame travels

Table 2 Case I

geometry	t_B, s	$\bar{S}_T$, ms^{-1}	$\bar{S}_T/u_o'$
rectangular	1.18×10^{-3}	21.18	67.0
cylindrical	2.18×10^{-3}	11.47	36.3
spherical	2.96×10^{-3}	8.45	26.7

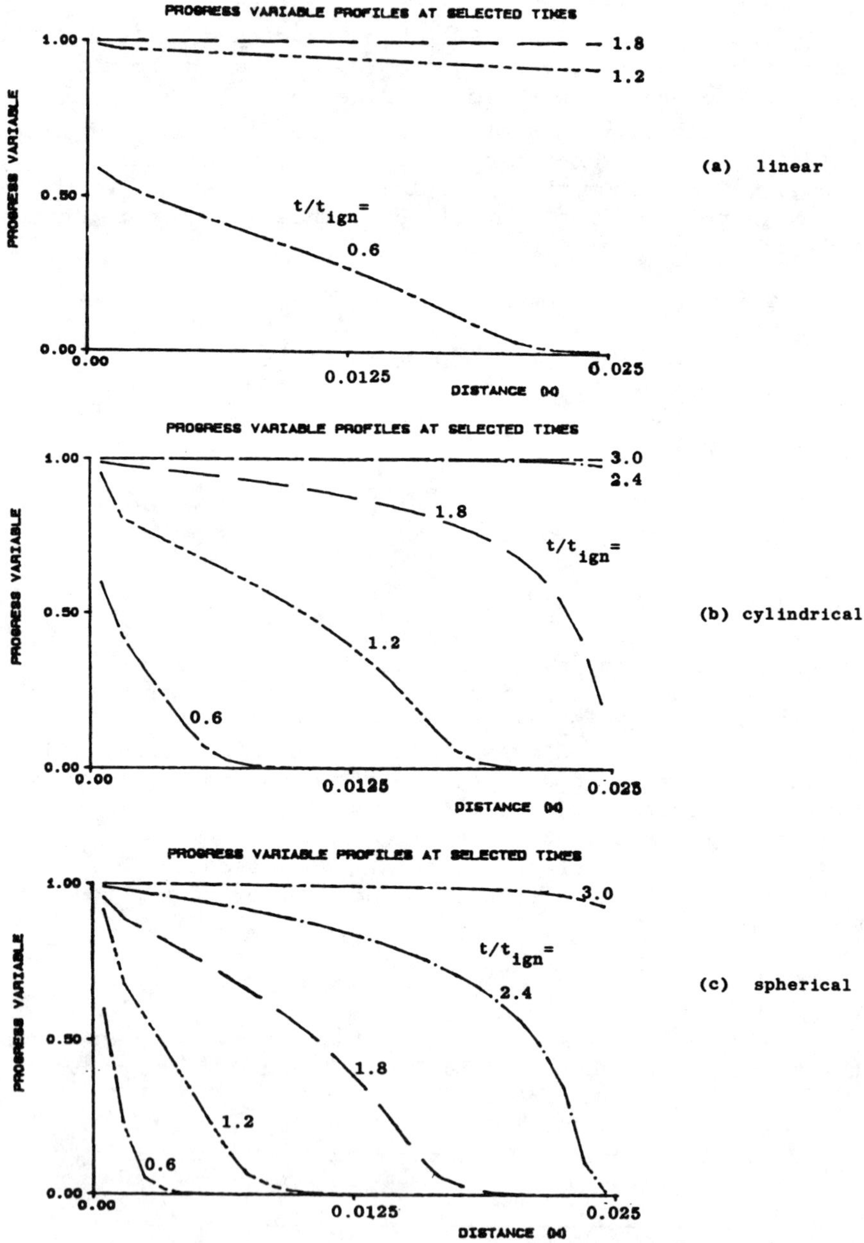

Fig. 1 Reaction progress variable $\tilde{c}$ vs time at selected times for: a) rectangular; b) cylindrical; and c) spherical geometry. Case I: $r_{in} = 0.1$, $\varepsilon_{in} = 20$.

Table 3 Case II

geometry	t_B, s	$\overline{S}_T$, ms^{-1}	$\overline{S}_T/u_o'$
rectangular	5.90×10^{-3}	4.24	3.47
cylindrical	10.90×10^{-3}	2.29	1.88
spherical	14.80×10^{-3}	1.69	1.39

from left to right, and after it reaches the right-hand end of the chamber the final stages of burning occur almost homogeneously. The rectangular geometry shows a faster burning rate than the cylindrical, which shows a faster burning rate than the spherical. Table 2 shows values of an overall burning velocity $\overline{S}_T$ calculated from

$$\overline{S}_T = L/t_B \tag{11}$$

where t_B is the time required to burn the complete charge, which in this case is determined mainly by the ignition time t_{ign}. Also shown in the table are the ratios of $\overline{S}_T/u_{in}'$. These ratios do not imply rapidly propagating turbulent flames because burning occurs almost homogeneously.

Figure 2 and Table 3 shows results of similar calculations using initial values of case II. With these initial values, a slowly propagating flame appears, which is thin compared with that of case I. Values of $\overline{S}_T/u_{in}'$ are now of the order of unity. Note that in contrast to case I, ε_{in} is much larger than ε_q, so the flamelets are initially very highly strained.

Contours of the mean reaction progress variable, $\tilde{c} = 0.1 - 0.9$ in intervals of 0.2, are plotted on a time-distance graph and shown in Figs. 3 and 4 for cases I and II, respectively. These results confirm the description presented previously.

V. Conclusions

Three geometries are studied in the present work: rectangular, cylindrical, and spherical for one-dimensional premixed propagating flames in closed vessels with fixed volume. The results show that the geometry can have a large effect on the turbulent flame thickness and burning velocity.

Further developments will follow by replacing the analytical flamelet model by a full strained laminar flamelet "library" of computed data.[9] Comparisons with experimental data will become available in the near future.

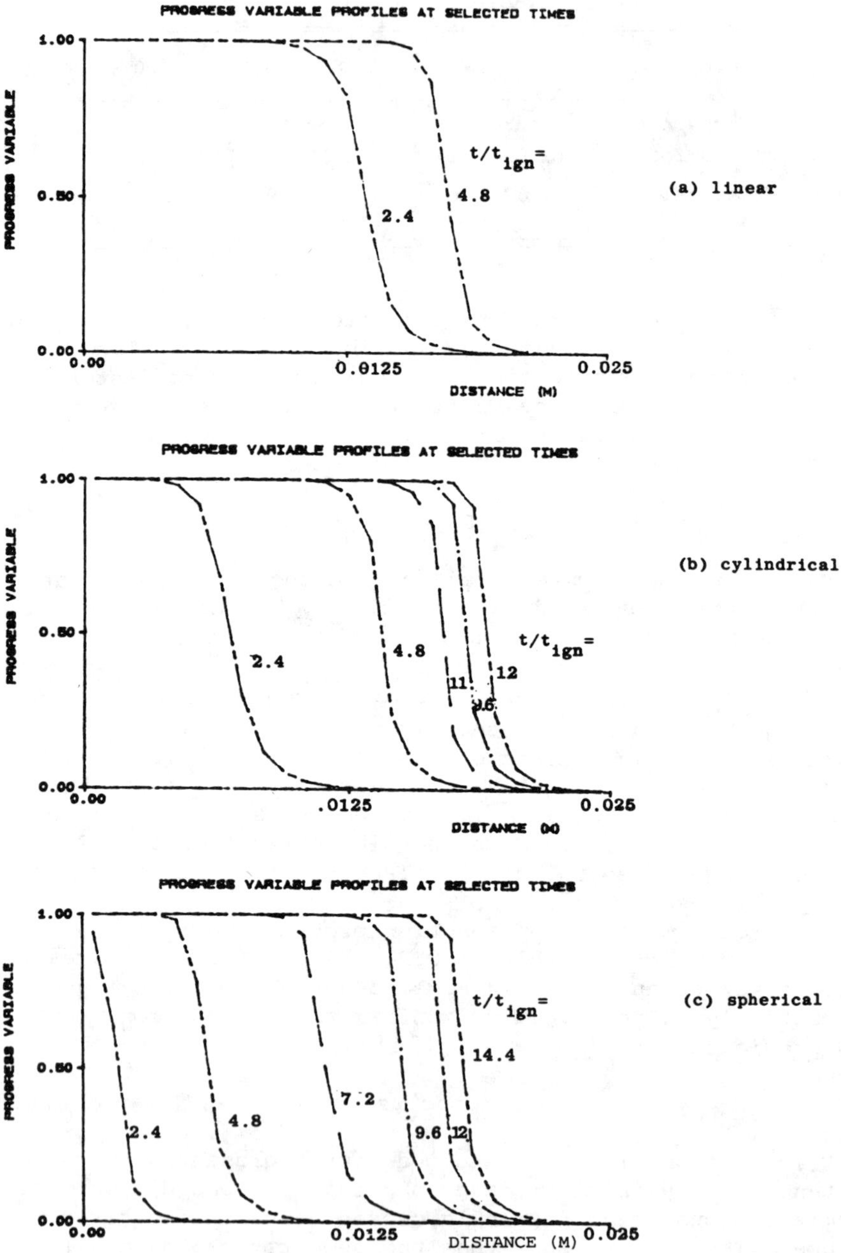

Fig. 2 Reaction progress variable $\tilde{c}$ vs time at selected times for: a) rectangular; b) cylindrical; and c) spherical geometry. Case II: r_{in} = 1.5, ε_{in} = 4000.

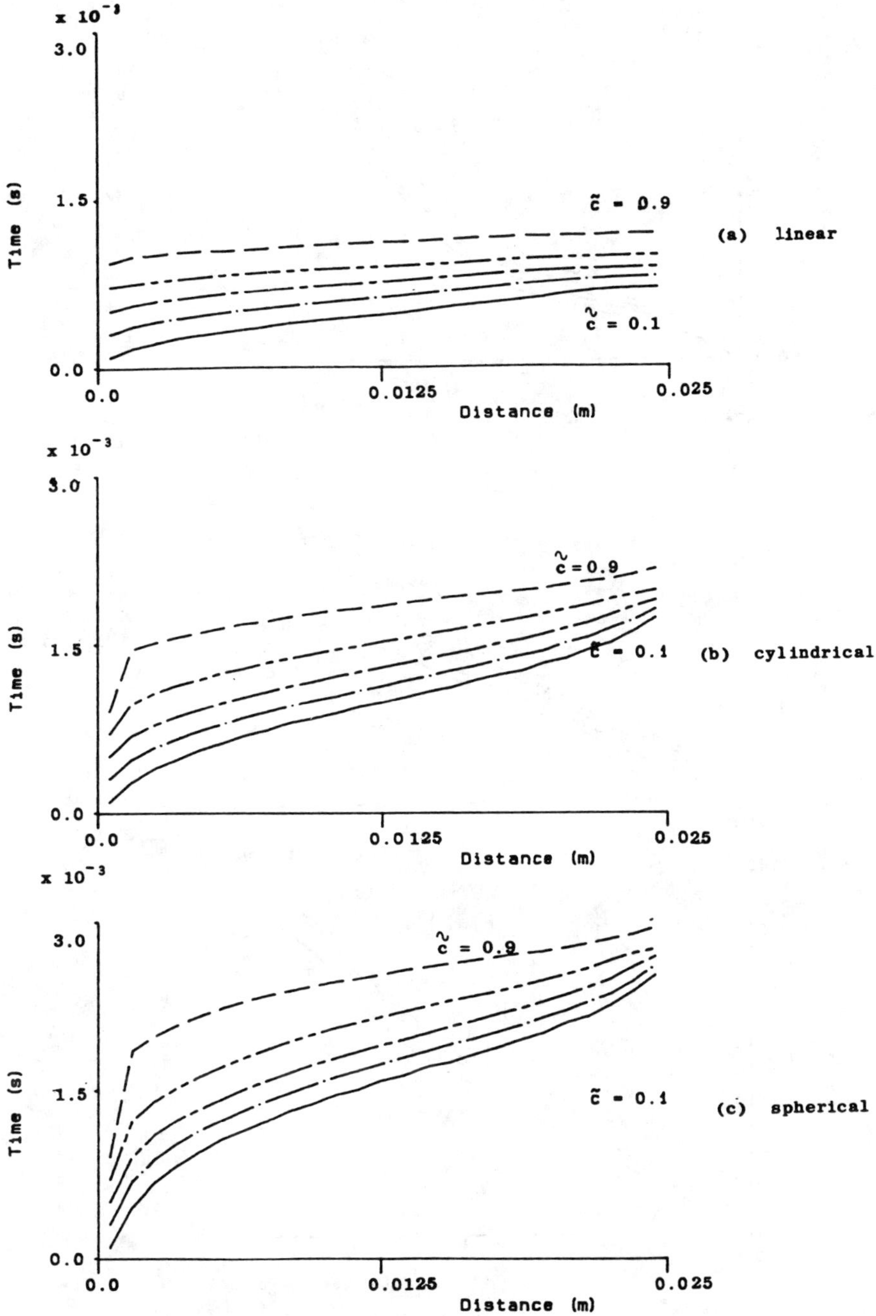

Fig. 3 Contours of reaction progress variable $\tilde{c}$ shown on a distance-time plot for: a) rectangular; b) cylindrical; and c) spherical geometry. Case I: $r_{in} = 0.1$, $\varepsilon_{in} = 20$.

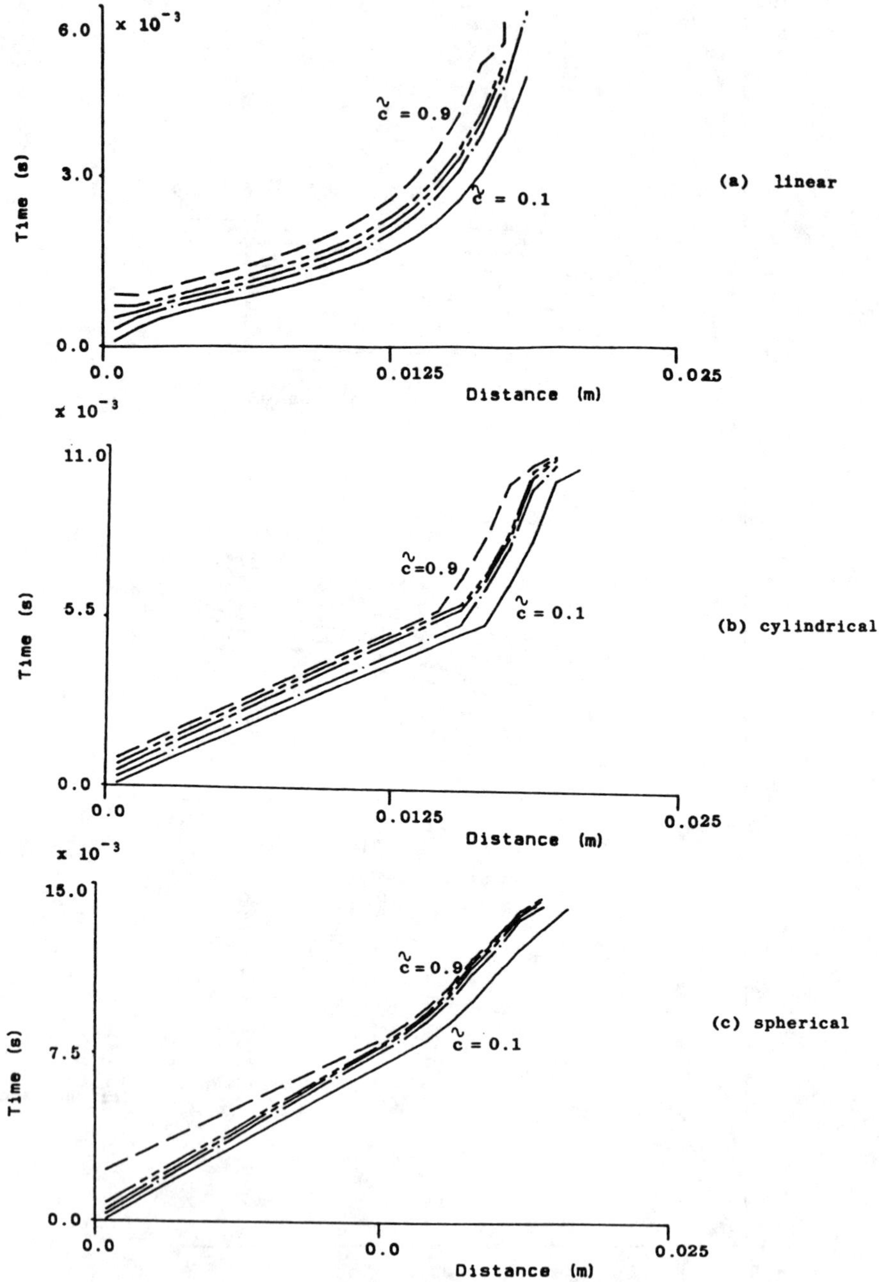

Fig. 4 Contours of reaction progress variable $\tilde{c}$ shown on a distance-time plot for: a) rectangular; b) cylindrical; and c) spherical geometry. Case II: $r_{in} = 1.5$, $\varepsilon_{in} = 4000$.

Acknowledgments

The research reported here is supported by the Science and Engineering Research Council, Shell Research Ltd., and the European Community.

References

[1]Cant, R.S., and Bray, K.N.C., "A Theoretical Model of Premixed Turbulent Combustion in Closed Vessels", Combustion and Flame, Vol. 76, Nos. 3,4, 1989, pp. 243-263.

[2]Cant, R.S., and Bray, K.N.C., "Strained Laminar Flamelet Calculations of Premixed Turbulent Combustion in a Closed Vessel," 22nd Symposium (International) on Combustion, The Combustion Institute, Seattle, WA, 1988, pp. 791-799.

[3]Mason, H.B., and Spalding, D.B., "Prediction of Reaction Rates in Turbulent Premixed Boundary Layer Flows", edited by F. J. Weinberg, The Combustion Institute European Symposium, Academic, New York, 1973, pp. 601-606.

[4]Bray, K.N.C., LIbby, P.A., and Moss, J.B., "Unified Modelling Approach for Premixed Turbulent Combustion - Part I: General Formulation," Combustion and Flame, Vol. 61, pp. 87-102.

[5]Cant, R. S., "Application of the PISO method to a Second-order Closure Model of Turbulent Combustion," (to be published).

[6]Bray, K.N.C., Libby, P.A., and Moss, J.B., "Scalar Length Scale Variation in Premixed Turbulent Flames", 20th Symposium (International) on Combustion, The Combustion Institute, Pittsburgh, Penn., 1984, pp. 421-427.

[7]Kolmogorov, A.N., "A Refinement of Previous Hypotheses Concerning the Local Structure of Turbulence of a Viscous Incompressible Fluid at High Reynolds Number," Journal of Fluid Mechanics, Vol. 13, No.1, 1962, pp. 82-85.

[8]Issa, R.I., Journal of Computational Physics, Vol. 62, 1986, pp.40-65.

[9]Cant, R.S., Rogg, B., and Bray, K.N.C., "On Laminar Flamelet Modelling of the Mean Reaction Rate in a Premixed Turbulent Flame," Combustion Science and Technology, Vol. 69, No. 1, 1990, pp.53-61.

Multivariate PDF Closure Applied to Oxidation of CO in a Turbulent Flow

E. Gutheil*
Technische Hochschule Darmstadt, Darmstadt, Germany

Abstract

To depart from the assumption of statistical independence, a multivariate probability density function (pdf) is formulated to average the chemical reaction rates in a turbulent diffusion flame. The oxidation of carbon monoxide (CO) at moderate Damköhler numbers has been evaluated to demonstrate applicability of the model. The turbulent flowfield is described using a conventional k-ϵ turbulence model; the chemical reaction is described adopting a global rate expression. The reaction model is closed employing a three dimensional clipped Gaussian distribution. The effect of turbulent fluctuations on the mean reaction rate is discussed in detail. In regions where high concentrations of CO prevail, the variances and covariances cause enhancement of the mean reaction rate. In regions with low concentrations of CO, the mixture fraction determines the concentration-dependent term of the kinetic rate expression, and a deceleration of the mean reaction rate is received. Calculated profiles of CO volume fractions and temperature are compared with experimental data available from the literature for this system. The temperature dependence of the oxidation of CO is reproduced clearly by the calculations. The agreement between measurements and calculations is reasonably good.

I. Introduction

Since most practical combustors, such as industrial furnaces and gas turbines involve turbulent, nonpremixed flames, research has focused on developing a fundamental understanding of such flames. A coupling of experiments and mathematical models is desirable to examine the complex processes taking place in turbulent dif-

---

*Wissenschaftliche Mitarbeiterin, Institut für Chemische Technologie; currently, Visiting Scholar, Department of Applied Mechanics and Engineering Sciences, University of California at San Diego, La Jolla, CA.

fusion flames. In practical systems, there are many chemical reactions occurring at the same time and place requiring complex modeling. Therefore, simple systems must be evaluated to resolve the problem.

In many turbulent diffusion flames, turbulent mixing is controlling combustion and a conserved scalar approach[1] is appropriate. For such flames, a conserved scalar approach in combination with a "laminar flamelet" structure[2–4] has been applied successfully.

Turbulent reacting flows at moderate Damköhler numbers are not controlled totally by turbulent mixing, but also by the chemical reaction. Therefore, it is necessary to consider in detail the finite rates of the chemical reactions. Featuring this problem, transport equations for the concentrations of the chemical species have to be solved as well as transport equations describing the turbulent flowfield. Following this approach, on the one hand, the higher moments, which appear in all Favre or Reynolds averaged equations, have to be modeled; conversely, the mean reaction rates of the chemical species have to be formulated with special regard to the turbulent structure.

There are several approaches to close the chemical reaction rate. Borghi[5] proposed a series expansion for the chemical source term, and another method is the formulation using a probability density function (pdf). Pope[6] evaluated the shape of the pdf from a separate transport equation; another way is to use an a priori distribution.[7,8]

Most often, the multivariate pdf, which depends on all variables entering the kinetic rate expression, is simplified to either a one dimensional pdf [8,9] or to a product of a number of one dimensional pdf's,[10] because sufficient information about joint pdf's is lacking. In the present study, these restrictions are relaxed, and a multivariate pdf is used to average the instantaneous reaction rates. The applicability of this model is discussed referring to the oxidation of carbon monoxide (CO) at moderate Damköhler numbers. This reaction can be described by a global rate expression depending on three variables; hence, a three dimensional joint Gaussian normal distribution is appropriate for this problem.

Experimental data for the oxidation of CO at moderate Damköhler numbers are available in Refs. 8 and 11.

The scope of this study is to discuss the sensitivity of the multivariate pdf closure with respect to the various quantities entering the averaging procedure of the reaction rate expressed by a global rate expression. However, the global rate expression is no restriction to this pdf approach, the basics of which are outlined in Ref. 14. Since the fundamentals of the method applied in this work are described in Refs. 14 and 21, the combustion model is sketched here only briefly.

II. Mathematical Modeling

The process considered is the oxidation of CO by air in a turbulent flow. A mixture of carbon monoxide and nitrogen is admixed axially into the completely burnt flue gas of a natural gas flame, which is operated with excess air. For details see Refs. 8 and 11.

A. Modeling the Flowfield

The flowfield is modeled by transport equations for a steady, swirl free, axisymmetric flow. The closure of the equations is achieved using a conventional eddy-viscosity gradient model of turbulent transport and a k-ϵ turbulence model. Thus, the description of the flowfield comprises the transport equations for total mass, momentum in the axial and radial directions and the turbulence quantities k and ϵ, which are used in their boundary-layer approximations (see, for example, Refs. 1,12,13). Reynolds averaging is used throughout.

The applicability of the turbulence model and the employed standard parameters for the experimental system is demonstrated in Ref. 14.

B. Modeling the Chemical Reaction

The turbulent character of the chemical reaction is modeled using a multivariate pdf closure. According to the principles of the pdf approach, the mean chemical reaction rate of a species i entering the transport equation for the mass fraction Y_i of species i is formulated as

$$\overline{\dot{w}_i} = \int_0^1 \cdots \int_0^1 \int_0^\infty \int_0^\infty \dot{w}_i P(\rho, T, Y_1, Y_2, \ldots, Y_N; x, r) \mathrm{d}\rho \mathrm{d}T \mathrm{d}Y_1 \ldots \mathrm{d}Y_N \quad (1)$$

where $\overline{\dot{w}_i}$ is the mean reaction rate of the component i, and $P(\ )$ is a joint pdf of all variables entering the kinetic rate expression $\dot{w}_i$.

As mentioned earlier, the oxidation of CO is described by a global rate expression[15]:

$$\dot{w}_{\mathrm{CO}} = -3.98 \times 10^{17} \rho^{1.75} Y_{\mathrm{CO}} \left(\frac{Y_{\mathrm{H_2O}}}{M_{\mathrm{H_2O}}}\right)^{0.5} \left(\frac{Y_{\mathrm{O_2}}}{M_{\mathrm{O_2}}}\right)^{0.25} \exp\left(\frac{-T_a}{T}\right) \quad (2)$$

which holds over a wide range of reaction conditions. According to Ref. 15, the activation temperature T_a equals 20,000 K.

Considering the global rate expression, water vapor can be treated as inert. Therefore, the dependency of the kinetic rate expression can be reduced to three variables: a normalized mixture fraction f, the mass fraction Y_{CO} of carbon monoxide, and enthalpy h. Hence, the application of Eq. (2) requires the formulation of a three dimensional pdf depending on the variables Y_{CO}, f, and enthalpy h.

In the present work, a three dimensional joint Gaussian distribution is adopted, the density of which is given, in general, by[16]

$$g(x_1, x_2, x_3) = \frac{1}{(2\pi)^{3/2}(\mid M \mid)^{1/2}} \times \exp\left[-\frac{1}{2\mid M\mid}\sum_{j,k=1}^{3} \mid M_{jk} \mid (x_j - \xi_j)(x_k - \xi_k)\right] \quad (3)$$

where $\mid M \mid \neq 0$ is the determinant of the matrix of the second moments of the variables x_i, and $\mid M_{jk} \mid$ denotes the algebraic complement of the element λ_{jk} in $\mid M \mid$. ξ_i is the first moment of the marginal distribution of variable x_i. The Gaussian distribution described by Eq. (3) has to be clipped, as the variables f and Y_{CO} are defined between 0 and 1 and enthalpy between the values of the incoming gas streams $h_{1,\mathrm{in}}$ and $h_{2,\mathrm{in}}$.

Therefore, the shape of the applied pdf $P(Y_{\mathrm{CO}}, f, h)$ must be evaluated from the following conditions for the first and second moments of f, Y_{CO} and h at each stage of the calculation:

$$\bar{\phi}_i = \frac{1}{Z}\int_0^1\int_0^1\int_{h_{1,\mathrm{in}}}^{h_{2,\mathrm{in}}} \phi_i \; P(Y_{\mathrm{CO}}, f, h)\mathrm{d}h \; \mathrm{d}f \; \mathrm{d}Y_{\mathrm{CO}} \quad (4)$$

and

$$\overline{\phi_i'\phi_j'} = \frac{1}{Z}\int_0^1\int_0^1\int_{h_{1,\mathrm{in}}}^{h_{2,\mathrm{in}}} (\phi_i\phi_j) \; P(Y_{\mathrm{CO}}, f, h)\mathrm{d}h \; \mathrm{d}f \; \mathrm{d}Y_{\mathrm{CO}} - \bar{\phi}_i\bar{\phi}_j \quad (5)$$

where Z is a normalization factor due to the density of the pdf used. For $i = j$, Eq. (5) gives the conditions for the variances of f, Y_{CO} or h; for $i \neq j$, Eq. (5) gives the conditions for the corresponding covariances.

Hence, $P(Y_{\mathrm{CO}}, f, h)$ is a function of the variables $\xi_{Y_{\mathrm{CO}}}, \xi_f, \xi_h$ and the corresponding second moments λ_{jk} with respect to conditions given by Eqs. (4) and (5). The shape of the clipped Gaussian distribution is determined from the actual values of the means, variances, and covariances of Y_{CO}, f, and h throughout the calculation domain.[14,21]

It should be mentioned that for all $\overline{\phi_i'\phi_j'} = 0$, with $i \neq j$, the multivariate Gaussian distribution passes into the product of the marginal distributions of f, h, and Y_{CO}, which is the condition for statistical independence for these variables.

Modeling of the variances and covariances is achieved by solving modeled transport equations assuming gradient transport and "dissipation hypothesis"[1,17] analogously to the modeling of the turbulent flowfield.

With the denotation of a turbulent parabolic transport operator L,

$$L(\bar{\phi}) \equiv \frac{\partial(\bar{\rho}\bar{u}\bar{\phi})}{\partial x} + \frac{1}{r}\frac{\partial(r\bar{\rho}\bar{v}\bar{\phi})}{\partial r} - \frac{1}{r}\frac{\partial}{\partial r}\left(r\frac{\mu_t}{\sigma_\phi}\frac{\partial\bar{\phi}}{\partial r}\right) \quad (6)$$

for any variable $\bar{\phi}$, the transport equations for the second moments are

$$L(\overline{\phi_i'\phi_j'}) = C_1\eta_t\left[\left(\frac{\partial\bar{\phi}_i}{\partial x}\frac{\partial\bar{\phi}_j}{\partial x}\right) + \left(\frac{\partial\bar{\phi}_i}{\partial r}\frac{\partial\bar{\phi}_j}{\partial r}\right)\right] - C_2\bar{\rho}\bar{\epsilon}\,\frac{\overline{\phi_i'\phi_j'}}{k} + \overline{\phi_i' S_j'} + \overline{S_i'\phi_j'} \qquad (7)$$

where ϕ_i and ϕ_j stand for f, Y_{CO}, or h, respectively. The constants C_1 and C_2 as well as the turbulent Prandtl numbers are standard parameters[18,19] – – C_1 equals 2.8 and C_2 equals 2.0, the Prandtl numbers for all of the above-mentioned variables equal 0.7.

According to the principles of pdf closure, the correlations appearing in Eq. (7) are calculated from

$$\overline{\phi_i' S_{\phi_j}'} = \frac{1}{Z}\int_0^1\int_0^1\int_{h_{1,\mathrm{in}}}^{h_{2,\mathrm{in}}} (\phi_i - \bar{\phi}_i)\,(S_{\phi_j} - \bar{S}_{\phi_j})P(Y_{CO}, f, h)\mathrm{d}h\,\mathrm{d}f\,\mathrm{d}Y_{CO} \qquad (8)$$

C. Numerical Solution Procedure

The resulting system of differential equations for an axisymmetric stationary turbulent flow has been solved with respect to the appropriate boundary conditions[21] using a finite-volume procedure developed by Rosten and Spalding.[20]

A nonuniform grid with 21 × 19 nodes was sufficient to produce numerically accurate results. The CPU time on an IBM 3090 VF not using the vector facility was about 20 min, including the integration over the three dimensional Gaussian distribution.

III. Results and Discussion

A. Calculation of Variances and Covariances

The application of the multivariate pdf requires the calculation of the variances of the mixture fraction, the mass fraction of CO and enthalpy, and the corresponding covariances.

Figure 1 exemplarily shows radial profiles of the mean fluctuations of the three variables for a test series at a distance of 0.16 m from the exit nozzle. The mean fluctuations of the mixture fraction and the mass fraction of CO show a radial decrease due to the radial increase of the gradients of the corresponding mean values, they attain high values within the turbulent reaction zone. The mean fluctuations of enthalpy increase again at higher distances from the axis. This is a consequence of the higher temperature of the oxidixing gas flow compared with the temperature of the CO/N_2 mixture.

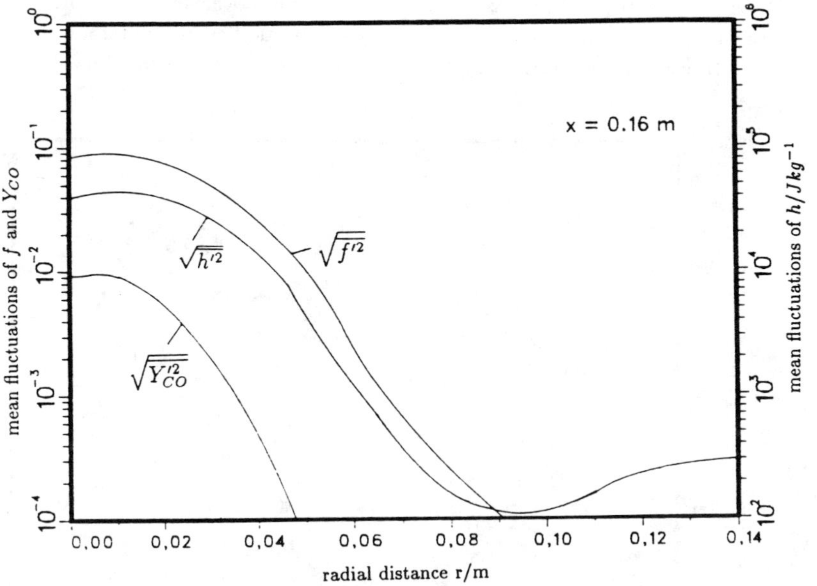

Fig. 1 Radial profiles of mean fluctuations. Experimental conditions: $Y_{\mathrm{CO,in}} = 0.09$, $T_{\mathrm{fu}} =$ 830 K, $Re_{\mathrm{fu}} = 6770$, $T_{\mathrm{ox}} = 1100$ K, $Re_{\mathrm{ox}} = 3920$.

The results of the calculations of the covariances are described in terms of a correlation coefficient $\gamma_{ij} = \overline{\phi'_i \phi'_j}/(\overline{\phi'^2_i}\ \overline{\phi'^2_j})^{1/2}$. Figure 2 shows radial profiles of the correlation coefficients of f, Y_{CO}, and h. The transport equation for enthalpy contains only a weak sink, so that enthalpy may be regarded as a quasiconserved scalar. Therefore, the correlation coefficient of f and h attains a value of approximately unity near the axis; it decreases at greater distances from the axis because of the finite value of enthalpy in the oxidizing gas flow, which causes different gradients of the mean values of f and h. The correlation coefficients $\gamma_{fY_{\mathrm{CO}}}$ and $\gamma_{Y_{\mathrm{CO}}h}$ exhibit a radial decrease due to chemical conversion, which causes a negative source in the corresponding transport equations, cf. Eq. (8). The correlation between enthalpy and the mass fraction of CO is lower than the correlation between f and Y_{CO}, because the gradients of the mean values of Y_{CO} are greater than the gradients of $\bar{f}$. Furthermore, the transport equation for the correlation coefficient of h and Y_{CO} contains an additional sink $\overline{Y'_{\mathrm{CO}} S'_h}$. For the present application, all correlation coefficients fulfill the condition $\mid \gamma_{ij} \mid \leq 1$. Reacting systems with higher gradients of the mean values or higher reaction rates, which cause higher fluctuations of the turbulent reaction rate and of the mass fractions of the chemical species and enthalpy, cause the calculation of high values of the correlations $\overline{\phi'_i \phi'_i}$ by Eq. (7) in connection with Eq. (8), so that the above-mentioned normalization condition for the correlation coefficient is no longer fulfilled. Therefore, the present model is restricted to turbulent combustion at moderate Damköhler numbers.

Using the calculated results of the variances and covariances of f, h, and Y_{CO}, the variance of temperature and the correlation coefficients of temperature and mixture fraction or mass fraction of CO may be evaluated from the following al-

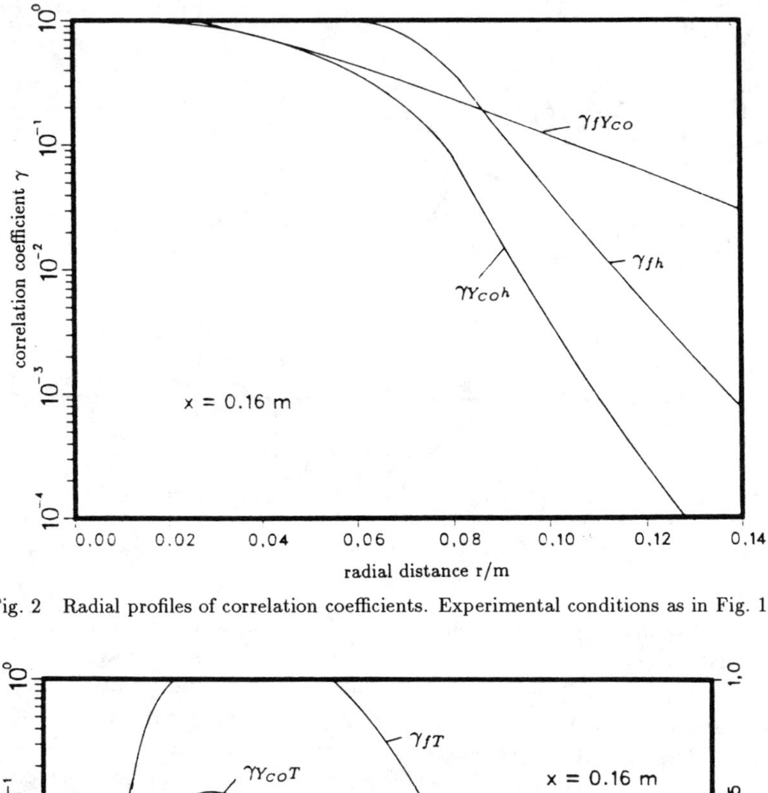

Fig. 2 Radial profiles of correlation coefficients. Experimental conditions as in Fig. 1.

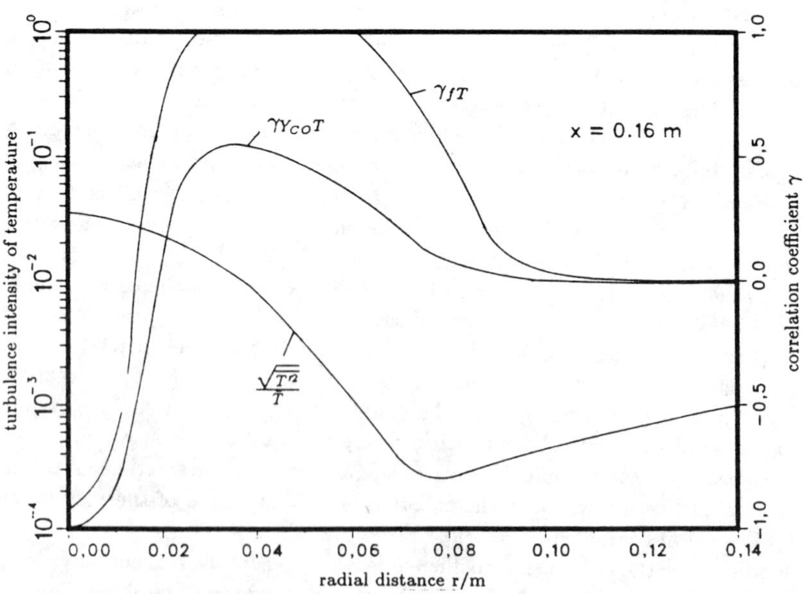

Fig. 3 Radial profiles of the turbulence intensity of temperature and correlation coefficients. Experimental conditions as in Fig. 1.

gebraic relations:

$$\overline{T'^2} = (\overline{h'^2} - 2H_{\mathrm{fu}}\overline{h'Y'_{\mathrm{CO}}} + H_{\mathrm{fu}}^2\overline{Y'^2_{\mathrm{CO}}})/\overline{c_P}^2 \tag{9}$$

$$\overline{f'T'} = (\overline{f'h'} - H_{\mathrm{fu}}\overline{f'Y'_{\mathrm{CO}}})/\overline{c_P} \tag{10}$$

and

$$\overline{Y'_{\mathrm{CO}}T'} = (\overline{Y'_{\mathrm{CO}}h'} - H_{\mathrm{fu}}\overline{Y'^2_{\mathrm{CO}}})/\overline{c_P} \tag{11}$$

Figure 3 displays radial profiles of the turbulence intensity of temperature and of the correlation coefficients γ_{fT} and $\gamma_{Y_{\mathrm{CO}}T}$. The turbulence intensity attains high values within the turbulent reaction zone and decreases at higher distances from the axis, where the chemical reaction slows down or is negligible. There is a slight increase of the turbulence intensity of temperature near the wall due to the higher level of temperature of the oxidizing gas flow compared with the temperature of the CO/N_2 mixture.

The correlation coefficients γ_{fT} and $\gamma_{Y_{\mathrm{CO}}T}$ attain negative values of nearly -1 near the axis, and increase to high positive values and fall to zero in regions where the gradients of the mean values of temperature show no substantial alteration, whereas the means of f and Y_{CO} change due to mixing or chemical conversion. Near the axis, the values of temperature increase and the values of the mixture fraction and mass fraction of CO decrease; they are negatively correlated. The correlation coefficients pass a value of zero where temperature and f respectively Y_{CO} are linearly independent; this is the place in the combustion chamber where the greatest gradients of the mean values of f and Y_{CO} occur while T reaches its maximum value (cf. Sec. III.C). In regions where the gradients of the mixture fraction decrease in a way similar to the temperature, γ_{fT} attains a value of approximately 1. The correlation coefficient $\gamma_{Y_{\mathrm{CO}}T}$ is smaller than γ_{fT}, as the gradients of $\overline{Y_{\mathrm{CO}}}$ decrease faster than the gradients of $\bar{f}$.

B. Effect of Turbulent Fluctuations on Mean Reaction Rate

The calculated variances and covariances of the previous paragraph have been used to choose the shape of the employed multivariate Gaussian distribution. The present section intends to investigate in detail the effect of turbulent fluctuations on the mean chemical reaction rate.

The first part of Fig. 4 demonstrates the influence of the fluctuations of a single variable on the reaction rate [cf. Eq. (2)] for two spot situations within the combustion chamber. The ratio of $\overline{\dot{w}_{\mathrm{CO}}(\phi_i)}/\dot{w}_{\mathrm{CO}}(\bar{\phi}_i)$ is shown in dependency of the turbulence intensity of variable ϕ_i, where $\overline{\dot{w}_{\mathrm{CO}}(\phi_i)}$ denotes the mean reaction rate

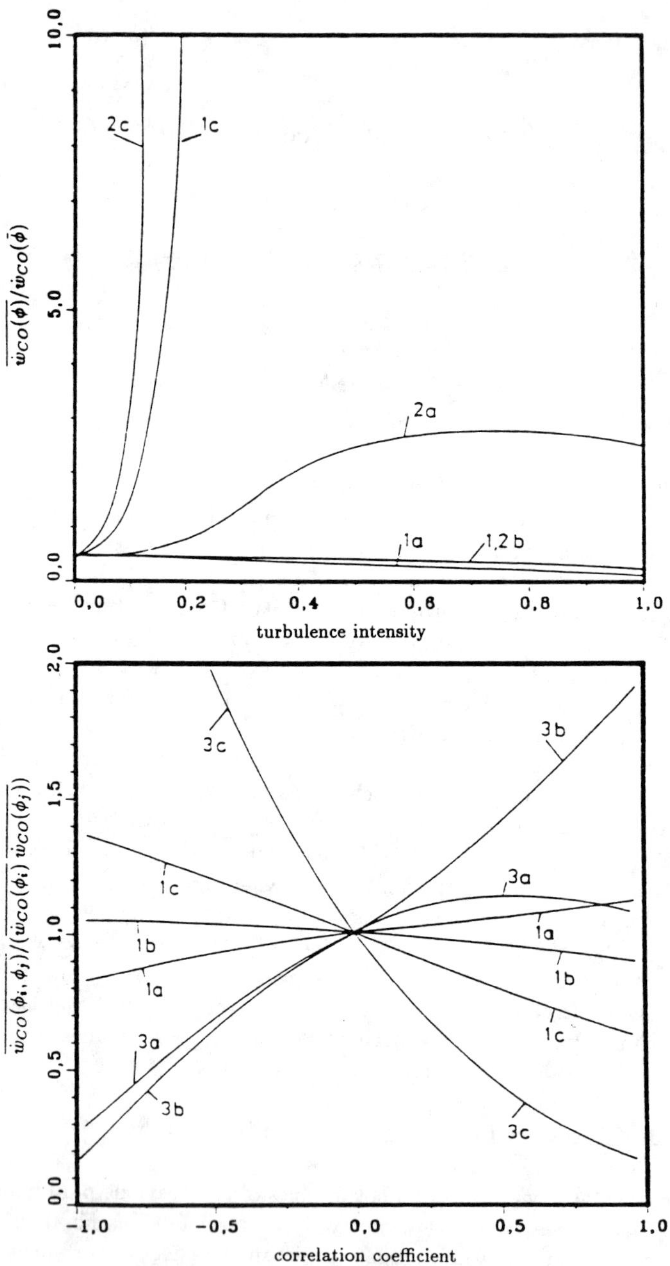

Fig. 4 Effect of turbulent fluctuations on mean reaction rate. Case 1: $\bar{f} = 0.323$, $\bar{h} = 1.5\ 10^6$ J/kg, $\bar{Y}_{CO} = 0.012$; case 2: $\bar{f} = 0.5$, $\bar{h} = 1.75\ 10^6$ J/kg, $\bar{Y}_{CO} = 0.05$; case 3: $\bar{f} = 0.7$, $\bar{h} = 1.6\ 10^6$ J/kg, $\bar{Y}_{CO} = 0.034$. a: $\phi = Y_{CO}$; b: $\phi = f$; c: $\phi = h$. Turbulence intensities for the lower part. Case 1: 30% for f, 3% for h, 60% for Y_{CO}; case 3: 40% for f, 5% for h, 60% for Y_{CO}. a: $\phi_i, \phi_j = Y_{CO}, h$; b: $\phi_i, \phi_j = f, Y_{CO}$; c: $\phi_i, \phi_j = f, h$.

obtained by weighting with a one dimensional Gaussian distribution depending on variable ϕ_i. Enthalpy fluctuations cause a pronounced acceleration of the chemical reaction, the power of which depends on the enthalpy level of the flow situation that is caused by the exponential temperature dependence of the kinetic rate expression. The fluctuations of the mixture fraction that enter with an exponent of less than unity into the kinetic rate expression cause a retardation of the reaction for both situations. The fluctuations of the mass fraction of CO contribute in two different ways to the chemical reaction rate: they cause an enhancement of the reaction for high values of $\bar{Y}_{CO}$ since they enter into the exponential term of the kinetic rate expression; at low values of $\bar{Y}_{CO}$ the concentration-dependent term of Eq. (2), where the mass fraction of CO enters with exponent 0.25 through the calculation of Y_{O_2}, determines the turbulent reaction rate.

The second part of Fig. 4 demonstrates the influence of covariances of the several variables. The ratio of a two dimensional joint Gaussian distribution and the product of the marginal distributions of the corresponding variables $\overline{\dot{w}_{CO}(\phi_i,\phi_j)}/(\overline{\dot{w}_{CO}(\phi_i)}\ \overline{\dot{w}_{CO}(\phi_j)})$ is shown in dependency of the correlation coefficient γ_{ij}. Again, the covariances substantially influence the mean chemical reaction rate depending on the absolute values of the mean values of the several variables.

Figure 5 displays the effect of a three dimensional Gaussian distribution on the chemical source term. Case 1 shows results obtained including only the mean values of f, Y_{CO}, and h, and case 2 displays reaction rates adopting the product of the marginal Gaussian distributions where all γ_{ij} equal zero presuming statistically independence of the three variables. Calculation 3 employs the three dimensional joint Gaussian distribution. Within the turbulent reaction zone at a distance of 0.075 m from the nozzle plane near the axis, where the CO mass fractions are not below 10% of the inlet mass fraction, turbulent fluctuations of f,h, and Y_{CO} enhance the chemical reaction in both cases, which regard the variables as statistically independent or positively correlated. At higher axial or radial distances, the CO mass fraction attains low values due to chemical conversion. In this region, the effect is inverted: the low CO mass fractions cause a low chemical reaction rate. The accelerating effect of enthalpy fluctuations is overweighted by the diminishing effect of the fluctuations of the mixture fraction, cf. Fig. 1. At locations with low CO mass fractions, enthalpy attains a low value; thus, the concentration-dependent term of the mean reaction rate determines the kinetic rate expression of Eq. (2).

C. Profiles of Volume Fractions of CO and Temperature

Figures 6 and 7 show radial profiles of CO volume fractions and temperature at a distance of 0.16 m from the nozzle plane for three different test series. The symbols represent results from the measurements, and the lines show the calculated profiles, where 1 to 3 refer to the three weighting procedures for the mean reaction rate discussed in the previous paragraph. The symbols refer to the three different test series.

Within the turbulent reaction zone, the different weighting procedures for the reaction rates markedly affect the temperature profiles. The three dimensional

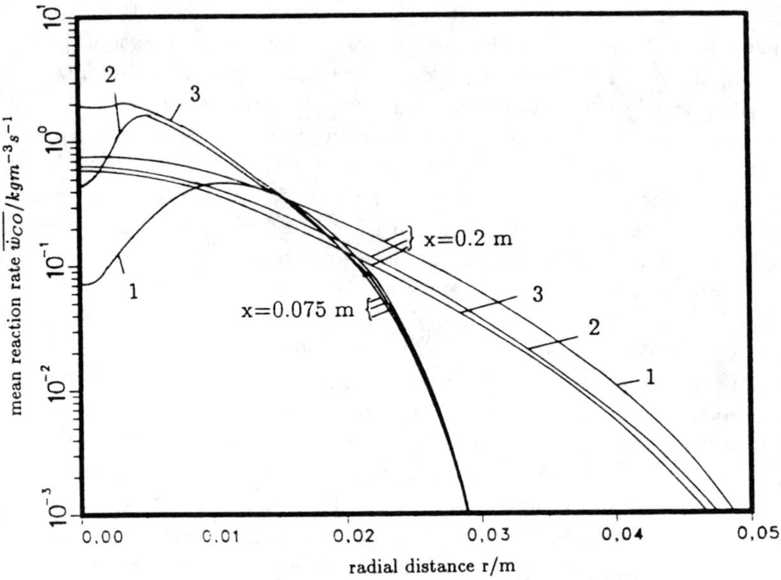

Fig. 5 Radial profiles of mean reaction rate. Experimental conditions as in Fig. 1.

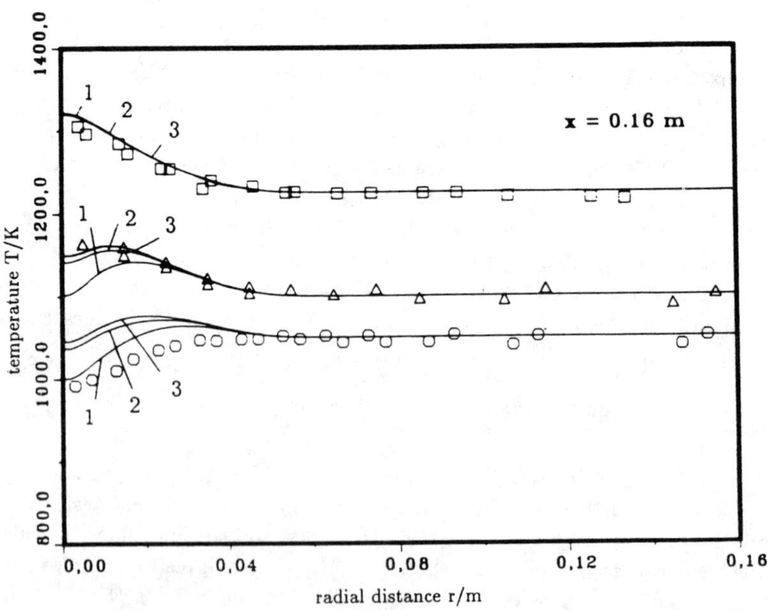

Fig. 6 Radial profiles of temperature for three test series. Experimental conditions: series 1 (○): $Y_{CO} = 0.09$, $T_{fu} = 790$ K, $Re_{fu} = 6970$, $T_{ox} = 1050$ K, $Re_{ox} = 3990$; series 2 (△) as in Fig. 1; series 3 (□): $Y_{CO} = 0.09$, $T_{fu} = 860$ K, $Re_{fu} = 6620$, $T_{ox} = 1225$ K, $Re_{ox} = 3710$.

joint Gaussian distribution causes higher temperature profiles than the product of the marginal distributions, which again produce higher values of temperature than the reaction rate involving no fluctuations. A comparison of the results of measured and calculated profiles shows good agreement for test series 2 and 3. The discrepancies for test series 1 may be due to the low temperature of the incoming fuel stream. For temperatures less than 800 K, the employed global rate expression is not longer valid.[15] Therefore, extremely high temperature profiles are predicted near the axis.

Considering Fig. 7, the predicted volume fractions of CO for cases 3 and 2 attain lower values than for case 1. This is a consequence of the higher mean reaction rates for cases 3 and 2 in comparison with case 1 within the turbulent reaction zone. The calculated profiles of the volume fractions of CO obtained adopting the three different weighting procedures are equalized by the turbulent flowfield. The differences of the three weighting procedures for test series 1 and 2 are more pronounced than for series 3 as the Reynolds numbers for series 1 and 2 are higher, producing higher turbulence intensities.

The agreement of measured and calculated CO profiles is reasonably good near the axis. At the edge of the fuel jet, where very low CO volume fractions prevail, there are substantial differences between the measured and calculated profiles. On one hand, this may be due to errors measuring very low volume fractions of carbon monoxide.[8,11] Conversely, these discrepancies hint of shortcomings of the mathematical model. The chemical reaction may be influenced by "coherent structures" near the edge of the fuel jet. The k-ϵ model is not suitable to predict

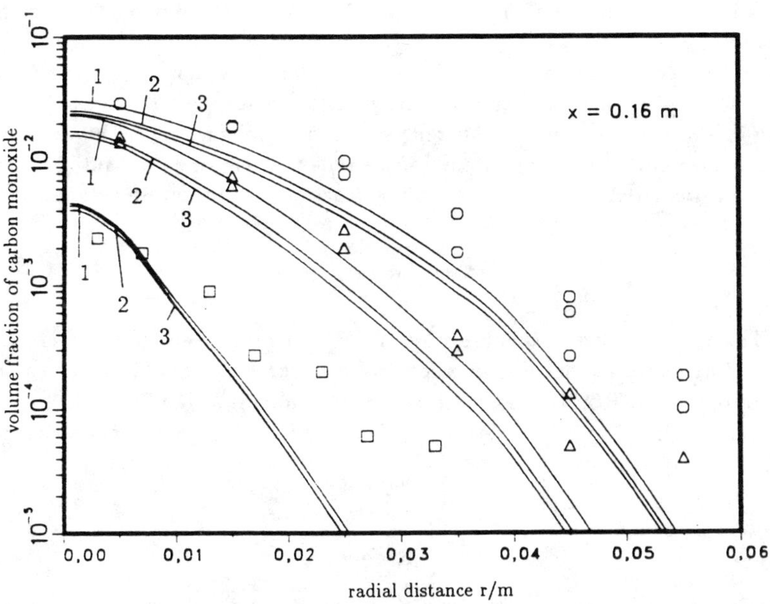

Fig. 7 Radial profiles of CO volume fractions. Experimental conditions as in Fig. 6.

these structures adequately. Furthermore, the variables determining the turbulent reaction rate are no longer Gaussian distributed in that region. Hence, the effect of turbulent fluctuations on the mean reaction rate is predicted to be too weak, leading to overestimated mean reaction rates. Furthermore, the adopted global rate expression does not include the effects of chemical nonequilibrium of the reaction.

IV. Conclusions

A multivariate probability density function (pdf) to close the chemical source in a turbulent diffusion flame has been derived and applied to the oxidation of carbon monoxide at moderate Damköhler numbers.

The model for the turbulent flow with chemical reaction employs the k-ϵ turbulence model and an eddy-viscosity gradient model, which is also used to link the chemical reaction model to the turbulent flowfield.

Accepting the description of the oxidation of carbon monoxide by a global rate expression, which may be reduced to a three-variable function, the pdf closure involves a three dimensional joint Gaussian distribution. The applied second-order pdf closure requires calculation of the variances and covariances of the dependent variables of the pdf, which has been achieved by solving modeled transport equations.

The results clearly demonstrate the effect of turbulent fluctuations on the mean chemical reaction rate. For the present application, they cause an acceleration of the chemical reaction at locations where high concentrations of carbon monoxide prevail. At locations with low concentrations of carbon monoxide, this effect is converted.

Calculated profiles of temperature and volume fractions of carbon monoxide are compared with measurements, showing reasonably good agreement. At the edge of the fuel jet, where "coherent structures" might occur, the reaction rate is no longer Gaussian distributed. In this region, there are substantial differences between predicted and measured volume fractions of carbon monoxide.

Acknowledgments

The author wishes to thank Professor F. Fetting for his sustained interest in this work. She expresses her appreciation to Professor D. B. Spalding for leaving the computer code PHOENICS 84 and to Dr. C. Aldham from CHAM Ltd., London, for much helpful advice in setting up PHOENICS 84. The financial support of Stiftung Volkswagenwerk is greatfully acknowledged.

References

[1]Bilger, R. W., Turbulent Flows with Nonpremixed Reactants, Turbulent Reacting Flows, edited by P. A. Libby and F. A. Williams, Springer, New York, 1980, p. 65.

[2]Williams, F. A., "Recent Advances in Theoretical Descriptions of Turbulent Diffusion Flames", Turbulent Mixing in Nonreactive and Reactive Flows, edited by S. N. B. Murthy, Plenum, New York, 1975, p. 189.

[3]Peters, N., Laminar Diffusion Flamelet Models in Non–Premixed Turbulent Combustion, Progress in Energy and Combustion Science, Vol. 10, 1984, p. 319.

[4]Liew, S. K., Bray, K. N. C. and Moss, J. B., A Stretched Laminar Flame Model of Turbulent Non–Premixed Combustion, Combustion and Flame, Vol. 56, 1984, p. 199.

[5]Borghi, R., Chemical Reactions Calculations in Turbulent Flows: Application to a CO–Containing Turbojet Plume, Advances in Geophysics, Vol. 18B, 1974, p. 349.

[6]Pope, S. B., Pdf Methods for Turbulent Reactive Flows, Progress in Energy and Combustion Science, Vol. 11, 1985, p. 119.

[7]Janicka, J. and Kollmann, W., A Two–Variables Formalism for the Treatment of Chemical Reactions in Turbulent H_2-Air Diffusion Flames, Seventeenth Symposium (International) on Combustion, The Combustion Inst., Pittsburgh, PA, 1979, p. 421.

[8] Bockhorn, H. and Lutz, G., The Application of Turbulent Reaction Models to the Oxidation of CO in a Turbulent Flow, Twentieth Symposium (International) on Combustion, The Combustion Inst., Pittsburgh, PA, 1985, p. 377.

[9]Pratt, D. T., Mixing and Chemical Reaction in Continuous Combustion, Progress in Energy and Combustion Science, Vol. 1, 1976, p. 367.

[10] Janicka, J., The Application of Partial Equilibrium Assumptions for the Prediction of Diffusion Flames, J. Non-Equilibrium Thermodynamics, Vol. 6, 1981, p. 367.

[11]Lutz, G., "Die Oxidation von Kohlenstoffmonoxid in turbulenten Verbrennungsprozessen", Dissertation, TH Darmstadt, Federal Republic of Germany, 1984.

[12]Libby, P. A. and Williams, F. A. (eds.), "Fundamental Aspects", Turbulent Reacting Flows, Springer, Heidelberg, 1980, p. 1.

[13]Launder, B. E. and Spalding, D. B., Mathematical Models of Turbulence, Academic, New York, 1972.

[14]Gutheil, E. and Bockhorn, H., The Effect of Multi Dimensional Pdf's on the Turbulent Reaction Rate in Turbulent Reacting Flows, PhysicoChemical Hydrodynamics, Vol. 9, No. 3/4, 1987, p. 525.

[15]Westbrook, C. K. and Dryer, F. L., Simplified Reaction Mechanism for the Oxidation of Hydrocarbon Fuels in Flames, Combustion Science Technology, Vol. 27, 1981, p. 31.

[16]Fisc, M., "Wahrscheinlichkeitstheorie und Mathematische Statistik", Deutscher Verlag der Wissenschaften, Berlin, German Democratic Republic, 1980.

[17]Spalding, D. B., Concentration Fluctuations in a Round Turbulent Free Jet, Chemical and Engineering Science, Vol. 26, 1971, p. 95.

[18]Kent, J. H. and Bilger, R. W., Turbulent Diffusion Flames, Fourteenth Symposium (International) on Combustion, The Combustion Inst., Pittsburgh, PA, 1973, p. 615.

[19]Launder, B. E. and Spalding, D. B., The Numerical Computation of Turbulent Flows, Computer Methods in Applied Mechanics and Engineering, Vol. 3, 1974, p. 269.

[20]Rosten, H. I. and Spalding, D. B., PHOENICS–Beginner's Guide and User's Manual, CHAM, TR/100, CHAM Ltd., London, United Kingdom, 1986.

[21]Gutheil, E., "Modellierung turbulenter CO/Luft–Diffusionsflammen", Dissertation, TH Darmstadt, Federal Republic of Germany, 1988.

Modeling Turbulent Reacting Flows: Detailed Chemical Reaction Mechanisms and Sensitivity Analysis

H. Bockhorn*
Technische Hochschule Darmstadt, Darmstadt, Germany

Abstract

The aim of the present work is twofold. Firstly, a turbulent combustion model without simplifying assumptions about the chemical reactions is developed. Secondly, a sensitivity analysis of the turbulent combustion model is performed showing concepts for the simplification of the closure for the mean chemical reaction rates. The turbulence model is conventional and based on gradient transport assumptions and an eddy viscosity concept. In addition to the equations of the turbulent flow field the species conservation equations are solved. For the system under consideration, the oxidation of carbon monoxide by air at high Reynolds numbers and low Damköhler numbers, the 15 chemical species N_2, O_2, CO, CO_2, H_2, H_2O, H_2O_2, H, O, OH, HO_2, CHO, CH_2O, N, and NO are involved. The chemical reactions are described by a detailed reaction mechanism comprising 72 elementary reactions. The source terms in the species conservation equations are treated by means of a presumed-shape pdf-closure relaxing the assumption of statistically independent variables. This treatment necessitates the computation of variances and covariances of the variables entering the chemical rate expression. The sensitivity analysis for the turbulent combustion model is performed in terms of the local first order sensitivity coefficients. The sensitivity analysis demonstrates the limitations of the turbulence model and the relative importance of single elementary steps of the reaction mechanism. From this some arguments for the simplification of the closure for the mean chemical reaction rates are derived.

---

* Professor, current address: Physikalisch-Chemisches Institut, Universität Heidelberg, Heidelberg, West-Germany

Turbulent Combustion Model

The Turbulent Flow Field

The turbulent mixing of a round fuel jet with oxidizer is described by the transport equations for steady, swirl-free, axisymmetric flow. The turbulent quantities are characterized by their Favre-mean values. Reynolds stresses and Reynolds fluxes in the averaged transport equations are directly modeled by means of gradient approaches utilizing the conventional eddy-viscosity concept. The governing equations for the turbulent flow field, thus, are given by

$$L(\tilde{\phi}) = \bar{S}_{\phi}, \quad \tilde{\phi} = \tilde{u}, \tilde{v}, \tilde{k}, \tilde{\epsilon} \tag{1}$$

where $L(\tilde{\phi})$ is a steady, axisymmetric, elliptic turbulent transport operator defined by

$$L(\tilde{\phi}) \equiv \frac{\partial(\bar{\rho}\tilde{u}\tilde{\phi})}{\partial x} + \frac{1}{r}\frac{\partial(r\bar{\rho}\tilde{v}\tilde{\phi})}{\partial r} - \frac{\partial}{\partial x}\left(\frac{\mu_t}{\sigma_{\phi\ eff}}\frac{\partial\tilde{\phi}}{\partial x}\right) - \frac{1}{r}\frac{\partial}{\partial r}\left(r\frac{\mu_t}{\sigma_{\phi\ eff}}\frac{\partial\tilde{\phi}}{\partial r}\right) \tag{2}$$

The turbulent viscosity is given by the k-ϵ model of turbulence

$$\mu_t = C_{\mu}\bar{\rho}\frac{\tilde{k}^2}{\tilde{\epsilon}} \tag{3}$$

In Eq. (3) C_{μ}, $C_{\mu} = 0.09$ is a model parameter. The eddy viscosity concept is transferred to the turbulent transport of other scalar quantities. In this way the turbulent transport coefficients of scalar quantities are derived from the turbulent viscosity by means of turbulent Schmidt or Prandtl numbers $\sigma_{\phi\ eff}$. In Eq. (1) the $\bar{S}_{\phi}$ are the source terms in the respective transport equations. Because the aim of this work is directed towards the inclusion of detailed chemical reaction mechanisms and the sensitivity analysis of the turbulent combustion model, first order closures for the Reynolds fluxes and Reynolds stresses are thought to be sufficient at the present state of analysis.

The turbulent flow field model thus contains equations for the mean velocities in x- and r- direction $\tilde{u}$ and $\tilde{v}$, with $\sigma_{u\ eff}, \sigma_{v\ eff} = 1$, for the mean turbulence energy $\tilde{k}$, with $\sigma_{k\ eff} = 1$, and the dissipation rate of turbulence energy $\tilde{\epsilon}$, with $\sigma_{\epsilon\ eff} = 1.3$. For details of these equations see Refs. 1 - 3. The system of equations is complemented by an equation for calculating the mean pressure $\bar{p}$ according to Refs. 4 and 5.

Linking of Chemical Reactions

The link between the turbulent flow field and chemical reactions is the mean density $\bar{\rho}$ which follows from the averaged thermal equation of state

$$\bar{\rho} = \frac{\bar{p}}{R} \sum_{i=1}^{N} \frac{M_i}{(\tilde{T}\tilde{Y}_i + \widetilde{Y_i''T''})} \tag{4}$$

Equation (4) contains the unknown mean temperatures $\tilde{T}$, the mean mass fractions of the chemical species $\tilde{Y}_i$, and the second moments $\widetilde{Y_i''T''}$.

The mean temperature may be derived from the caloric equation of state which, upon expressing the specific enthalpies of the chemical species as temperature polynomials, $h_i = \sum_{k=1}^{P} a_{ik}T^k$, takes the form

$$\tilde{h} = \sum_{i=1}^{N} \left[\sum_{k=0}^{P} a_{ik}\tilde{Y}_i\tilde{T}^k + \sum_{k=2}^{P} a_{ik}\frac{k(k-1)}{2}\tilde{Y}_i\tilde{T}^{k-2}\widetilde{T''^2} + \sum_{k=1}^{P} a_{ik}k\widetilde{Y_i''T''}\tilde{T}^{k-1} \ldots \right] \tag{5}$$

The mean specific enthalpy of the mixture is computed from the transport equation

$$L(\tilde{h}) = 0 \ , \sigma_{h\ eff} = 0.7 \tag{6}$$

Equation (5) again contains the second moments $\widetilde{Y_i''T''}$ and $\widetilde{T''^2}$. Higher moments that arise from the temperature polynomials of the species specific enthalpies are omitted. This is indicated by the dots in Eq. (5).

The mean mass fractions of the chemical species are given by the solution of the transport equations

$$L(\tilde{Y}_i) = \bar{w}_i, i = 1, \ldots, N-1, \ \sigma_{Y_i\ eff} = 0.7 \tag{7}$$

where $\bar{w}_i$ is the mean chemical reaction rate of species i. The latter follows from the instantaneous reaction rate

$$w_i = \sum_{j=1}^{R} w_{ij} = \sum_{j=1}^{R} M_i(\nu_{ij}'' - \nu_{ij}')k_j \prod_{i=1}^{N} (\frac{\rho Y_i}{M_i})^{\nu'_{ij}} \tag{8}$$

by weighting with a more-dimensional probability density function

$$\overline{w_i} = \sum_{j=1}^{R} \int \cdots \int w_{ij}(\rho, T, Y_{mj}, \ldots, Y_{oj}) \\ \delta(\rho - \bar{\rho}) P(T, Y_{mj}, \ldots, Y_{oj}; \mathbf{x}) d\rho dT dY_{mj} \ldots dY_{oj} \tag{9}$$

In Eq. (8) M_i is the molar mass of the chemical species i, and the k_j are the rate coefficients of the elementary reaction j where the species i is involved in. The rate coefficients are functions of temperature, $k_j = A_j T^{\alpha_j} exp(-\frac{E_{aj}}{RT})$. Furthermore, in Eq. (9) δ is the Dirac-delta function which indicates that the model is applied for systems with small density fluctuations. The subscripts $mj,\ldots,oj$ in Eq. (9) apply for those chemical species which are involved in the jth elementary reaction, i.e., the stochiometric coefficients ν'_{ij} of which are different from 0. In general, up to three chemical species are involved in elementary reactions. The probability density function $P(T, Y_{mj}, \ldots, Y_{oj}; \mathbf{x})$, therefore, is a joint probability density function for up to four variables.

The averaging procedure given by Eq. (9) requires the specification of this joint probability density function. In this work the probability density function is specified by a more-dimensional normal distribution. Temperature and mass fractions of the chemical species are defined in finite ranges. For the temperature this range is limited by the temperature of the inlets and the adiabatic flame temperature corresponding to the local mixture composition. The mass fractions of the chemical species that are subject to production as well as consumption range between zero and a limit which follows from the conservation of elements in the reacting mixture. Hence, the permitted domain of the pdf is complicated and not simply a square or cube. The multi-dimensional normal distribution has to be clipped properly to this permitted domain. The averaging procedure according to Eq. (9) gives the mean reaction rate $\overline{w_i}$ of the species i by

$$\overline{w_i} = \sum_{j=1}^{R} M_i(\nu''_{ij} - \nu'_{ij}) A_j \tilde{T}^{\alpha_j} exp(-\frac{E_{aj}}{R\tilde{T}}) \prod_{i=1}^{N} (\frac{\bar{\rho}\tilde{Y}_i}{M_i})^{\nu'_{ij}} F_1\ F_2\ F_3 \ldots \tag{10}$$

where the $F_1, F_2, F_3,\ldots$ are polynomials of the form

$$lnF_1 = a_{F_1 1} \left[T u_T (\frac{E_{aj}}{R\tilde{T}}) \right] + a_{F_1 2} \left[T u_T (\frac{E_{aj}}{R\tilde{T}}) \right]^2 \tag{11}$$

$$lnF_2 = 0,\ \text{for}\nu'_{ij} = 1, \quad lnF_2 = a_{F_21}\,(Tu_{Y_i}) + a_{F_22}\,(Tu_{Y_i})^2,\ \text{for}\ \nu'_{ij} = 2, \text{etc.} \tag{12}$$

and

$$lnF_3 = \{a_{F_31}\,(Tu_{Y_i}) + a_{F_32}\,(Tu_{Y_i})^2 \ + a_{F_33}\left[Tu_T(\frac{E_{aj}}{R\tilde{T}})\right] + a_{F_34}\left[Tu_T(\frac{E_{aj}}{R\tilde{T}})\right]^2 \gamma_{T\,Y_i} \tag{13}$$

respectively, for details see Refs. 6 and 7. These polynomials reflect the weighting procedure and the effect of turbulent fluctuations of temperature and mass fractions and their correlations on the mean chemical reaction rate. In Eqs. (11-13) Tu_T and Tu_{Y_i} mean the turbulence intensities of temperature and mass fractions and $\gamma_{T\,Y_i}$ are the correlation coefficients for temperature and the respective mass fractions. The coefficients a_{F_il} of the polynomials (11-13) can be computed by averaging the instantaneous reaction rates according to Eq. (9) for a large number of sets of predefined mean values, variances and covariances of mass fractions and temperatures as well as activation energies and stochiometric coefficients. [6,7]

Following Eq. (10) the averaging procedure to calculate the mean reaction rates $\overline{w_i}$ may be interpreted as applying rate expressions of the form of Eq. (8) employing the mean values of mass fractions and temperatures and correction factors F_l. The latter may be combined with the preexponential factors of the rate coefficients to give ''effective'' rate coefficients. The above discussed closure for the mean chemical reaction rates requires the turbulence intensities of temperature and mass fractions and correlation coefficients to determine the shape of the more-dimensional probability density function and, hence, the values of the polynomials $F_1, F_2, F_3, \ldots$. They can be obtained from the second moments which are computed by solving the transport equations

$$L(\widetilde{\phi''\psi''}) = C_{\epsilon_1\phi\psi}\mu_t\left[\frac{\partial\tilde{\phi}}{\partial x}\frac{\partial\tilde{\psi}}{\partial x} + \frac{\partial\tilde{\phi}}{\partial r}\frac{\partial\tilde{\psi}}{\partial r}\right] - 2\,C_{\epsilon_2\phi\psi}\bar{\rho}\frac{\tilde{\epsilon}}{\tilde{k}}\widetilde{\phi''\psi''} + \overline{\phi'' S_\psi} + \overline{\psi'' S_\phi},$$
$$\tilde{\phi} = \tilde{h}, \tilde{Y}_i, i = 1,\ldots,N,\ \tilde{\psi} = \tilde{Y}_i, i = 1,\ldots N,\ \sigma_{\phi\,\psi\ eff} = 0.7 \tag{14}$$

with $C_{\epsilon_1\phi\psi} = 2/\sigma_{\phi\psi,eff}$ and $C_{\epsilon_2\phi\psi} = 1$. The second moments $\widetilde{Y_i''T''}$ and $\widetilde{T''^2}$ which are also contained in the thermal equation of state and the caloric equation of state, respectively, follow from $\widetilde{\phi''\psi''}$ with $\tilde{\phi}, \tilde{\psi} = \tilde{h}$ or $\tilde{\phi} = \tilde{h},\ \tilde{\psi} = Y_i$ applying the caloric equation of state. The last two expressions on the right hand side of Eq. (14) are obtained by pdf-closure as well.

Numerical Method, Sensitivity Analysis

Numerical Method

The system of differential equations, because of the large number of equations, is solved by an ADI-technique.[8,9] Numerical stability and convergence is achieved by applying linear relaxation and linearization of the source terms by $\bar{S}_\phi = \bar{S}u_\phi + \bar{S}p_\phi\,\tilde{\phi}$. A nonuniform grid is used with higher concentration of grid lines in the region of steep gradients of the variables. Independency of the computed results on the grid has been checked. The boundary conditions are given as usual for elliptic problems, compare Refs. 8 and 9.

Sensitivity Analysis

The system of transport equations in discrete form can be rewritten as

$$F(\tilde{\Phi}) = 0 \tag{15}$$

In Eq. (15) $\tilde{\Phi}$ is the vector of the discrete variables at the single grid nodes. The system of equations $F(\tilde{\Phi})$ contains the discrete transport equations at the $L \cdot M$ grid nodes. It is obvious from the transport Eqs. (1), (6), (7), and (14) that Eq. (15) depends explicitly and, by the various couplings, implicitly on numerous model parameters. Equation (15), therefore, can be written as

$$F(\tilde{\Phi}(\beta), \beta) = 0 \tag{16}$$

where β is the vector of model parameters. The sensitivity of the model is given in terms of the local first order sensitivity coefficients $\partial\tilde{\Phi}/\partial\beta_i$ which, in general, depend on the local values of the variables. The local first order sensitivity coefficients are computed from Eq. (16) by

$$\frac{\partial F}{\partial \tilde{\Phi}} \frac{\partial \tilde{\Phi}}{\partial \beta_i} + \frac{\partial F}{\partial \beta_i} = 0, \quad i = 1, \ldots B \tag{17}$$

The resulting system of linear equations, because of the large number of equations, has again to be solved by an iterative ADI-technique.

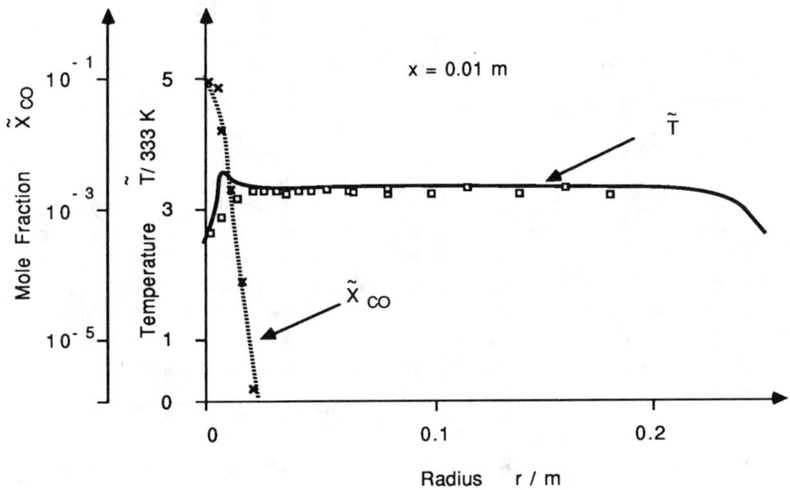

Fig. 1 Radial profiles of mean temperature and carbon monoxide mole fraction at $x = 0.01$ m distance from the fuel nozzle. Symbols refer to measurements. Lines give results of calculations. Experimental conditions: $\tilde{X}_{CO\,0} = 0.0909$; $\tilde{T}_{0\,Fu} = 830K$; $Re_{Fu} = 6770$; $\tilde{X}_{O_2\,0\,Ox} = 0.122$; $\tilde{T}_{0\,Ox} = 1100K$; $Re_{Ox} = 3920$.

In the following example the sensitivities of the model are given as relative sensitivity coefficients $Sr_{\phi\beta_i} = (\frac{\partial\tilde{\phi}}{\tilde{\phi}})/(\frac{\partial\beta_i}{\beta_i})$ or semirelative sensitivity coefficients $Sa_{\phi\beta_i} = \partial\tilde{\phi}/(\frac{\partial\beta_i}{\beta_i})$.

Results

The system under consideration is the oxidation of carbon monoxide by hot air at moderate Damköhler numbers and high Reynolds numbers. The concept of the experiment underlying the validation of the model is simple: carbon monoxide diluted by nitrogen is injected through a nozzle of 10 mm diameter into hot coflowing air-which is produced by an air rich natural gas flame. In the downstream region from the fuel nozzle mean velocities in axial direction, mean temperatures and mean composition of the chemically reacting mixture have been measured. Details of the experimental setup and results of the experiments which have been performed in a combustor of 5 m length and 0.5 m diameter are given in Refs. 10 and 11.

The reaction mechanism of the oxidation of carbon monoxide is constituted by the reactions between the 15 chemical species CO, CO_2, CHO, CH_2O, H_2, H_2O, H, OH, O_2, O, H_2O_2, HO_2, N_2, N, NO summarized in Table 1, see Ref. 12. Table

Table 1 Reaction mechanism for the $CO/H_2/O_2/N_2$ -system[12]

Reaction	A_j	α_j	E_{aj}	j
H_2–O_2–reactions				
$O_2 + H \rightleftarrows OH + O$	$2.0 \cdot 10^{14}$	0.0	70.3	(1,2)
$H_2 + O \rightleftarrows OH + H$	$5.06 \cdot 10^{4}$	2.67	26.3	(3,4)
$H_2 + OH \rightleftarrows H_2O + H$	$1.0 \cdot 10^{8}$	1.6	13.8	(5,6)
$OH + OH \rightleftarrows H_2O + O$	$1.5 \cdot 10^{9}$	1.14	0.42	(7,8)
$H + H + M \rightleftarrows H_2 + M$	$1.8 \cdot 10^{18}$	-1.0	0.0	(9,10)
$H + OH + M \rightleftarrows H_2O + M$	$2.2 \cdot 10^{22}$	-2.0	0.0	(11,12)
$O + O + M \rightleftarrows O_2 + M$	$2.9 \cdot 10^{17}$	-1.0	0.0	(13,14)
$H + O_2 + M' \rightleftarrows HO_2 + M'$	$2.3 \cdot 10^{18}$	-0.8	0.0	(15,16)
$HO_2 + H \rightleftarrows OH + OH$	$1.5 \cdot 10^{14}$	0.0	4.2	(17,18)
$HO_2 + H \rightleftarrows H_2 + O_2$	$2.5 \cdot 10^{13}$	0.0	2.9	(19,20)
$HO_2 + O \rightleftarrows OH + O_2$	$1.8 \cdot 10^{13}$	0.0	-1.7	(21,22)
$HO_2 + OH \rightleftarrows H_2O + O_2$	$6.0 \cdot 10^{13}$	0.0	0.0	(23,24)
$HO_2 + HO_2 \longrightarrow H_2O_2 + O_2$	$2.5 \cdot 10^{11}$	0.0	-5.2	(25)
$OH + OH + M' \rightleftarrows H_2O_2 + M'$	$3.25 \cdot 10^{22}$	-2.0	0.0	(27,28)
$H_2O_2 + H \rightleftarrows H_2 + HO_2$	$1.7 \cdot 10^{12}$	0.0	15.7	(29,30)
$H_2O_2 + H \rightleftarrows H_2O + OH$	$1.0 \cdot 10^{13}$	0.0	15.0	(31,32)
$H_2O_2 + O \rightleftarrows OH + HO_2$	$2.8 \cdot 10^{13}$	0.0	26.8	(33,34)
$H_2O_2 + OH \rightleftarrows H_2O + HO_2$	$5.4 \cdot 10^{12}$	0.0	4.2	(35,36)
CO–CO_2–reactions				
$CO + OH \rightleftarrows CO_2 + H$	$4.4 \cdot 10^{6}$	1.5	-3.1	(37,38)
$CO + HO_2 \rightleftarrows CO_2 + OH$	$1.5 \cdot 10^{14}$	0.0	98.7	(39,40)
$CO + O + M' \rightleftarrows CO_2 + M'$	$7.1 \cdot 10^{13}$	0.0	-19.0	(41,42)
$CO + O_2 \rightleftarrows CO_2 + O$	$2.5 \cdot 10^{12}$	0.0	200.0	(43,44)
$CHO + H \rightleftarrows CO + H_2$	$2.0 \cdot 10^{14}$	0.0	0.0	(45,46)
$CHO + O \rightleftarrows CO + OH$	$3.0 \cdot 10^{13}$	0.0	0.0	(47,48)
$CHO + O \rightleftarrows CO_2 + H$	$3.0 \cdot 10^{13}$	0.0	0.0	(49,50)
$CHO + OH \rightleftarrows CO + H_2O$	$1.0 \cdot 10^{14}$	0.0	0.0	(51,52)
$CHO + O_2 \rightleftarrows CO + HO_2$	$3.0 \cdot 10^{12}$	0.0	0.0	(53,54)
$CHO + M' \rightleftarrows CO + H + M'$	$7.1 \cdot 10^{14}$	0.0	70.3	(55,56)
$CH_2O + H \rightleftarrows CHO + H_2$	$2.5 \cdot 10^{13}$	0.0	16.7	(57,58)
$CH_2O + O \rightleftarrows CHO + OH$	$3.5 \cdot 10^{13}$	0.0	14.6	(59,60)
$CH_2O + OH \rightleftarrows CHO + H_2O$	$3.0 \cdot 10^{13}$	0.0	5.0	(61,62)
$CH_2O + HO_2 \rightleftarrows CHO + H_2O_2$	$1.0 \cdot 10^{12}$	0.0	33.5	(63,64)
$CH_2O + M' \rightleftarrows CHO + H + M'$	$1.4 \cdot 10^{17}$	0.0	320.0	(65,66)
NO–reactions				
$N + OH \rightleftarrows NO + H$	$3.0 \cdot 10^{13}$	0.0	0.0	(67,68)
$N + O_2 \rightleftarrows NO + O$	$6.4 \cdot 10^{9}$	1.0	26.1	(69,70)
$N_2 + O \rightleftarrows NO + N$	$1.8 \cdot 10^{14}$	0.0	319.0	(71,72)

$$M = [H_2] + 0.4\,[O_2] + 0.4\,[N_2] + 6.5\,[H_2O]$$

$$M' = [H_2] + [O_2] + [N_2] + [H_2O]$$

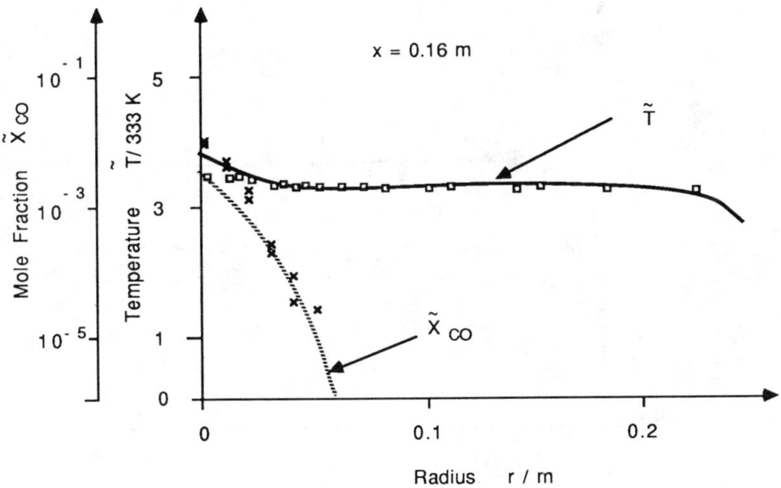

Fig. 2 Radial profiles of mean temperature and carbon monoxide mole fraction at $x = 0.16$ m distance from the fuel nozzle. Symbols refer to measurements. Lines give results of calculations. Experimental conditions see Fig. 1.

1 contains the rate coefficients for the forward reactions in the form $k_j = A_j T^{\alpha_j} exp(-\frac{E_{aj}}{RT})$. The units for A_j are mol, cm^3 and s. E_{aj} is given in units of $kJmol^{-1}$.

The oxidation of carbon monoxide mainly occurs through the reaction $CO + OH \rightarrow CO_2 + H\,(37)$. The direct oxidation by oxygen $CO + O_2 \rightarrow CO_2 + O\,(43)$ is strongly temperature dependent and, therefore, meaningless under the prevailing conditions. The oxidation by oxygen atoms $CO + O + M' \rightarrow CO_2 + M'\,(41)$ the reaction rate of which decreases with increasing temperature is the other main channel for the formation of carbon dioxide. The OH radicals which are necessary for the oxidation of carbon monoxide according to reaction (37) are formed in the chain branching and propagation reactions of the $H_2/O_2/N_2$ system, reactions (1-8). The $H_2/O_2/N_2$ system is driven by the strongly temperature dependent shuffle reactions (1-8) and attains equilibrium by the less temperature dependent recombination reactions (9-13). The mechanism in Table 1 also contains the formation of nitrogen oxide through the thermal Zeldovich mechanism.

Figures 1 and 2 contain the comparison of the radial profiles of measured and calculated mean temperatures and mean mole fractions of carbon monoxide for two cross sections of the combustor. The experimental conditions are

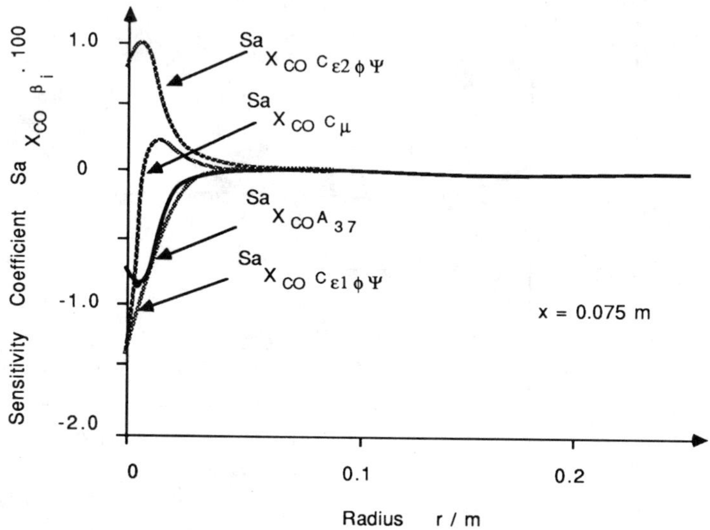

Fig. 3 Radial profiles of sensitivity coefficients of carbon monoxide mole fraction with respect to C_μ, $C_{\epsilon_{1\phi\psi}}$, $C_{\epsilon_{2\phi\psi}}$ and A_{37}. Experimental conditions see Fig. 1.

given in the caption of the figures. The figures exhibit very good agreement between measurements and predictions.

Because one aim of this work is the sensitivity analysis of the above discussed turbulent combustion model, some results of the sensitivity analysis are discussed in the following in more detail. Other results, in particular results for the turbulent fluctuations and their influence on the local turbulent reaction rates, are given elsewhere.[6,13] The most important reaction for the oxidation of carbon monoxide is the reaction $CO + OH \rightarrow CO_2 + H$ (37). Figure 3 gives radial profiles of the local first order sensitivity coefficients of the carbon monoxide mole fraction with respect to the preexponential factor of this reaction in comparison with the sensitivity with respect to the model parameters of the turbulence model. The profiles are computed for an axial distance of 0.075 m where the rate of consumption of carbon monoxide on the axis has a maximum. In the central region of the jet the sensitivity coefficients $\partial\tilde{X}_{CO}/(\frac{\partial A_{37}}{A_{37}})$ are negativ. This is obvious, because an acceleration of reaction (37) lowers the carbon monoxide mole fraction. Near the jet boundary, where the reaction rate for the consumption of carbon monoxide tends to very small values, the sensitivity of the carbon monoxide mole fraction with respect to that reaction vanishes.

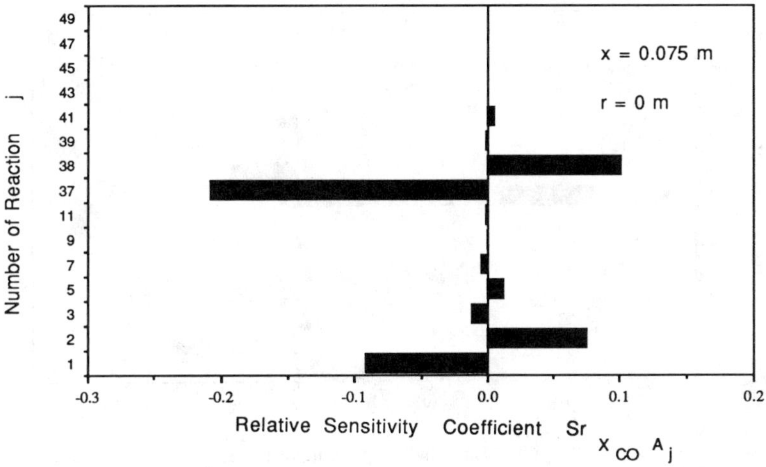

Fig. 4 Sensitivity coefficients of carbon monoxide mole fraction with respect to the preexponential factors of the rate coefficients from Table 1 at $x = 0.075$ m. Experimental conditions see Fig. 1.

The rate coefficient of the reaction $CO + OH \rightarrow CO_2 + H$ (37) influences the carbon monoxide mole fraction by the same order of magnitude as the parameters of the turbulence model C_μ, $C_{\epsilon_{1\phi\psi}}$, or $C_{\epsilon_{2\phi\psi}}$. Again, the sensitivity of the carbon monoxide mole fraction with respect to these parameters vanishes in the region of the jet boundary. This is a consequence of the low reaction rates near the jet boundary and of the applied gradient transport concept, which predicts production of turbulent fluctuations only in the region with large gradients of the mean variables. In this way, the second moments, the production and dissipation of which are modeled via the parameters $C_{\epsilon_{1\phi\psi}}$ and $C_{\epsilon_{2\phi\psi}}$, influence only weekly the reaction rates for carbon monoxide near the jet boundary. Near the center of the jet where the reaction rates are high and the production and dissipation of turbulent fluctuations due to higher gradients of the mean values of the variables are large, the influence of the model parameters for modeling the second moments under the prevailing conditions is comparable to the influence of the rate coefficient of the main carbon monoxide consuming reaction.

The carbon monoxide mole fraction is essentially affected only by the fuel specific reactions (37),(38), and the nonfuel specific reactions (1) and (2), respectively. This can be seen from Figs. 4 and 5 where the sensitivity of the carbon monoxide mole fraction with respect to the rate coefficients of some reactions from Table 1 is given. The

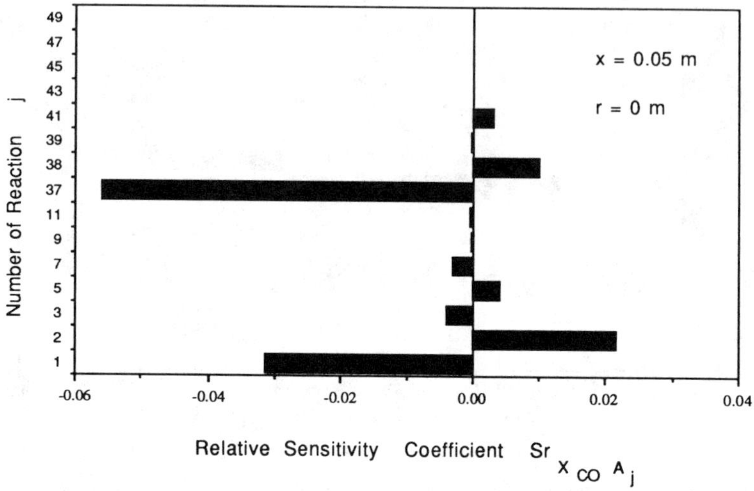

Fig. 5 Sensitivity coefficients of carbon monoxide mole fraction with respect to the preexponential factors of the rate coefficients from Table 1 at $x = 0.05$ m. Experimental conditions see Fig. 1.

sensitivity coefficients are displayed for the axial position with the maximum rate of consumption of carbon monoxide, $x = 0.075$ m, and for the position $x = 0.05$ m downstream from the fuel nozzle. Similar sensitivities can be found in laminar flames,[14] and the fact that nonfuel specific reactions affect the mole fraction of the fuel is known as well from other fuels. The sensitivity analysis reflects the importance of OH radicals for the oxidation of carbon containing fuels. Figures 4 and 5 also demonstrate a change in the mechanism depending on the local conditions. At the location where the rate of carbon monoxide consumption is maximum, the relative importance of reaction (37) is larger than at lower consumption rates.

The sensitivity analysis identifies the reactions, the rate coefficients of which affect the properties of the turbulent reacting system. For the computation of, e.g., the carbon monoxide mole fraction, the most important reactions are the reactions $CO + OH \rightleftharpoons CO_2 + H$ (37,38) and $O_2 + H \rightleftharpoons OH + O$ (1,2). For the prediction of, e.g., the carbon monoxide mole fraction, the rate of these reactions have to be determined exactly. Hence, the influence of turbulent fluctuations on the rate of these reactions must be modeled in a proper way. The reaction rates of the other reactions do not affect the calculated mole fractions of carbon monoxide very much. Hence, the rate coefficients or ''effective'' rate coefficients of these reactions need not to be calculated exactly, i.e., the correction factors F_l in the rate expression (10) for these reactions may be set to 1 without affecting the computed mole fractions of carbon monoxide noticeably. By this, a lot of computational effort may be saved.

References

[1] Libby, P.A., and Williams, F.A., ''Fundamental Aspects'', Turbulent Reacting Flows, edited by P.A. Libby and F.A. Williams, Springer-Verlag, Berlin, Heidelberg, 1980,p.1.

[2] Bilger, R.W., ''Turbulent Flows with Nonpremixed Reactants'', Turbulent Reacting Flows, edited by P.A. Libby and F.A. Williams, Springer-Verlag, Berlin, Heidelberg, 1980,p.65.

[3] Behrendt, F., Bockhorn, H., Rogg, B., and Warnatz, J., ''Modelling of Turbulent CO/Air Diffusion Flames with Detailed Chemistry'', Complex Chemical Reaction Systems, Mathematical Modelling and Simulation, edited by J.Warnatz and W.Jäger, Springer-Verlag, Berlin, Heidelberg, 1987, p. 376.

[4] Spalding, D.B., ''Basic Equations of Fluid Mechanics and Heat and Mass Transfer'', Imperial College London, Mechanical Engineering Department, Rep. HTS /76/6, London, 1976.

[5] Patankar, S.V., Numerical Heat Transfer and Fluid Flow, Hemisphere, New York, Washington, Philadelphia, London, 1980, p.113.

[6] Bockhorn, H., Zur Struktur Turbulenter Flammen, Habilitationsschrift, T.H. Darmstadt, Darmstadt, 1989, p. 36.

[7] Bockhorn, H., ''Modeling of Turbulent Diffusion Flames with Detailed Chemistry'', Mathematical Modeling in Combustion and Related Topics, edited by C.M. Brauner and C. Schmidt-Laine, Martinus Nijhoff, Dordrecht, 1988, p. 411.

[8] Gosman, A.D., Pun, W.M., Runchal, A.K., Spalding, D.B., and Wolfshtein, M., Heat and Mass Transfer in Recirculating Flows, Academic, London, 1969.

[9] Srivatsa, S.K., ''CORA2 - A Computer Code for Axi-Symmetrical Combustion Chambers'', CHAM, Technical Rep. TR 201/1, London, 1977.

[10] Bockhorn, H., and Lutz, G., ''The Application of Turbulent Reaction Models to the Oxidation of CO in a Turbulent Flow'', Twentieth Symposium (International) on Combustion, The Combustion Institute, Pittsburgh, PA, 1984, p. 377.

[11] Bockhorn, H., and Lutz, G., ''Reaction Models for Simulation of the Oxidation of Carbonmonoxide in Turbulent Diffusion Flames'', Chemical Engineering and Technology, Vol. 10, 1987, p. 43.

[12] Warnatz, J.,''Rate Coefficients in the C/H/O System'', Combustion Chemistry, edited by W.C. Gardiner Jr., Springer-Verlag, New York, 1984, p. 196.

[13] Bockhorn, H., ''Simulation der Verbrennung von CO in turbulenten Diffusionsflammen unter Berücksichtigung komplexer Reaktionsmechanismen'', VDI Berichte Nr. 645, VDI-Verlag, Düsseldorf, 1987, p. 499.

[14] Maas, U., and Warnatz, J., ''Mathematische Modellierung von Selbstzündung und Zündgrenzen im H_2-O_2-CO-System'', VDI Berichte Nr. 645, VDI-Verlag, Düsseldorf, 1987, p. 509.

Numerical Simulation of a Premixed Flame in a Turbulent Boundary Layer

B. Rogg*

University of Cambridge, Cambridge, England, United Kingdom

Abstract

The objective of the present paper is to model and to numerically simulate premixed turbulent combustion in a boundary layer which arises from the injection of cold reactants into a hot stream of corresponding combustion products. A laminar-flamelet approach is adopted with attention being focused on the chemical-source-term closure. For various injection rates of the fresh mixture, a sensitivity analysis is performed to investigate the sensitivity of selected quantities, such as the location of the flame, with respect to the various model constants. The numerical results are compared to experimental data available in the literature.

Introduction

At the present time, a general description of the physics and chemistry in turbulent flames is impossible and, therefore, models are needed - and used - to describe, to analyse and to predict turbulent reactive flows. If numerical predictions of such flows are to serve seriously as a basis for the design of equipment, the diagnosis of malfunction problems or the development of modifications for im-

*Professor, Department of Engineering.

provement of performance, the predictions must be trustworthy. As a consequence, hand in hand with adequate numerical-solution methods and computers, models of turbulent reacting flows are needed that are able not only to predict the flowfield and global flame properties such as the flame length or the overall heat release, but that allow equally well for an accurate and confident prediction of the temperature and the species concentrations throughout the flame.

Much of current world-wide combustion research is devoted to *premixed* turbulent combustion which generally is preferred to the nonpremixed mode if intense combustion in small volumes is desired such as in Otto-engines, in domestic burners, or for gas welding. A particularly important regime of premixed turbulent combustion is the laminar-flamelet regime, the most prominent representative of which is combustion in Otto-engines.[1] To describe premixed turbulent combustion in this regime, relatively comprehensive models are available such as the "BMCL-model" developed by Bray, Moss , Champion and Libby,[2–5] and the "Stochastic Flamelet Model" developed by Pope.[6,7] In contrast to other models, these two models have been successful in accounting for some of the experimental observations related to counter-gradient diffusion and flame-generated turbulence.[8,9] Within the framework of flamelet-models of premixed turbulent combustion, an important topic of current research is the prediction of the local flame-surface area per unit volume σ, which is also referred to as the local flame-surface density, a quantity which is needed to calculate the local turbulent-burning rate in the flame. For the Stochastic Flamelet Model first evolution equations have been given from which a local mean value for σ can be calculated.[7] Equivalently, to supplement the BMCL-model, Bray, Champion and Libby have recently derived a transport equation for the probability of flamelet-burning.[10] As an alternative to the latter equation, a modelled transport equation for the mean of the flame-surface density σ has been proposed by Peters.[11,12] For convenience of notation, subsequently the latter equation will be referred to as the "σ-transport equation".

The objective of the present paper is to model and to numerically simulate premixed turbulent combustion in a boundary layer which arises from the injection of cold reactants into a hot stream of corresponding combustion products. A laminar-flamelet approach is adopted with attention being focused on the chemical-source-term closure, for the purpose of which a modified version of the σ-transport equation proposed by Peters[11,12] is employed. For various injection rates of the fresh mixture, a sensitivity analysis is performed to investigate the sensitivity of selected quantities, such as the absolute values and the location of the maximum mean flame-surface density, with respect to the various model constants, particularly to those appearing in the σ-transport equation. The numerical results are compared to experimental data available in the literature.

Modeling

Turbulent premixed combustion in a boundary layer is studied that arises from the injection of cold reactants into a hot stream of corresponding combustion products. A schematic of this flow geometry is shown in Fig. 1. The constant-enthalpy case is considered, i.e., a situation in which the enthalpies of the reactants and the products are the same. Since previous work[4,5] showed that for constant-enthalpy combustion in turbulent boundary layers both counter-gradient transport and dissipation through flamelets are unlikely to occur, and because herein the focus of interest is the chemical-source-term closure, an early version of the BMCL-model[13]

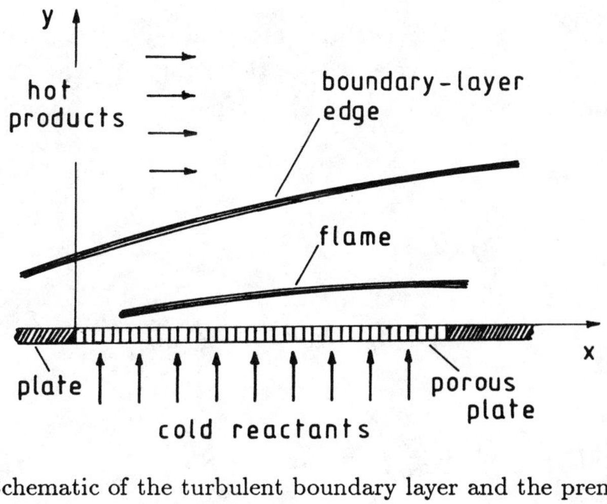

Fig. 1 Schematic of the turbulent boundary layer and the premixed flame.

is adopted which is based on a conventional eddy-viscosity gradient model of turbulent transport together with the k-ϵ turbulence model and the aforementioned σ-transport equation.

It should be noted here that with respect to the flowfield model that will be presented below, the formulation is similar to that of Meunier et al.[14] who investigated experimentally and numerically the same type of flow. However, to close the mean chemical-reaction term these authors employed a direct-closure approach using a presumed univariate two-parameter probability-density function for the reaction-progress variable. Since the emphasis of their work was on the comparison between the numerical and experimental results, this approach was adequate. The general objective of the present work is different as we are interested in investigating in a turbulent reactive boundary layer a recently developed, more universal approach to closure of the mean chemical source term. A specific aim of this study is the validation of model constants in turbulent reactive boundary layers, particularly of those appearing in the σ-transport equation. For the latter purpose a rigorous sensitivity analysis is performed.

The Flowfield Model

Upon introducing a parabolic turbulent-transport operator L,

$$L(\tilde{\phi}) \equiv \frac{\partial(\bar{\rho}\tilde{u}\tilde{\phi})}{\partial x} + \frac{\partial(\bar{\rho}\tilde{v}\tilde{\phi})}{\partial y} - \frac{\partial}{\partial y}\left(\frac{\mu_t}{\sigma_\phi}\frac{\partial\tilde{\phi}}{\partial y}\right)$$

for any variable $\tilde{\phi}$, for the turbulent boundary-layer problem under consideration, the conservation equations for mass and momentum and the equations of the k-ϵ turbulence model can be written as

$$L(1) = 0, \tag{1}$$

$$L(\tilde{u}) = -d\bar{p}/dx, \tag{2}$$

$$L(\tilde{k}) = \mu_t\left(\frac{\partial\tilde{u}}{\partial y}\right)^2 - \bar{\rho}\tilde{\epsilon} - \overline{u''}\frac{d\bar{p}}{dx} - C_k\bar{\rho}\tilde{k}\left(\frac{\partial\tilde{u}}{\partial x} + \frac{\partial\tilde{v}}{\partial y}\right), \tag{3}$$

$$L(\tilde{\epsilon}) = \frac{\epsilon}{k}\left[C_{\epsilon 1}\left(\frac{\partial\tilde{u}}{\partial y}\right)^2 - C_{\epsilon 2}\bar{\rho}\epsilon\right]. \tag{4}$$

In deriving Eqs. (1) to (4) the Favre-averaging procedure has been used; tildes identify Favre-averages, overbars time averages; x and y denote the longitudinal and transversal coordinates, respectively, and $\tilde{u}$ and $\tilde{v}$ the respective flow-velocity components; $\bar{\rho}$ is the density, μ_t the turbulent diffusivity, $\tilde{k}$ the turbulence kinetic energy, $\tilde{\epsilon}$ the turbulence dissipation, and the σ_ϕ's and $C_{\epsilon i}$'s are model constants for which alternatively two slightly different sets have been employed, viz., $\sigma_u = \sigma_k = 1.0, \quad \sigma_\epsilon = 1.3, \quad C_{\epsilon 1} = 1.45, \quad C_{\epsilon 2} = 1.9$ according to Ref. 15 and $\sigma_u = \sigma_k = 1.0, \quad \sigma_\epsilon = 1.3, \quad C_{\epsilon 1} = 1.57, \quad C_{\epsilon 2} = 2.0$ according to Ref. 16. In Eq. (3), the term involving C_k describes the change of the kinetic turbulence energy due to the heat release, and herein is modelled according to Ref. 14, as is the term involving $\overline{u''}$. Equations (1) to (4) govern the variations of $\tilde{u}$, $\tilde{v}$, $\tilde{k}$ and $\tilde{\epsilon}$. Note that, as a consequence of the boundary-layer approximation, the gradient of the mean pressure appearing in Eq. (2), $d\bar{p}/dx$, is the external pressure gradient that is imposed on the boundary layer. Apart from appropriate boundary conditions the solution of these equations requires information about the variations of the density $\bar{\rho}$ and the turbulent viscosity μ_t. For μ_t the relationship

$$\mu_t = C_\mu \bar{\rho} k^2 / \epsilon \tag{5}$$

is adopted, with a constant value of 0.09 for C_μ, see Refs. 15 and 16. The mode of providing information about $\bar{\rho}$ depends on the approach to closure of the chemical source term and is briefly discussed in the following subsection.

The Laminar – Flamelet Model

The basic idea of the flamelet concept of premixed turbulent combustion is to describe the turbulent flame locally and instantaneously in terms of laminar flames, the so-called flamelets; this concept is applicable if the thickness of the flamelets is small compared to the smallest length scale of turbulence, the Kolmogorov scale. The flamelets are subjected to strain and develop curvature as consequences of the turbulent velocity fluctuations. Effects of both influences, combined with effects of differential diffusion of chemical species and heat, cause the inner structure of the flamelets to respond, thereby leading to a change in burning velocity and in some cases to extinction[18,19] In the simplest cases the structure of a flamelet is conveniently described by a single reaction-progress variable c which takes values between zero in the cold reactants and unity in the burnt gas; for the problem considered herein c is chosen to be a normalised product mass fraction or, equivalently, the normalised temperature. It is straightforward to derive a transport equation for the mean value of c,[13] viz,

$$L(\tilde{c}) = \overline{\omega}, \tag{6}$$

where ω is a suitably defined chemical source term and $\overline{\omega}$ its mean; the latter can be interpreted as a local mean turbulent-burning rate. Assuming that the chemical reactions in the turbulent flame proceed due to local laminar-flame propagation normal to the instantaneous turbulent flame-surface, it is easy to show that

$$\overline{\omega} = S_L \widetilde{\sigma} \tilde{c}(1 - \tilde{c}). \tag{7}$$

Here S_L denotes the laminar-flame speed, which for simplicity has been assumed to be constant, and $\widetilde{\sigma}$ is the local Favre-averaged flame-surface density. Obviously, closure of Eq. (7) requires information on $\widetilde{\sigma}$. As an advanced alternative to algebraic closure models for Eq. (7), such as that proposed by Gouldin, Bray and Chen,[17] herein closure is achieved by supplementing the governing partial differential equations by a transport equation for the mean flame-surface density which has been proposed recently by Peters.[11,12] Using the Favre-averaging procedure, this equation can be cast into a form which is compatible with Eqs. (1) - (4) and (6). In particular, the σ-transport equation contains terms describing the production of mean specific flame-surface area $\widetilde{\sigma}$ due to strain and dissipation terms that characterise shrinking and annihilation of $\widetilde{\sigma}$ e.g. due to local flamelet extinction as a consequence of excessive strain, see Refs. 18 and 19.

According to the BMCL-model, the mean density $\bar{\rho}$ which is still needed for closure of the set of governing equations is a simple function of the mean progress variable,

$$\frac{\bar{\rho}}{\rho_b} = \frac{1+\tau}{1+\tau\tilde{c}}, \tag{8}$$

where the heat-release parameter τ is defined as

$$\tau = (T_b - T_u)/T_u. \tag{9}$$

In Eqs. (8) and (9), and below, the subscripts u and b identify conditions in the unburnt and burnt mixture, respectively. Recall that the model employed herein implies that the burnt mixture is in chemical equilibrium at the adiabatic flame temperature. For details of the derivation of Eq. (8), as well as for relationships that enable the local means of other thermochemical variable of interest to be calculated, Ref. /13/ may be consulted.

Initial and Boundary Conditions; Numerical Method

At the wall the thermochemical state of the cold reactants and their rate of injection, and hence $\tilde{v}$, are specified; $\tilde{u}$ there is zero. The boundary conditions for $\tilde{k}$, $\tilde{\epsilon}$ and μ_t at the wall are calculated using the wall functions described in Ref 20. At the boundary-layer edge the thermochemical state of the hot gas is that of chemical equilibrium at the adiabatic flame temperature, and $\tilde{u}$, $\tilde{k}$, and $\tilde{\epsilon}$ there are specified; $\tilde{v}$ there is zero. At $x = 0$ physically and mathematically consistent profiles are specified for all dependent variables.

The set of parabolic governing equations is solved by a space-marching method with respect to x starting at x=0. Successively, in each x-plane x^n, n=0, 1, $\cdots$, the governing equations are discretized. The discretization with respect to x is of the backward-Euler type, that with respect to y is performed on a nonuniform mesh $\mathbf{M}$ of grid points,

$$\mathbf{M}^n = \{0 = y_1^n < y_2^n < ... < y_m^n = y_{max}\}. \tag{10}$$

In the y-direction a static adaptive-gridding procedure[21] combined with a piecewise monotonic cubic Hermite interpolation method[22] is employed that concentrates grid points in regions where the profiles of the dependent variables exhibit steep gradients and/or strong curvature. Therefore, the number of gridpoints, i.e. the subscript m in Eq. (10), may vary with x, but y_{max} is kept constant and is selected sufficiently large as to ensure that the conditions at the boundary-layer edge are asymptotically satisfied.

The marching scheme advances from one x-plane, x^n, to the next, x^{n+1}, in each plane solving the nonlinear system of difference equations

$$F(U^{n+1}, U^n; C) \equiv f(U^{n+1}; C) - (U^{n+1} - U^n)/h^{n+1} = 0 \qquad (11)$$

by a modified Newton method. In Eq. (11), and below, h^{n+1} stands for x^{n+1} - x^n, n=0,1,$\cdots$; U^i denotes the solution of the nonlinear system in plane x^i, i=n and $n+1$, viz.,

$$\begin{aligned} U^i = (&\tilde{u}_1^i, \tilde{v}_1^i, \tilde{k}_1^i, \tilde{\epsilon}_1^i, \tilde{c}_1^i, \tilde{\sigma}_1^i, \tilde{u}_2^i, \tilde{v}_2^i, \tilde{k}_2^i, \tilde{\epsilon}_2^i, \tilde{c}_2^i, \tilde{\sigma}_2^i, \\ &\cdots \tilde{u}_m^i, \tilde{v}_m^i, \tilde{k}_m^i, \tilde{\epsilon}_m^i, \tilde{c}_m^i, \tilde{\sigma}_m^i), \end{aligned} \qquad (12)$$

and C is the parameter M-vector of the M model constants that appear in the governing equations,

$$C = (C_{\epsilon_1}, C_{\epsilon_2}, C_{\epsilon_k}, C_\mu, \cdots). \qquad (13)$$

Thus, at the $k+1^{st}$ Newton-iteration step, k=0, 1, $\cdots$, the linear system

$$\begin{aligned} &\Big(J(U_k^{n+1}) - I/h^{n+1}\Big)(U_{k+1}^{n+1} - U_k^{n+1}) \\ &\quad = -\omega_k^{n+1}\Big(f(U_k^{n+1}; C) - (U_k^{n+1} - U^n)/h^{n+1}\Big), \end{aligned} \qquad (14)$$

$k = 0, 1, ...,$ is solved. Here U_k^{n+1} denotes the solution after k iterations, ω_k^{n+1} is the damping parameter, and $J(U_k^{n+1})$ and I are the Jacobian based on U_k^{n+1} and the identity matrix, respectively, The damping strategy[23,24] allows the Jacobians to be re-evaluated only periodically.[24]

Sensitivity Analysis

For the reactive boundary-layer problem under consideration, sensitivity analysis is a useful tool to investigate quantitatively the dependence of the numerical predictions on the particular choice of the numerical values that have been assigned to the individual model constants appearing in the governing equations. Since the model constants of the k-ϵ model are relatively well established, in the present paper attention is focused on the σ-transport equation. In particular, we are interested in the sensitivity on the σ-equation model constants of the predicted absolute values and locations of the maximum mean flame-surface density, quantities that are a direct measure for the turbulent-burning rate.

In general, to predict the effect of the variation of the model constants contained in C on the dependent variables contained in U, where U and C are defined in Eqs. (12) and (13), respectively, the first-order sensitivity coefficients $\partial U/\partial C_\alpha$, α=1,...,M, must be calculated.[25] The appropriate equation for the latter quantities is obtained by differentiating Eq. (11) with respect to C, viz.,

$$\frac{d}{dC_\alpha}\Big(F(U^{n+1},U^n;C)\Big) = \frac{\partial F}{\partial U^{n+1}}\frac{\partial U^{n+1}}{\partial C_\alpha} + \frac{\partial F}{\partial U^n}\frac{\partial U^n}{\partial C_\alpha} + \frac{\partial F}{\partial C_\alpha} = 0, \tag{15}$$

$\alpha = 1,...,M$. Equation (15) can be rewritten as

$$\Big(J(U^{n+1}) - \frac{I}{h^{n+1}}\Big)\frac{\partial U^{n+1}}{\partial C_\alpha} = -\Big(\frac{\partial F}{\partial C_\alpha} + \frac{I}{h^{n+1}}\frac{\partial U^n}{\partial C_\alpha}\Big) \tag{16}$$

Since the first-order sensitivity coefficients at plane x^n, $\partial U^n/\partial C_\alpha$, are known, those at plane $x^{(n+1)}$, $\partial U^{n+1}/\partial C_\alpha$, can readily be calculated from Eq. (16) by solving a linear system of equations.

Results and Discussion

Due to space limitations in the present volume, only a few results can be presented and discussed. These are results for one of the flames investigated by Meunier[14] et al., viz., a nitrogen-diluted, slightly lean propane/air flame, the reactants being injected with a rate of F=1.05 percent. The data are: fuel-equivalence ratio 0.97, nitrogen mass fraction in the diluted reactive mixture 0.28, initial wall temperature (at x=5cm) 1000 K, initial freestream temperature 1500 K and initial freestream velocity 75.3 m/s. The conditions are such that wall temperature decreases and freestream velocity increases slightly (approximately linearly between x=5 cm and x = 19 cm by 10 percent); the freestream temperature remains constant. Under these conditions it was found that a flame stabilizes, and experimental data are reported in Ref. 14 for the two axial locations. Herein the data at x=19 cm are used for comparison with the numerical predictions based on the σ-transport equation.

Shown in Fig. 2 are the measured and computed velocity profiles. It is seen that the agreement is good. Shown in Fig. 3 are the measured and calculated temperature profiles. The agreement, which is seen to be excellent, has been obtained for the following set of parameters: $C_{\sigma 1}$= 7.5, $C_{\sigma 2}$= 0.075 and σ_σ=1 . It is interesting to note that the ratio

$$r = C_{\sigma 2}/C_{\sigma 1} = 0.01 \tag{16}$$

is small, as was observed by Peters for the one-dimensional, constant-

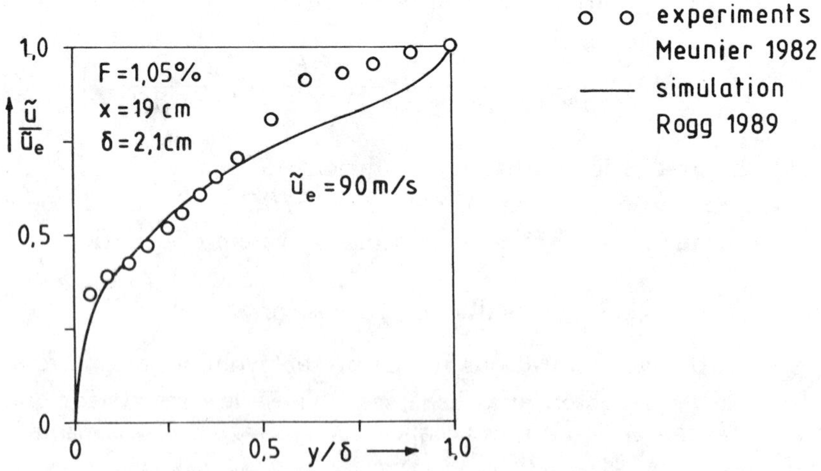

Fig. 2 Comparison between measured and predicted mean-velocity profiles.

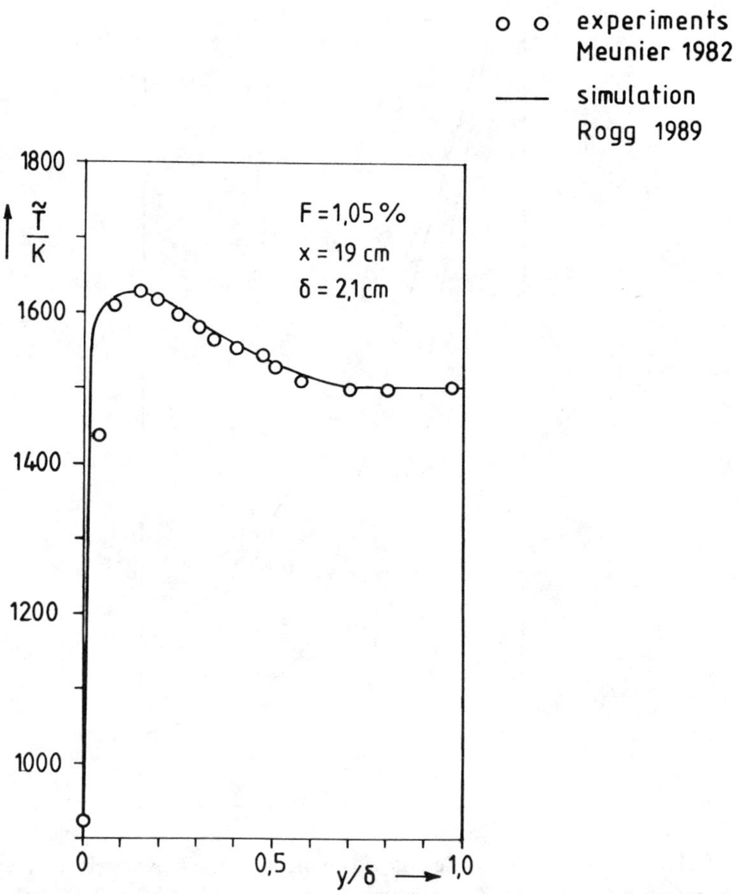

Fig. 3 Comparison between measured and predicted mean-temperature profiles.

property case.

Shown in Figs. 4 to 6 are some results obtained from the sensitivity analysis. The abscissa in these figures is normalized such that $x^\star \equiv (x - 5cm)/(19cm - 5cm)$. It is important to note that the relative sensitivity coefficients, which in the figures are denoted by

$$\beta_{i,j} \equiv \frac{C_j}{U_i}\frac{\partial U_i}{\partial C_j} \approx \frac{\Delta U_i/U_i}{\Delta C_j/C_j}, \tag{17}$$

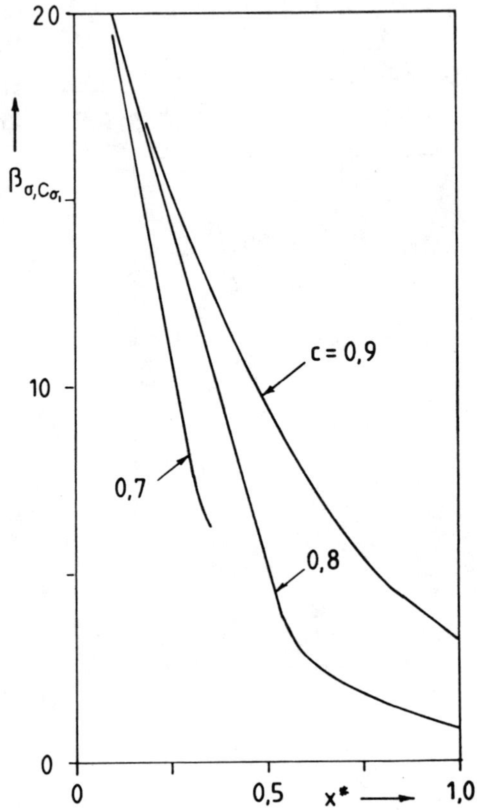

Fig. 4 Relative sensitivity coefficient of the mean flame-surface density with respect to $C_{\sigma 1}$ along the contours where $\tilde{c}$=0.7, 0.8, and 0.9, respectively.

are local quantities, i.e., quantities that vary through the flame. Shown in Figs. 4 and 5 are the sensitivities of the mean flame-surface density with respect to the model parameter $C_{\sigma 1}$ and the ratio $r = C_{\sigma 2}/C_{\sigma 1}$ for the three contour lines where $\tilde{c}$ is 0.7, 0.8 and 0.9, respectively. The sensitivity of σ with respect to $C_{\sigma 1}$ is remarkably high for small to intermediate values of $x^\star$. Inspection of the temperature field (not shown here) reveals that these high sensitivities are related to the ignition process which takes place in the early parts of the boundary layer. The sensitivity of σ with respect to r (or $C_{\sigma 2}$) is seen to be fairly constant throughout the length of the boundary layer.

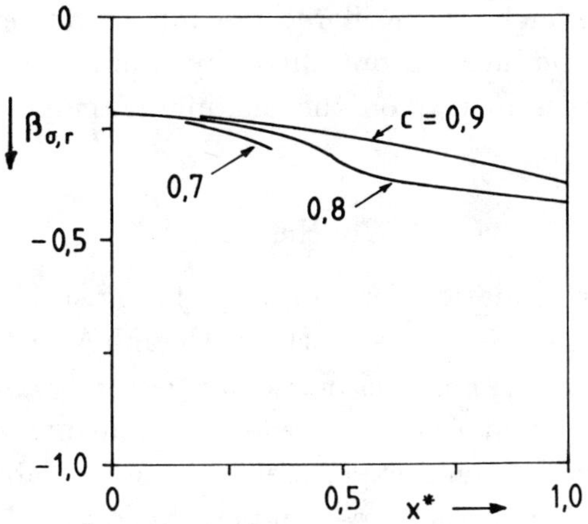

Fig. 5 As Fig. 4, but sensitivity with respect to the ratio $C_{\sigma 2}/C_{\sigma 1}$.

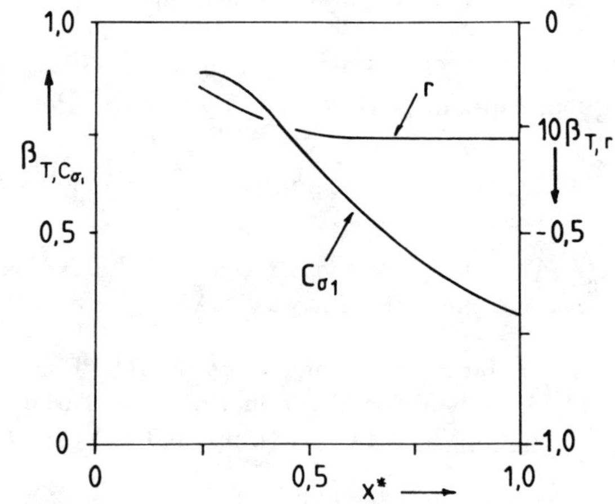

Fig. 6 Relative sensitivity coefficients of the mean temperature with respect to $C_{\sigma 1}$ and the ratio $C_{\sigma 2}/C_{\sigma 1}$ along the contours where $\widetilde{T}$ attains its maximum.

Shown in Fig. 6 is the sensitivity of the mean temperature *along the contour of maximum temperature* with respect to $C_{\sigma 1}$ and r. It is seen that throughout the flame the maximum temperature can be expected to be affected only little by changes in r, whereas the impact of changes in $C_{\sigma 1}$ on the maximum temperature increases with increasing x.

Conclusions

A transport equation for the mean flame surface-to-area ratio has been tested for premixed turbulent combustion in a boundary layer with injection. For this particular case model constants have been determined that lead to excellent agreement between numerical model predictions and experimental data available in the literature. Sensitivity analyses have been performed to indicate particular sensitivities of the predictions with respect to the empirical constants. Further computations for different geometries and different conditions and comparisons to corresponding experimental data are needed to further verify and establish the new transport equation.

Acknowledgments

The author appreciates stimulating discussions of this research with colleagues, especially K.N.C. Bray and N. Peters.

References

[1]Abraham, J., Williams, F.A. and Bracco, F.V., "A Discussion of Turbulent Flame Structure in Premixed Charges," SAE paper 850345, 1985.

[2]Bray, K. N. C., Libby, P. A. and Moss, J. B., "Flamelet Crossing Frequencies and Mean Reaction Rates in Premixed Turbulent Combustion," Combust. Sci. and Tech., Vol 41, 1984, pp. 143-172.

[3]Bray, K. N. C., Libby, P. A. and Moss, J. B., "Scalar Length Scale Variations in Premixed Turbulent" Proceedings of the Twentieth Symposium (International) on Combustion, The Combustion Institute, Pittsburgh, PA, 1984, pp. 421-427.

[4]Champion, M. and Libby, P.A., "Turbulent Premixed Combustion in a Boundary Layer," Combust. Flame, Vol 38, 1984, pp. 267-291.

[5]Bray, K. N. C., Champion, M., Dave, N. and Libby, P. A., "On the Thermochemistry of Premixed Turbulent Combustion in Variable Enthalpy Systems," Combust. Sci. and Tech., Vol 46, 1986, pp. 31-43.

[6]Pope, S. B., "Turbulent Premixed Flames," Annual Review of Fluid Mechanics, Vol. 19, 1987, pp. 237-270.

[7]Pope, S. B. and Cheng, W.K., "Stochastic Flamelet Model of Turbulent Premixed Combustion," Proceedings of the Twentysecond Symposium (International) on Combustion, The Combustion Institute, Pittsburgh, PA, 1988, pp. 781-789.

[8]Libby, P. A., "Theory of Normal Premixed Turbulent Flames Revisited," Progress in Energy and Combustion Science, Vol. 11, 1985 pp. 83-96.

[9]Anand, M.S. and Pope, S.B., "Calculations of Premixed Turbulent Flames by PDF Methods," Combust. Flame, Vol 67, 1987, pp. 127-142.

[10]Bray, K. N. C., Champion, M., Libby, P. A., private communication, 1988.

[11]Franke, Ch. and Peters N., "Flammenoberfläche und Brenngeschwindigkeit von turbulenten Vormischflammen,", Proceedings of Sonderforschungsbereich 224, RWTH Aachen, October 1988, pp. 126-147.

[12]Peters N., private communication, 1989.

[13]Bray, K.N.C., "Turbulent Flows with Premixed Reactants," Turbulent Reacting Flows, edited by P.A. Libby and F.A. Williams, Springer, Berlin, 1980, pp. 115-183.

[14]Meunier, S., Champion, M. and Bellet, J.C., "Premixed Combustion in a Turbulent Boundary Layer with Injection," Combust. Flame, Vol. 50, 1983, pp. 231-238.

[15]Pope, S.B., "An Explanation of the Turbulent Round-Jet/Plane-Jet Anomaly," AIAA Journal, Vol. 16, 1978, pp. 279-281.

[16]Kuo, K.K. and Keswani, S.T., "A Comprehensive Theoretical Model for Carbon-Carbon Composite Nozzle Recession," Combust. Sci. and Tech., Vol 42, 1986, pp. 145-164.

[17]Gouldin, F.C., Bray, K.N.C. and Chen, J.-Y, "Chemical Closure Model for Fractal Flamelets," Combust. Flame, Vol. 77, 1989, pp. 2241-259.

[18]Rogg, B., "Response and Flamelet Structure of Stretched Premixed Methane -Air Flames," Combust. Flame, Vol. 73, 1988, pp. 45-65.

[19]Rogg, B., "Numerical Analysis of Strained Premixed CH_4- Air Flames with Detailed Chemistry," Numerical and Applied Mathematics, edited by W.F. Ames, Baltzer, Switzerland, 1989, pp. 159-167.

[20]Srivatsa, S.K., "Cham Computer Code 201, CORA2 – A Computer Code for Axi-Symmetrical Combustion Chambers," CHAM Technical Report 201/1, 1977.

[21]Giovangigli, V. and Smooke, M. D., "Extinction of Strained Premixed Laminar Flames with Complex Chemistry," Combust. Sci. and Tech., Vol 53, 1987, p. 23.

[22]Fritsch, F.N. and Butland, J., "A Method for Constructing Local Monotone Piecewise Cubic Interpolants," SIAM Journal of Scientific and Statistic Computing, Vol. 5, 1984, pp. 300-304.

[23]Deuflhard, P., "A Modified Newton Method for the Solution of Ill-Conditioned Systems of Nonlinear Equations with Application to Multiple Shooting," Numerische Mathematik, Vol. 22, 1974, p. 289.

[24]Smooke, M. D., "An Error Estimate for the Modified Newton Method with Application to the Solution of Nonlinear Two-Point Boundary Value Problems," Journal Opt. Theory and Applications, Vol. 39, 1983, p. 489.

[25]Reuven, Y., Smooke, M.D. and Rabitz, H., "Sensitivity Analysis of One-Dimensional Mixed Initial-Boundary Value Problems: Application to Freely Propagating Premixed Laminar Flames," Yale University, Report ME-102-87, 1987.

Chapter V. Flame Propagation in Combustion Engines

Flame Imaging Studies of Flame Development in a SI Four-Stroke Engine

S. C. Bates*

General Motors Research Laboratories, Warren, Michigan

Abstract

The progressive stages of flame development in a spark-ignition (SI) four-stroke engine have been recorded in stereo flame luminosity images taken through a transparent liner and piston-top window. Flame evolution in each cycle is measured by superimposing early and later flame images.

Flame images are shown for the early, middle, and late stages of combustion. Image data are presented for three lean equivalence ratios, at 500 and 1000 rpm, and for one-and two-port intake, and average flame speeds are measured. The early flame kernel is shown to vary greatly in size, shape, and location, and this kernel is seen to have a major effect on combustion development by setting the basic flame shape throughout the cycle. The flame images permit various inferences to be drawn about the in-cylinder flow. Variable swirl from cycle to cycle, and a large scale cross-flow for one-port intake flow have major effects on flame growth.

Flame shapes are correlated with mass burn fraction curves derived from cylinder pressure histories. Mass burn history is found to be determined by the early flame kernel size and shape together with the large-scale midcycle flame shape and its interaction with the combustion chamber shape.

Introduction

The study of combustion in spark-ignition four-stroke engines has been going on for over a hundred years. Because of the difficulty of performing high-speed diagnostics, engine

*Engine Research Department

modeling and experiments have been based primarily on global variables and cycle-averaged quantities. Current trends,[1] however, involve the use of advanced diagnostics to examine individual cycles or sets of cycles and how they affect overall engine performance.

Turbulent combustion in engines is a special case of both turbulence and combustion. The entire process takes place in a closed volume, with the piston as a rigid time-varying boundary that tends to drive the entire process at its own frequency. The turbulence in the cylinder decays after intake valve closing and is complicated by a variety of mean motions (swirl or tumble) and secondary flows that can be stable or unstable. All of these flows are affected by the time-varying volume; some in different ways than others. Furthermore, combustion characteristics that are deemed desirable can be quite different at different engine operating conditions, especially idle vs high speed.

As external constraints such as pollution controls and fuel economy standards increase, there is greater need for detailed understanding of combustion and improved design of the chamber. Answers may be supplied by optical diagnostics that can investigate the physical processes taking place in the cylinder in each cycle.

Flame imaging in engines has been done for over 50 years.[2] Its past use as a combustion diagnostic has been limited by restricted optical access and the many experimental inadequacies of using and analyzing film; these obstacles are now fading. In the work described here, light-amplified flame imaging in a realistic engine with nearly total optical access below the head[3,4] provides a means for investigating the details of flame development and overall combustion behavior of the engine.

Equipment and Procedure

Since turbulent combustion in engines is dependent on many factors in unknown ways, as many engine and system parameters as possible should be specified to define adequately the physical context of the problem. Subtle details often have a major influence on combustion progress for some operating conditions.

Engine Characteristics

The flame image data of this work were taken from a single-cylinder visualization engine described in detail elsewhere.[3,5] The major characteristics of this engine are

Table 1 Engine parameters

Bore & stroke:	92.0 mm	Piston:	Extended Flat-topped
Connecting rod length:	191.6 mm	Fuel:	Propane
Compression ratio, C_r:	9:1	Spark:	Duration 1.5 ms; Energy 75 mj
rpm:	4000 max	Head:	Three-valve, pent-roof

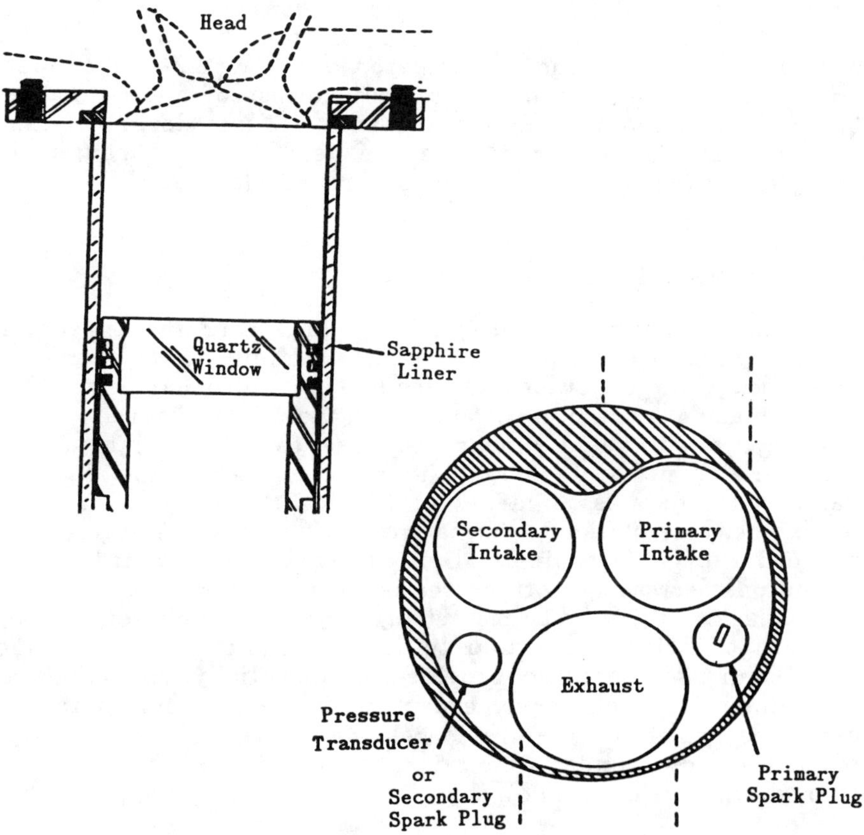

Fig. 1 Engine geometry: a) side view; b) head geometry, bottom view.

representative of a modern, four-stroke, spark-ignition internal-combustion engine. Global parameters for the engine are given in Table 1. The engine has a sapphire liner with a flat quartz piston-top window for optical access.

The engine is instrumented with a manifold absolute pressure (MAP) sensor, a piezoelectric cylinder pressure sensor, and a shaft angle encoder with 0.1-deg accuracy. Premixed propane and air fuels combustion for 50 cycles as the engine speed is held constant by a dynamometer.

Shown in Fig. 1a and 1b are the pent-roof head geometry and the transparent liner that extends to the bottom of the head. For the data presented in this work, the secondary spark plug has been replaced with a pressure transducer. The peripheral spark plug is oriented vertically (parallel to the cylinder axis) and is electrically driven by a standard High Energy Ignition (HEI) system. The spark gap is 8-9 mm above the bottom of the head. The piston window has a 69.8-mm smallest diameter compared with the 92-mm cylinder bore. Combustion chamber crevices consist of a 0.6-mm gap 5.0 mm deep around the piston, and a 3-mm-deep, 0.1-mm-wide gap at the bore radius in the head. Flow through the secondary valve is controlled by a throttle plate in the head.

Flame Imaging Diagnostic

Flame behavior is studied by analyzing instántaneous stereo images of flame luminosity taken by two image-intensified Charge-Coupled Device (CCD) video cameras. The flame imaging system is described in detail elsewhere.[3,5] The imaging system has a linear response to visible light and a gain variable up to 10,000. The basic format of the digitized flame image data is an array of 512 x 512 pixels, with an 8-bit intensity resolution. Spatial resolution of the images is on the order of a few pixels, limited by the image intensifier resolution, decreasing with increasing gain.

Camera optics and lens settings have been chosen to give a depth of field comparable to the dimensions of the flame. The 5-mm-thick sapphire provides an essentially distortion-free side view of the chamber, except within a few millimeters of each side, which is compressed relative to its real extent. The outer 1 cm of cylinder radius is blocked from the bottom view through the piston window by the piston shell surrounding the window.

Within the framework of 30 frames/s standard video framing rates, flame development is recorded during a single cycle at normal engine speeds by making two short exposures separated in time on one frame. Single-image exposures as short as a few microseconds are obtained by gating the image intensifier. Since the flame front radiation is much greater than the comparatively uniform and weak radiation from the burned gases, the early and later images can be easily separated in most cases.

Experiment Operation

The flame imaging results reported here were taken from sets of firing runs. The engine first reaches a motoring equilibrium at the desired speed; then the tank containing a calibrated propane-air mixture is switched into the intake manifold line with a pair of valves, and the gas is drawn though the engine. The video recorders that store the flame imaging data are turned on before the run, and the computer is triggered to take 50 sequential cycles of pressure data at the start of the run. The intake tank is evacuated and refilled between runs.

Data were taken with both valves operating normally but with the secondary port throttled off at the entrance to the head, giving a mean swirl ratio of 1. Operation referred to as "two-port" differs from this only in that this throttle plate is opened, resulting in very little swirl.

The start and the duration of the camera exposures are measured in degrees crank-angle (deg-CA) to an accuracy of 0.1 deg. This timing is preset by a thumbwheel controller at the start of each run.

Data Processing

The engine cylinder pressure and liner temperature are monitored by an IBM PC/AT. Pressure data from each run are stored and postprocessed to give mass burned progress using techniques similar to those described in Rassweiler and Withrow.[2]

Flame images are analyzed by a commercial image processing package on a high-resolution (1024 x 1280), 256-color IBM PC/AT monitor. Simple contrast enhancement is used for maximum detection of intensity details in the processed flame image. Threshholding allows the location of the flame front to be found automatically from a series of images. Many images can be manipulated at one time as arrays fixed in space. This permits superposition of the engine geometry outline onto flame images and the direct comparison of flame images from different engine cycles. A graphics editor is used to form accurate line drawings from reference images, and these line images are added to the flame images of a sequence of cycles without any adjustment of their relative location.

Experimental Advances

A number of factors contribute to the ability to provide new data on flame development in engines.

Extensive optical access is the most important capability. Easy simultaneous stereo imaging gives a way to decipher the complex three-dimensional behavior of the flow and combustion in the cylinder.

Next, the use of video makes possible the screening of large amounts of data that reflect both the turbulent nature of combustion processes and the large number of interacting parameters that control engine operation. The speed and power of inexpensive image processing and personal computers then permit manipulation of a significant part of these data.

Digital imaging also provides accurate and repeatable analysis of data to place the images easily and accurately in their spatial context relative to a prerecorded fixed engine geometry and to explore image intensity distribution for structure and relative magnitude.

Finally, flame radiation amplification permits low-light-level lean and very lean flames to be examined, an important operating regime not readily available for study previously.

Flame Development Physics in SI Engines

A spark-ignition engine places the naturally complex process of turbulent combustion in a closed, time-varying volume, changing both the turbulence and the combustion. Combustion progress depends on the temporal and spatial interaction of many geometric, flow, chemical, and thermal parameters. Flame images give direct information about many of these processes, but some, such as flame chemistry, can be studied only indirectly.

One framework for examining flame development in an engine is to study the appropriate processes in the context of combustion progress regimes.[4] These are: 1) spark initiation and very early flame kernel growth, 2) early flame kernel growth, 3) flame expansion, and 4) quasistationary combustion. Different physical effects are dominant in the different regimes, and these effects serve as a focus for exploring the dependence of flame development on the various parameters.

There are also dominant input and process variables that determine flame development in the engine. A flowchart of these variables is shown in Fig. 2. The blocks are arranged so that causality flows left to right as indicated by arrows. For example, the gas expansion component of the flame propagation velocity is fixed by the properties of the unburned mixture in front of the flame. Variables set externally during

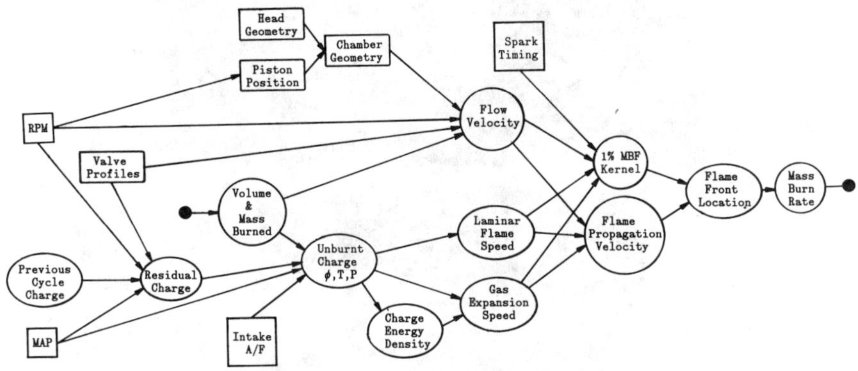

Fig. 2 Combustion physics flowchart.

operation are indicated by squares, whereas engine geometry variables (more difficult to change) are shown as horizontal rectangles. Primary process variables that describe flame development during the cycle are indicated by circles. Dependent or subsidiary parameters that partially control the development of these engine variables are shown as ellipses.

Flame development is determined by the temporal and spatial history of the propagation velocity of the flame. The progress of the flame is measured in terms of the amount of mixture burned and described in terms of volume or mass. The propagation velocity has contributions from the gas expansion behind the flame, the laminar flame speed, the turbulent flame speed and, in the fixed reference frame of the engine cylinder, the convection of the flame by the large-scale flow. The complexity of the turbulent combustion arises from the interaction between all of these components.

The block diagram illustrates the physical processes by labeling the controlling parameters. The solid dot indicates where the process connects as a closed loop. The most important basic variables for flame development are the state of the unburned charge and the in-cylinder flowfield, in the context of the instantaneous combustion chamber geometry. Some parameters are shown because of the importance of their effect on flame development or because a single characteristic determines the flame behavior beyond a certain stage. An example of such a characteristic is the inclusion of a 1% mass burned fraction (MBF) flame kernel category. This kernel sets a critical junction for the future development of the flame and the combustion itself, as discussed later in this paper and in a previous work on cyclic variation.[4]

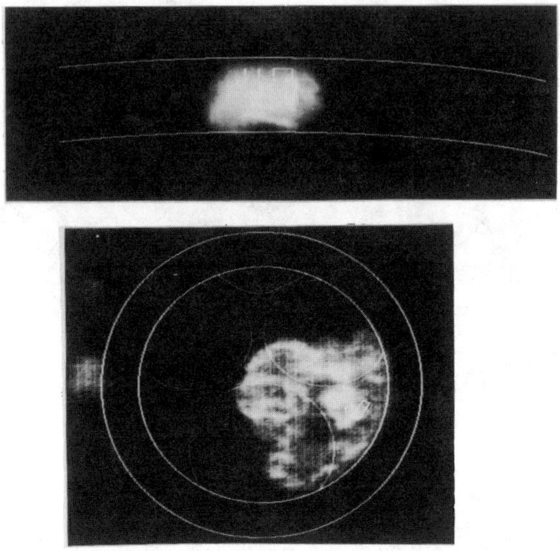

Fig. 3 Very lean early flame kernel growth without flow convection, one-port flow, ϕ_i = 0.60, spark at 60 deg-CA BTDC, MAP = 0.75 atm, 500 rpm: a) side view at 57 deg-CA BTDC and 20 deg-CA BTDC; b) bottom view at 20 deg-CA BTDC and TDC.

Two geometric effects dominate the behavior in this engine: the peripheral spark plug and the swirl that comes from the port configuration in use. The peripheral spark implies a long burn across the chamber and a flame geometry that is determined by only the side of the early flame kernel facing the center of the chamber. Also, swirl velocities are higher near the wall and larger relative to the flame speed so that, at this location the spark is more sensitive to swirl. The low swirl number for the configurations discussed in this work implies that the in-cylinder flow will, in general, contain variable large-scale motions rather than an overall rotation of the flow and flame from swirl.

An additional complexity that certainly exists in real engines is mixture nonuniformity. This has been eliminated in the present work by using premixed propane and air and will not be considered.

A velocity diagnostic has also been developed specifically for this engine to measure the in-cylinder flow. This diagnostic measures velocity by detecting the displacement of marked lines in the flow and is described in detail elsewhere.[6] It has not been used in combination with firing experiments at this writing.

Flame Image Data

The data and discussion that follow are organized in parallel with the combustion progress regimes given earlier. As combustion proceeds from ignition to extinction, flame history can proceed along many paths. These paths can diverge at different times because of separate effects, or can recombine. For instance, a flame may initially grow rapidly as a result of flow past the spark plug during ignition, then slow down as a result of large-scale counterflow in the rest of the chamber. For this reason, it is more important to examine the separate processes that affect flame development rather than trace the history of one flame through one cycle. The interwoven paths also lead to a great deal of confusion when one attempts to correlate cause and effect using overall engine statistical behavior.

Individual samples have been chosen from a large number of engine tests to provide evidence of different effects. These results are examples representative of a type of behavior seen repeatably in the engine. The dispersion in flame development paths has been examined in a complementary paper on cyclic variation in this engine.[4] This paper will be used as a reference for facts about flame development that have emerged from the study of entire sets of data.

It must be emphasized that the flame images represent line-of-sight integrals of the radiation from the flame. Thick flame fronts in the images do not necessarily imply a thick reaction zone. However, as the piston passes through the top-dead-center (TDC) position, the height of the chamber is much less than its width, forcing major variations in the mid- and late-cycle flame shape to be primarily radial rather than axial. This is not true of an engine flame in general; the early kernel, a different location of the spark plug, or a different combustion chamber shape will show three-dimensional flame growth.

Most of the flame images have line drawings of the engine derived from reference images that are superimposed onto the data appropriate to the camera view. For example, in views through the piston window such as Fig. 3b, the outer circle represents the cylinder bore, whereas the inner circle shows the extent of the actual image as defined by the piston window. At the top of the view through the window are shown the intake valves on one slope of the pent-roof and, at the bottom, the exhaust valve on the other slope. The position of the spark plug ground strap is shown by the outline at the lower right, with the electrode at its lower end. The protruding shape between the intake valves is the bottom flat of the head

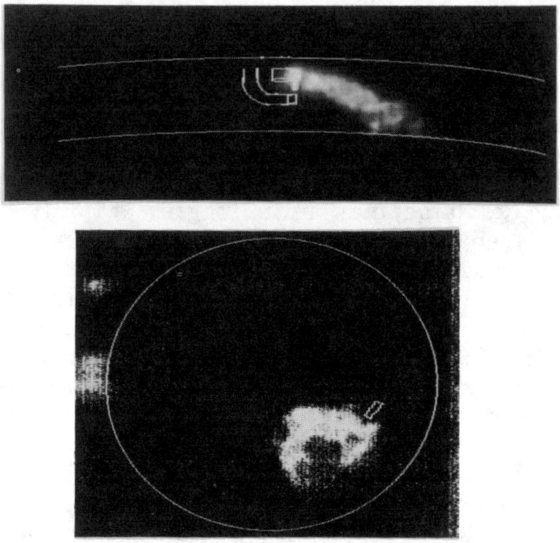

Fig. 4 Very lean early flame kernel growth, an example of flameholding, one-port flow, ϕ_i = 0.60, spark at 60 deg-CA BTDC, MAP = 0.75 atm, 500 rpm: a) side view at 57 deg-CA BTDC and 20 deg-CA BTDC; b) bottom view at 20 deg-CA BTDC and TDC.

protruding into the window view. The head pattern is identical to that given in the schematic of Fig. 1b.

Flame development in one cycle is examined by taking two short exposures during the interval of one video frame. As shown in the bottom view of Fig. 3b, the early exposure is the bright section around the spark plug, whereas the later exposure captures the flame halfway across the chamber. Behind the later flame, the burned-gas radiation is low and uniform, allowing the earlier flame image to be easily examined separately through image processing.

Spark and Early Flame Kernel

The beginning of turbulent combustion appears to be the most crucial of the stages of flame development.[7,8] In this stage, the small laminar flame initiated by the spark becomes a fully turbulent flame.[9,10] The basic physical process of interest is the interaction of the local flow with the flame.

For most current spark-ignition automotive engines (and the engine used in this work), combustion is initiated by a 1-2-ms-duration, high-voltage inductive discharge across a spark gap flush-mounted in the cylinder head. The spark plug consists of a central electrode and a narrow ground strap welded at the circumference and ending just beyond the plug

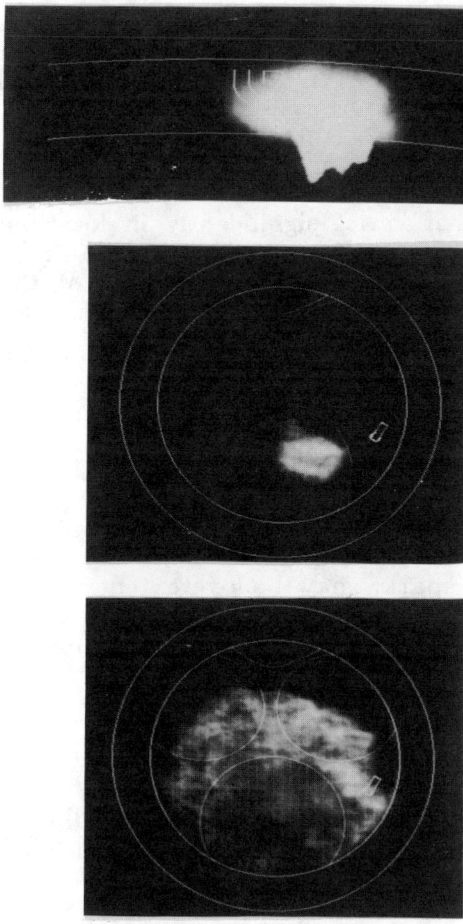

Fig. 5 Very lean early flame kernel growth showing blowoff due to convection in swirl direction, one-port flow, $\phi_i = 0.60$, spark at 60 deg-CA BTDC, MAP = 0.75 atm, 500 rpm: a) side view at 20 deg-CA BTDC; b) bottom view at 20 deg-CA BTDC; c) bottom view at TDC.

axis. Many other methods of ignition exist and can result in very different flame development.

The important characteristics of spark ignition are its energy, duration, and geometry.[9] It has been shown that the energy of the spark does not have a major impact on combustion if that energy is well beyond a known threshhold value[9]; current spark systems meet this requirement. Spark duration has effects on flame development that arise both from the total length of time that initiation energy exists and from the time that spark-flow interaction takes place. Spark-plug geometry is important with respect to its overall location in the chamber[11,12] as the center of the expanding flame in the context of the whole combustion chamber shape. It is also important through the orientation of the ground strap, which can act as a flameholder or a flow obstruction, depending on the flow direction.

The spark itself varies significantly in location at the plug both from cycle to cycle and during the discharge, with smaller variations in duration and energy. Without flow, the spark moves around the full area of the electrode from cycle to cycle,[4] resulting in important variations in the distance from the discharge to the ground strap base. This in turn causes a change in the interaction between the strap wake and flame propagation.

For flow velocities typical of current engines, flow through the gap often carries the low-inertia ionized gas column away from its original location to form a loop in the direction of the flow.[4] Flow in the engine is such that these loops may occur during one spark, may be overextended so that the arc restrikes, or may not occur at all, depending on the local flow at the time of the spark. Thus, the ignition source for the flames is often varying in space and time.

The early flame kernel grows from the spark to become a fully turbulent flame. During this growth, it changes its smallest dimension by over a factor of 10, from a length comparable to the smallest turbulence scale to a length that begins to be a significant fraction of the cylinder bore. This growth is determined by the laminar flame speed and the flow in the vicinity of the flame. At the present time, changes in the early flame kernel are thought to account for most combustion variation between cycles in spark-ignition engine combustion.[7,8] The three-dimensional shape and location of the early kernel are both important in this process.

The interaction of the flow and flame is most obvious at very lean mixtures, where the laminar flame speed is low enough so that the flame speed and flow speed are comparable. For properly phased combustion, these conditions in this engine are close to the misfire limit. Although misfires are a common and important engine phenomenon, they will not be considered here, where the emphasis is on flame development.

In an engine with low swirl (such as in this work), the flow at the spark plug at the time of the spark contains fine-scale turbulence and a series of larger-scale motions that are convective on the scale of the spark plug. The initial flame can grow while convective flows approach the plug in different directions or when there happens to be no convective flow at all.

Figures 3-6 illustrate early flow-flame interaction. Data were taken at an intake equivalence ratio of approximately 0.6, with a spark initiated at 60 deg-CA before TDC (BTDC). The manifold absolute pressure (MAP) was 0.75 atm at 500 rpm with one-port flow. The equivalence ratio is not constant from cycle to cycle because of varying amounts of residual fraction after partial burns or misfires. Slant side-view exposures were taken at 57 deg-CA BTDC for a duration of 0.1 deg-CA and at 20 deg-CA BTDC for a duration of 2.0 deg-CA, whereas bottom-view exposures were taken at 20 deg-CA BTDC for a duration of 2.0 deg-CA and at TDC for a duration of 1.0 deg-CA. The earlier image in the slant side view shows the spark at approximately the middle of its duration.

Orientation of the spark plug is indicated by the outline of the ground strap in each bottom view; the plug was inserted so that the open end of the strap is at the lower end in the images. Since the spark is 1.5 ms long, it lasts for approximately 4.5 deg-ca CA at 500 rpm. The flame images have been enhanced for maximum contrast, and the background cylinder patterns have been superimposed. The bottom line in the side view indicates piston position at 20 deg-CA BTDC.

The case of minimal convective flow is shown in Fig. 3a and 3b. The first exposures in both views show the early kernel to be localized at the spark plug and expanding from there. Some flow at a later time is indicated by the distorted flame shape at TDC in Fig. 3b. As expected, this case is the exception; there is usually a convective flow of some sort at the plug.

Figures 4a and 4b show a case in which the convective flow is aligned with the ground strap. Here the flame is able to propagate back into the stagnation region behind the ground strap connection, which then acts as a flameholder. The flame is shown stretching out from the plug, with the flow as continuous light from the spark to the piston top, expanding sidewise at the local turbulent flame speed. At higher flow velocities (high swirl), the few-millisecond spark permits continuous flame ignition and a similar extended flame.[13]

Two different degrees of crossflow are seen in Figs. 5 and 6. Partial crossflow is shown in Figs. 5a-5c, where the early flame kernel has blown off the spark plug into a region of low turbulence so that initially it grows slowly. Later, at TDC, the flame area is much larger than the other examples because of the higher equivalence ratio in this cycle. The richer mixture is also the cause of the saturated exposure in Fig. 5a. For this type of camera, image saturation causes vertical

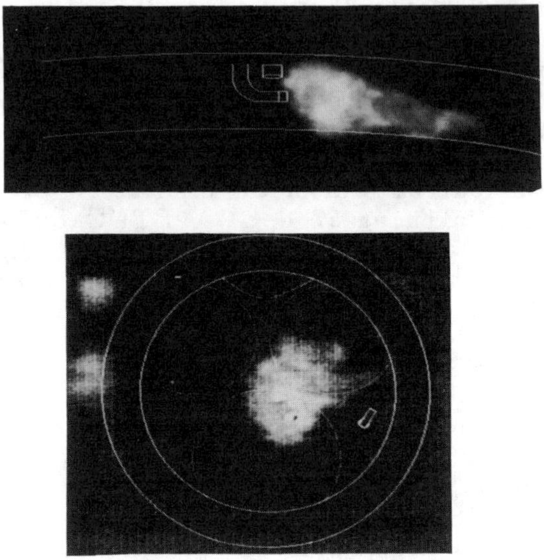

Fig. 6 Very lean early flame kernel growth showing blowoff from convection in toward the cylinder center, one-port flow, $\phi_i = 0.60$, spark at 60 deg-CA BTDC, MAP = 0.75 atm, 500 rpm: a) side view at 20 deg-CA BTDC; b) bottom view at 20 deg-CA BTDC and TDC.

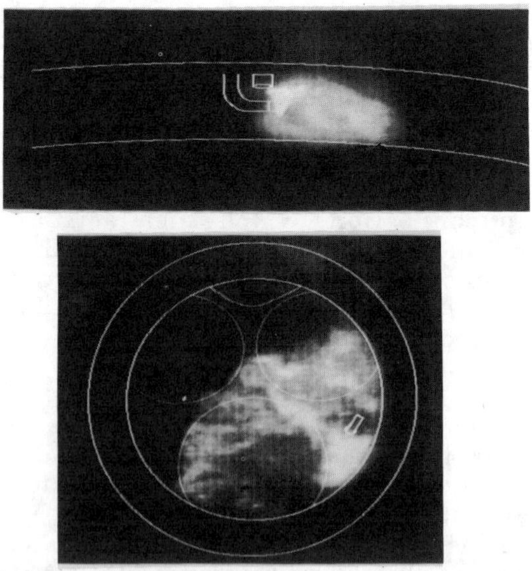

Fig. 7 Baseline flame behavior, one-port flow, $\phi_i = 0.75$, spark at 40 deg-CA BTDC, MAP = 0.75 atm, 500 rpm: a) side view at 37 deg-CA BTDC and 20 deg-CA BTDC; b) bottom view at 20 deg-CA BTDC and TDC.

blooming, - the white area extending below the piston line. The bottom view exposures of Figs. 5b and 5c happened to fall on separate video frames. Figures 6a and 6b show a case in which the crossflow is at nearly a right angle to the ground strap, leading to a growing kernel at the center of the chamber. The spark is not shown in Fig. 6a.

The effect of the flow on flame development decreases as the equivalence ratio increases, but it remains important. As the mixture becomes progressively richer, both the laminar flame speed and gas expansion speed increase, while the

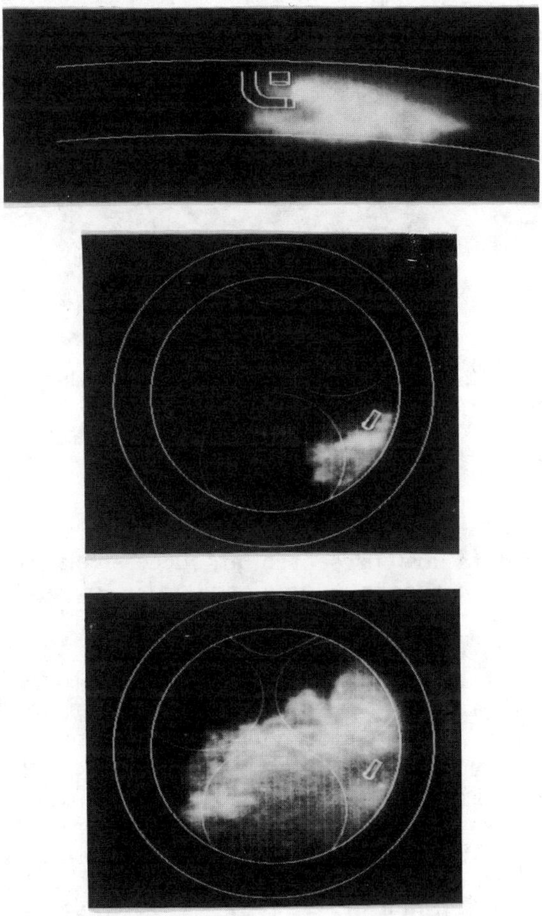

Fig. 8 Images of flow causing a spark loop and a large early flame kernel, one-port flow, spark at 40 deg-CA BTDC, $\phi_i = 0.75$, 500 rpm: a) side view at 37 deg-CA BTDC and 20 deg-CA BTDC; b) bottom view at 20 deg-CA BTDC; c) bottom view at TDC.

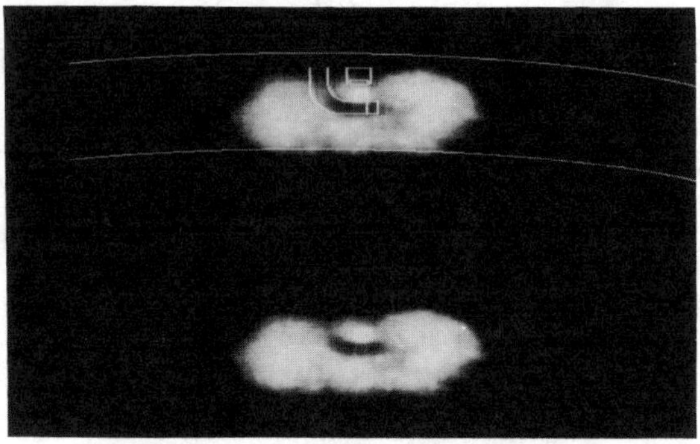

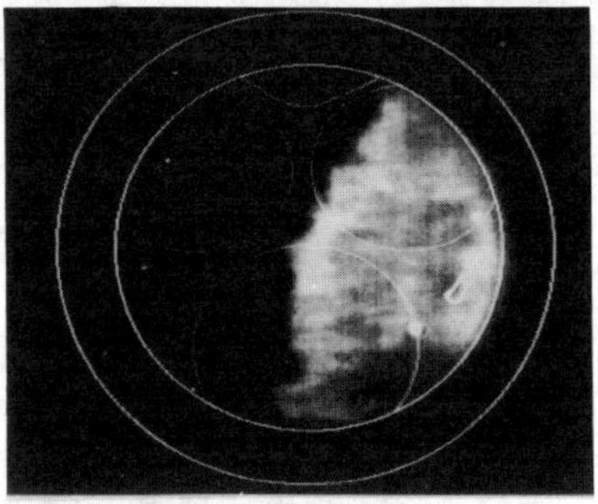

Fig. 9 Flame image data of flame blown toward the outer wall, one-port flow, spark at 40 deg-CA BTDC, ϕ_i = 0.75, 500 rpm: a) side view at 37 deg-CA BTDC and 20 deg-CA BTDC; b) bottom view at 20 deg-CA BTDC and TDC.

contribution of turbulence remains the same. To explore this dependence, data are presented at successively richer mixtures.

Data for an intake equivalence ratio of 0.75 are presented in Figs. 7-10. Spark timing was 40 deg-CA BTDC, which is MBT (mean spark advance for best torque) timing as indicated by a MBF of approximately 60% at 10 deg-CA after TDC (ATDC). Again, the MAP was 0.75 atm at 500 rpm with one-port flow. In the side view, the image consists of a 0.1-deg-CA long exposure 37 deg-CA BTDC, together with one at 20 deg-CA BTDC for a duration of 2.0 deg-CA. Superimposed exposures for the bottom view were taken at 20 deg-CA BTDC for a duration of 2.0 deg-CA and at TDC for a duration of 1.0 deg-CA.

Figure 7 shows data from a typical cycle. There is indication of some clockwise swirl, as shown by the early kernel displaced to the right of the plug in Fig. 7a. Fig. 7b shows continuing swirl at TDC from the rotation of the burned region with respect to the spark plug. This is

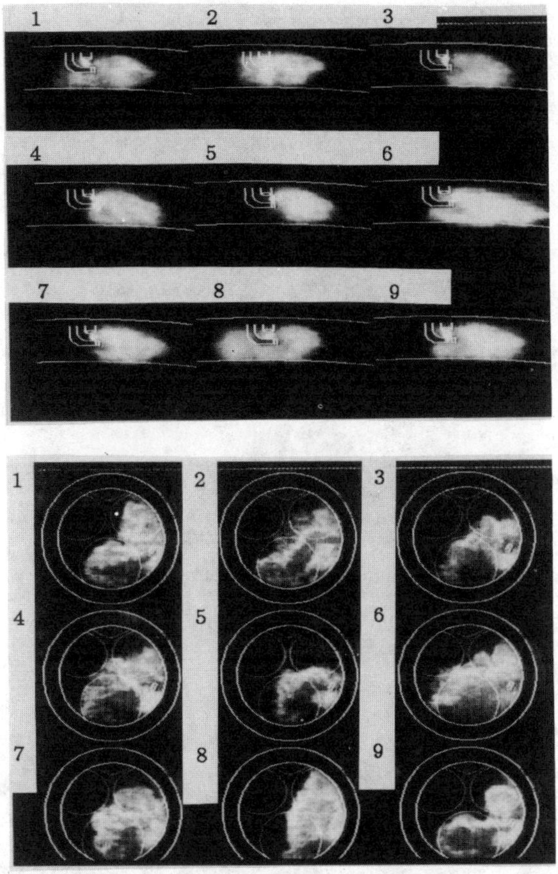

Fig. 10 Flame variation resulting from flow-flame interaction in nine consecutive cycles with background lines superimposed, one-port flow, spark at 40 deg-CA BTDC, ϕ_i = 0.75, 500 rpm: a) side view at 37 deg-CA BTDC and 20 deg-CA BTDC; b) bottom view at 20 deg-CA BTDC and TDC.

contrasted with the data of Fig. 8, where there is a much greater swirl at the time of the spark. The spark is blown out into a loop, as discussed earlier, and the flame is elongated as the spark plug apparently acts as a flame-holder to generate a much larger flame kernel in the same interval as the flame in Fig. 7.

Figure 9 shows a much more unusual case, in which the flame kernel has been blown toward the wall. Figure 9a shows the same side view of the early flame kernel with and without the line drawing, to emphasize the shadow cast by the ground strap. This shadow indicates that the flame is behind

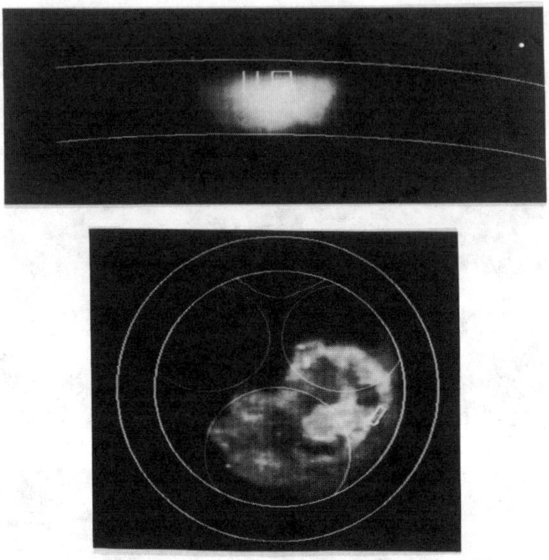

Fig. 11 Baseline flame behavior, two-port flow, spark at 40 deg-CA BTDC, $\phi_i = 0.75$, MAP = 0.75 atm, 500 rpm: a) side view at 37 deg-CA BTDC and 20 deg-CA BTDC; b) bottom view at 20 deg-CA BTDC and TDC.

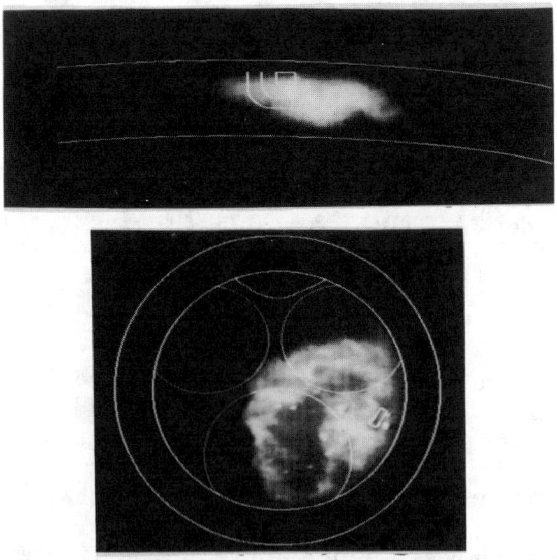

Fig. 12 Flame structure example for two-port flow, spark at 40 deg-CA BTDC, $\phi_i = 0.75$, MAP = 0.75 atm, 500 rpm: a) side view at 37 deg-CA BTDC and 20 deg-CA BTDC; b) bottom view at 20 deg-CA BTDC and TDC.

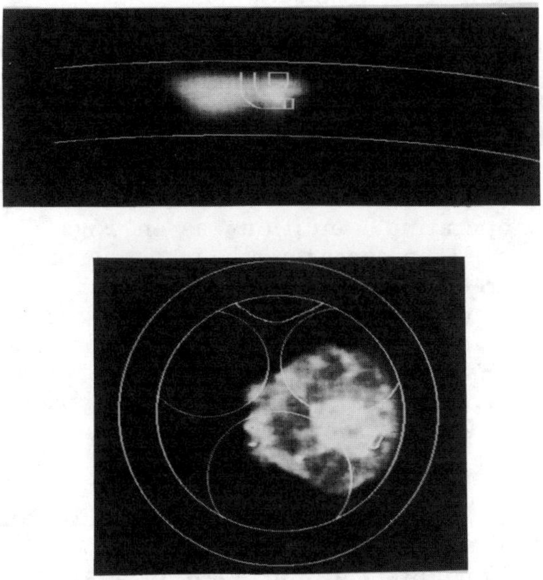

Fig. 13 Flame structure example for two-port flow, spark at 40 deg-CA BTDC, ϕ_i = 0.75, MAP = 0.75 atm, 500 rpm: a) side view at 37 deg-CA BTDC and 20 deg-CA BTDC; b) bottom view at 20 deg-CA BTDC and TDC.

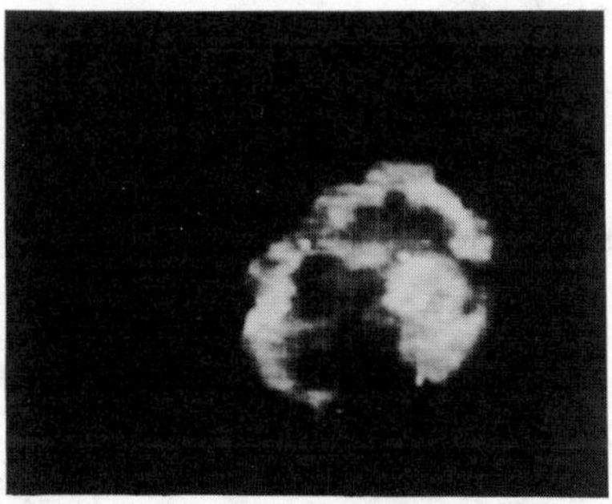

Fig. 14 Flame behavior for two-port flow, bottom view at 20 deg-CA BTDC and TDC, showing spark plug shadow, spark at 40 deg-CA BTDC, ϕ_i = 0.75, MAP = 0.75 atm, 500 rpm.

the strap and between it and the wall. For a delayed spark and consequently decreased clearance height at the time of spark, this location of the early flame kernel leads to a delay in flame propagation as the flame goes past the ground strap again and to a distortion in the flame throughout the cycle.[4]

A set of flame images from nine consecutive cycles from this one-port run is shown in Fig. 10 to indicate the typical shape, location, and structure of the flames. The data of Figs. 7-9 are taken from cycles 16, 18, and 20, respectively. Most of the cycles show the expected average swirl, seen as the rotation of the apparent center of flame expansion in the early and later bottom-view images.

The range of early kernel behavior for two-port operation is shown in Figs. 11-16 with the same camera settings and lean engine operating conditions as in Figs. 7-9. Figure 11 shows a typical case characteristic of the lack of swirl for this intake configuration; the early kernel is centered on the spark plug and expands from there. Figure 12a shows a kernel with a shape that indicates the interaction of the flame with intermediate-scale turbulence. Figure 13a shows a spark and flame kernel blown back into the ground strap. The bottom-view image of Fig. 14 shows that the kernel often develops in the top of the head, not obscuring the ground strap with radiation as in Fig. 11b. A similar dark area at the location of the ground strap is shown in Figs. 12b and 13b but is obscured by the lines indicating the strap.

Figure 15 shows the side view of nine consecutive cycles of the early flame kernel for the two-port configuration to show how the varying flow from cycle to cycle changes the flame. The kernels vary in size by more than a factor of 2 and can have very convoluted shapes. Figure 16 shows these flames with background patterns superimposed. Note that the early kernel motion is never toward the wall; there is never a shadow of the ground strap.

To show the behavior of flames closer to a stoichiometric mixture, Figs. 17 and 18 show flame images for one-port flow and an equivalence ratio of 0.90. The spark was moved to 30 deg-CA BTDC, and the spark exposure to 27 deg-CA, but otherwise conditions are the same as before.

Figure 17 shows a typical pair of flame images. The early flame kernel is the same size as before, indicating a much higher flame speed since the interval between the spark and the exposure has been halved. The same nine consecutive cycles are shown with and without background patterns in Figs. 18 and 19 at these conditions. The early kernel shape has diverged from the spark with minor shape perturbations, and there has been some movement of the center both radially (strap shadow) and circumferentially.

Figures 20-22 show flame image data for two-port intake flow for $\phi_i = 0.90$, while other parameters remain the same. Although a typical flame, as shown in Fig. 20, is very similar to the one-port case, the consecutive cycle images of Figs. 21a and 22a show a greater variation in flame location, shape, and size. These early flame kernels have not grown axially enough to eliminate the shadow of the ground strap, as shown in the

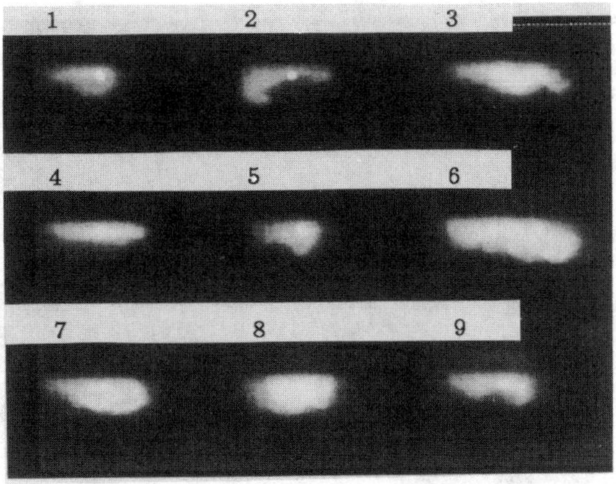

Fig. 15 Early flame kernel; variation due to flow-flame interaction, side view for nine consecutive cycles, two-port flow, spark at 40 deg-CA BTDC, exposures at 37 deg-CA BTDC and 20 deg-CA BTDC, $\phi_i = 0.75$, 500 rpm.

early bottom-view exposures in Fig. 21b, in the same manner as the $\phi = 0.75$ data.

Another major parameter that has an impact on the early flow-flame interaction is engine speed. Again, to emphasize flow effects, data were taken at $\phi_i = 0.75$ and conditions similar to those giving the data on Figs. 7-10. Engine speed was increased from 500 to 1000 rpm.

Figures 23-26 show one-port data for these flames. A large early kernel similar to that of Fig. 8 at 500 rpm is shown in Fig. 23a. This demonstrates that the early flow-flame interaction remains similar while time scales are halved and flow velocities are roughly doubled. Fig. 24a shows a case in which the flame image shows scale variations also very similar to those at 500 rpm. The consecutive cycle images shown in Figs. 25 and 26 confirm the similarity.

Figures 27-30 show the same similarity of behavior at different engine speeds for the two-port configuration. A large early kernel is shown in Fig. 27a, and Fig. 28a shows kernel scales at 1000 rpm. Consecutive cycle images shown in Figs. 29 and 30 again show the basic similarity of the images at the different speeds

Midcycle Burn

The midcycle burn stage includes most of the volume burned in the cylinder, while much of the mass remains unburned near the wall. The flame image data taken at TDC

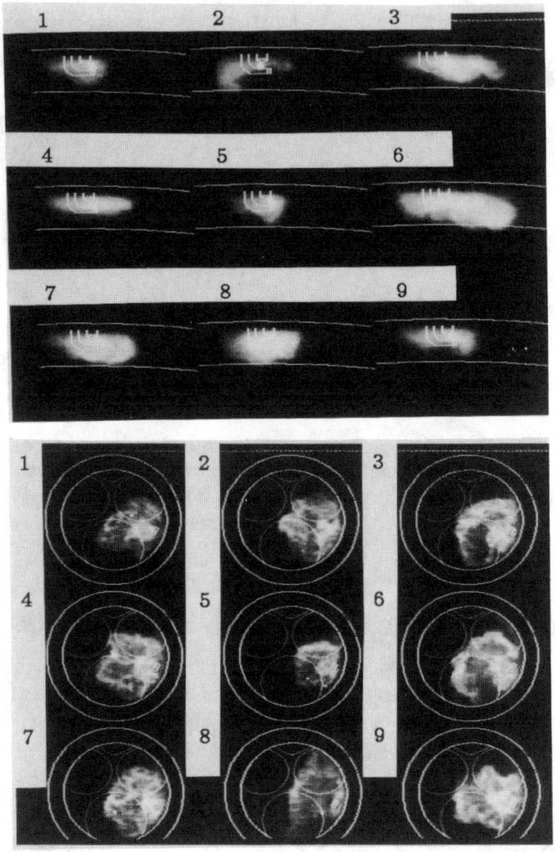

Fig. 16 Flame variation due to flow-flame interaction for nine consecutive cycles with background lines superimposed, two-port flow, spark at 40 deg-CA BTDC, ϕ_i = 0.75, 500 rpm: a) side view at 37 deg-CA BTDC and 20 deg-CA BTDC; b) bottom view at 20 deg-CA BTDC and TDC.

represents only 20-40% mass burned, depending on the equivalence ratio. This burn stage includes the major interactions between the flame and the combustion chamber shape, as well as between the flame and flows that are on the scale of the cylinder size.

The primary characteristic of the midcycle burn is the shape of the flame front. This shape has a mean large-scale curvature with superimposed convolutions representative of turbulence of progressively smaller scale down to the Kolmogorov length scale.

The mean large-scale curvature of the flame is set by the shape of the side of the early kernel facing the chamber,

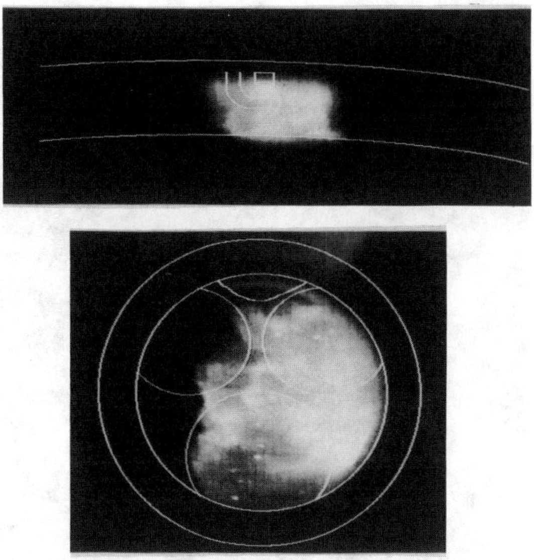

Fig. 17 Baseline flame behavior, one-port flow, spark at 30 deg-CA BTDC, ϕ_i = 0.90, MAP = 0.75 atm, 500 rpm: a) side view at 27 deg-CA BTDC and 20 deg-CA BTDC; b) bottom view at 20 deg-CA BTDC and TDC.

together with interaction with large-scale flows in the cylinder.[3,4] This overall shape similarity is also shown in the bottom-view figures presented here (e.g. Figs. 7-9, 11-13), which compare the early and later flame shapes. The cause is gas expansion behind the flame, which dominates the turbulent flame speed in the midcycle burn stage. The data show this to be true with changes in flow configuration, equivalence ratio, and engine speed. Turbulence and progression of the flame normal to its surface tend to smooth minor convex and concave distortions in the flame front.

The large-scale midcycle flame shapes are characteristic of the intake configuration. As discussed in the previous work on cyclic variation[4] and seen again in the data here in Fig. 10b, the one-port configuration flames are round, indented, or of an intermediate shape. The two-port flames are basically round (Fig. 16b) but often elongated along the pent-roof axis (Fig. 16, cycle 1).

Much structure can be seen in the light intensity variations in all of the TDC bottom-view exposures in this work, characteristic of the flow in each cycle. The bright and dark patterns arise from varying orientation of the flame front. A front aligned with the camera direction appears bright relative to a front normal to the camera view, giving much detailed flame front geometry information.

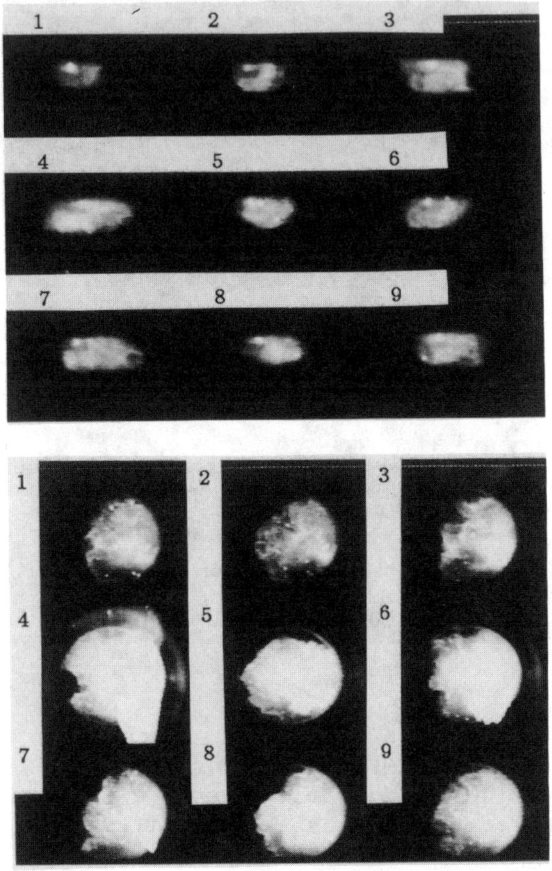

Fig. 18 Flame variation due to flow-flame interaction for nine consecutive cycles, one-port flow, spark at 30 deg-CA BTDC, ϕ_i = 0.90, 500 rpm: a) side view at 27 deg-CA BTDC and 20 deg-CA BTDC; b) bottom view at 20 deg-CA BTDC and TDC.

Overall rotation of the flame relative to the spark plug is indicative of swirl in the cycle. The one-port swirl number of 1 and the lack of swirl in the two-port configuration are shown consistently in the data. There is considerable variation in swirl from cycle to cycle, however.[4] This variable swirl is often the cause of the variations in the size of the early flame kernel; Fig. 8a shows a longer kernel in the swirl direction, and Fig. 8b shows continued rotation of the flame in the clockwise direction.

Combustion Completion

The final stage of combustion in a modern fast-burn engine takes place near the wall. The need for the flame to conform to the combustion chamber to achieve fast burns is well known.[11,14] This conformity is usually accomplished with a central spark plug and assumes spherically expanding flames. The interaction of variable flame location and shape with chamber shape to give extended burns is also known.[15,16] Two-plug chambers or less efficient chambers burn the end gases across a volume rather than through a thin shell. In all these cases, much of the mass burned occurs at constant or decreasing pressure. The flame propagation speed is then dominated by the turbulent flame speed rather than the gas expansion speed.

With a transparent liner, combustion completion close to the wall can be directly examined. Figures 31 and 32 show data from one-port and two-port lean firing runs. For these cycles, the manifold absolute pressure was 0.75 atm, and the intake equivalence ratio was 0.70 at 500 rpm. Slant side-view exposures were taken at 30 deg-CA BTDC for a duration of 2.0 deg-CA and at 50 deg-CA ATDC for a duration of 1.5 deg-CA, whereas bottom-view exposures were taken at 20 deg-CA BTDC for a duration of 2.0 deg-CA and at TDC for a duration of 1.0 deg-CA. The upper left corner of the view through the liner is blocked by a surface temperature probe, indicated by the superimposed lines. Spark timing was 40 deg-CA BTDC for the one-port configuration, and 50 deg-CA BTDC for the two-port configuration. The images have been enhanced for maximum contrast and have had the background cylinder patterns superimposed. The line drawing in the side view indicates piston position at 50 deg-CA ATDC.

As discussed elsewhere, the flame variation characteristic of operation at 500 rpm in this engine is shown in the bottom-view images. Combustion in cycles for the two-port case often result in partial burns resulting from the slow flame propagation rate.

The side-view images of combustion in each cycle show the location of continuing combustion at the wall. Fully burned regions result in uniform intensity, whereas radiation from the burning turbulent flame gives regions of highly structured intensity. The faint striations aligned with the piston-top line are caused by compression ring debris on the inner wall.

Figure 31b shows combustion completion for the one-port configuration, whereas Fig. 32b shows the same for the two-port case. These cycles were chosen for the extensive burning along the wall that shows flame structure. It is notable that the intensity structure shown in Figs. 31b and 32b is very similar. This is true for other cycles as well. These flame fronts are also almost all vertical.[4]

Almost all of the side-view images indicate that the flame front is vertical, supporting the earlier assumption that the bottom-view images are good representations of flame shape, even though they are axial averages.

Mass Burned Development

Mass burned history is the usual measure of combustion progress in an engine. However, both mass burned and pressure increase are integral parameters that result from the combined details of flame development in the cycle. Although these integral quantities are the measurable outputs of the process, and those most important for the engine, the fundamental process of interest is the flame development.

Flame areas have often been used to compute mass burned fractions (MBF), which can also be derived from the cylinder pressure.[2,15] Burning progress itself has been classified into various intervals. The interval of 0-1% MBF is classically known as ignition delay,[17] whereas 10-90% has been used as an indicator of primary combustion rate.[17]

In the previous work,[4] which classified the MBF curves relative to an average at 1%, 50%, and 90%, correlations were found between flame shape and mass burned rate. Fast early burns (short ignition delay) were found to correlate with large, very early kernels (before significant mass burned). Slower midcycle burns were found to correlate with flames that are flat on the average, relative to round flame fronts. The large-scale shape was found to explain MBF changes between cycles at one intake configuration and speed, even though changes in smaller-scale turbulence might be expected to have a major effect. This is consistent with the turbulence discussion that follows, showing that the intermediate scales do not change between cycles or even between operating conditions.

Mass burned fraction curves for the flames shown in Figs. 7-9 are presented in Fig. 33. The cycle in Fig. 7 is an average burn, whereas that shown in Fig. 8 is a fast early burn caused by the large early kernel, as shown in Fig. 8a.

The cycle shown in Fig. 9 and its corresponding MBF curve represent an unusually slow burn. This appears to be due to a series of effects. First, the spark-plug structure delays the flame that must pass it after having been blown toward the wall. The initial kernel also begins on the axis of the pent-roof and expands along this axis, as indicated by Fig. 9b. Most of the early kernels expand from the side of the pent-roof as a result of the swirl, so that expansion stretches

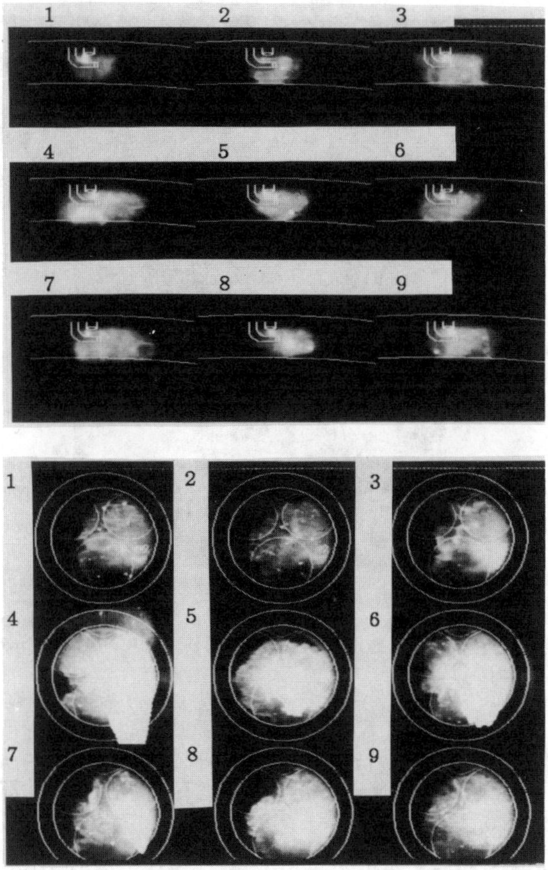

Fig. 19 Flame variation due to flow-flame interaction for nine consecutive cycles with background lines superimposed, one-port flow, spark at 30 deg-CA BTDC, $\phi_i = 0.90$, 500 rpm: a) side view at 27 deg-CA BTDC and 20 deg-CA BTDC; b) bottom view at 20 deg-CA BTDC and TDC.

the kernel relative to one in the center. The difference is one of a kernel expanding into either a contracting or expanding chamber. An expanding chamber volume accelerates growth by enlarging the flame surface. The case in Fig. 9 will be relatively slow because of this effect. Other more important lengthening effects on combustion are the flat flame front and the poor match of the flame shape to the chamber shape. All these effects combine to give a very slow mass burned, as indicated in Fig. 33.

Together, the effects just discussed explain all of the major divergences of the MBF from the average cycle, in a set of cycles. Adding the in-cylinder flow effects discussed later

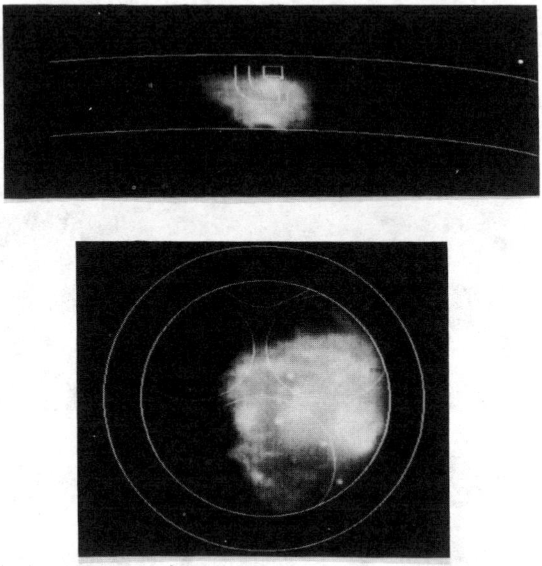

Fig. 20 Baseline flame behavior, two-port flow, spark at 30 deg-CA BTDC, $\phi_i = 0.90$, MAP = 0.75 atm, 500 rpm: a) side view at 27 deg-CA BTDC and 20 deg-CA BTDC; b) bottom view at 20 deg-CA BTDC and TDC.

and the effects of partial burns and misfires, an apparently complete picture is created of the major effects causing combustion progress in this engine.

These different effects mix to give various mass burned histories in a cycle. For instance, a large early kernel may cause a flat flame that gives an average overall burn length. Such effects would be extremely difficult to discover from an analysis of cycle pressure histories. Not included in this discussion are the effects of high swirl, or squish, which are not present in this engine. However, from the nature of the effects described here, high swirl and squish could be included through their effects on the phenomena already discussed. A more complicated effect also not considered is mixture nonuniformity.

Flame Speed Analysis

Flame speeds vary significantly for the operating conditions discussed previously. The flame propagation speed S_p increases in proportion to engine speed and turbulence level. Although the images presented capture the flame front at widely separated times, some useful flame speed calculations

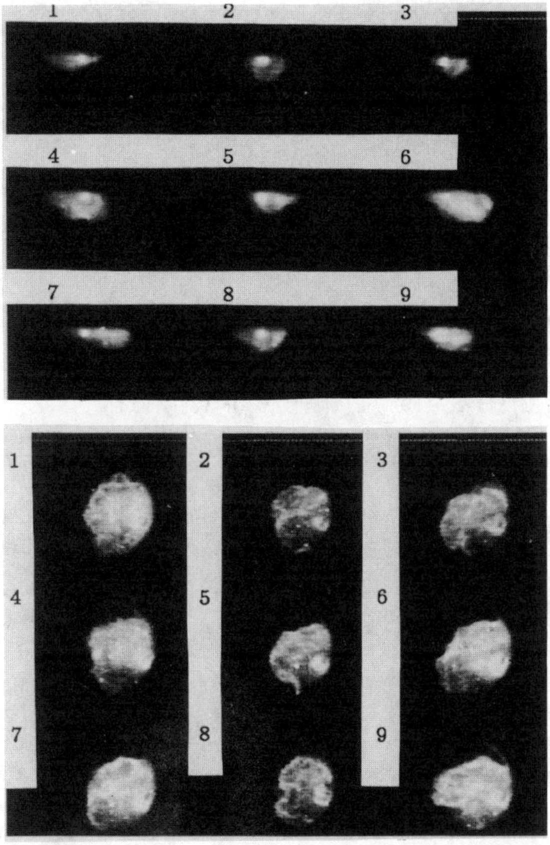

Fig. 21 Flame variation due to flow-flame interaction for nine consecutive cycles, two-port flow, spark at 30 deg-CA BTDC, ϕ_i = 0.90, 500 rpm: a) side view at 27 deg-CA BTDC and 20 deg-CA BTDC; b) bottom view at 20 deg-CA BTDC and TDC.

can be made taking advantage of the similarity in shape of the early and later flame kernels. This shape similarity permits approximate location of the same point in the flame front for the separate images and, therefore, measurement of its average displacement. Dividing by the appropriate time interval gives an average flame propagation speed.

Calibrating for image scale gives the results listed in Table 2. The measurements in Table 2 are averages from multiple points along the flame front in each image. They are approximately constant for an entire run and are accurate to 10-20%. They apply to only the middle portion of the burn.

The flame propagation speed has contributions from the laminar flame speed S_L, the gas expansion speed S_e, and the turbulent flame speed S_t.

Table 2 Derived flame propagation speeds

ϕ_i	Speed	S_p (one-port)	S_p (two-port)
0.75	500 rpm	5.3 m/s	2.5 m/s
0.75	1000	8.7	6.1
0.90	500	5.6	4.6

The laminar flame speed S_L changes according to the relation,[18,19]

$$S_L = S_{L\phi}(T_u/T_\phi)^{\alpha}(p/p_\phi)^{\beta}(1-2.1f) \quad (1)$$

where α and β are constants, T is the gas temperature, p is the pressure, 0 refers to atmospheric conditions, u refers to the unburned gas, and f is the diluent mass fraction. For lean propane-air mixtures, $\alpha = 2.2$ and $\beta = -0.23$. The dependence of $S_{L\phi}$ on equivalence ratio (ϕ) is given by

$$S_{L\phi} = 38.31 + 24.84(\phi-1) - 153(\phi-1)^2 \quad (2)$$

Changing ϕ from 0.70 to 0.90 implies an increase in S_{L0} from 32 to 39 cm/s, or 20%, and the larger pressure release due to an increase in fuel density implies a larger increase in S_L, depending on the time during the cycle. The increase in laminar flame speed is most important during early flame kernel growth and if the turbulence level is small enough. The empirical relationship between S_t and S_L during the middle of the burn[11] as a function of the fluctuating velocity u' is

$$S_t = S_L + 4u' \quad (3)$$

The gas expansion velocity results from the density decrease from the reactants to the products. It represents an amplification factor of local flame speed, since it is proportional to the mass entrained by the local flame front throughout the flame. S_e is reduced as the cycle progresses by increasing resistance from the compressed end gases.

The importance of S_L for the early flame kernel is shown by the fact that the images show that the flame kernels at 20 deg-CA BTDC are the same size for the $\phi_i = 0.75$ and the $\phi_i = 0.90$ cases although the time between the spark and the image has decreased from 20 deg-CA to 10 deg-CA. Since the turbulence is unchanged, the implied doubling of early flame

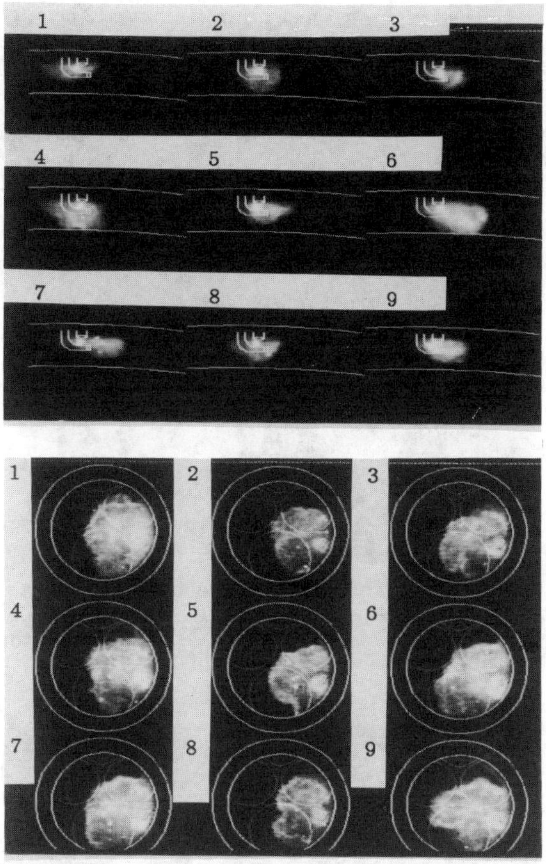

Fig. 22 Flame variation due to flow-flame interaction for nine consecutive cycles with background lines superimposed, two-port flow, spark at 30 deg-CA BTDC, ϕ_i = 0.90, 500 rpm: a) side view at 27 deg-CA BTDC and 20 deg-CA BTDC; b) bottom view at 20 deg-CA BTDC and TDC.

speed must be due primarily to the increase in S_L, amplified by the expansion velocity. It is known that the laminar flame speed dominates early flame kernel growth.[9]

The changes in the one-port results can be explained by the contributions from S_t and S_e during the main part of the volume burned. Comparing the $\phi_i = 0.75$ to the $\phi_i = 0.90$ flame speed measurements shows an increase of only 5-10%. S_e increases slightly in response to the increase in product gas temperature and appears to account for the small change. The turbulence remains unchanged at a large enough value so that the laminar flame speed contribution is not important.

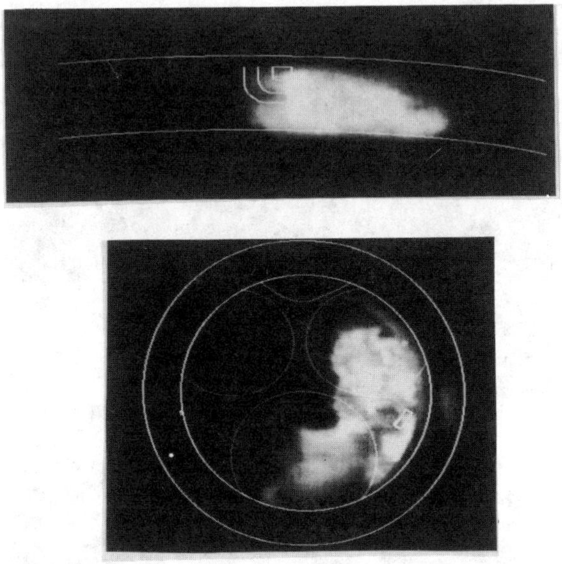

Fig. 23 Example of a cycle with a large flame kernel at 1000 rpm, one-port flow, spark at 40 deg-CA BTDC, ϕ_i = 0.75: a) side view at 37 deg-CA BTDC and 20 deg-CA BTDC; b) bottom view at 20 deg-CA BTDC and TDC.

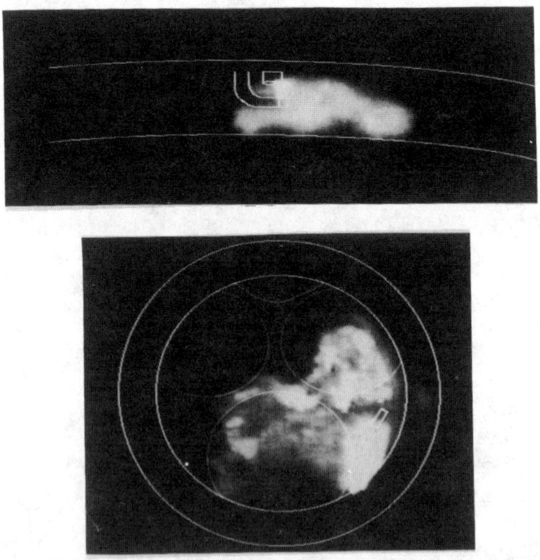

Fig. 24 Flame structure at 1000 rpm, one-port flow, spark at 40 deg-CA BTDC, ϕ_i = 0.75: a) side view at 37 deg-CA BTDC and 20 deg-CA BTDC; b) bottom view at 20 deg-CA BTDC and TDC.

Comparing the 500- to 1000-rpm data shows that doubling the engine speed almost doubles the propagation speed. This is consistent with an increase in turbulent velocities in accordance with the dependence of S_t on u'. More importantly, the smallest turbulence scale is halved, doubling the wrinkling. As will be shown later, changes in S_t appear to be caused by changes in the smallest scales of turbulence.

The two-port data display similar trends over the changes in operating parameters. However, the increase in equivalence ratio has a much larger effect compared with the one-port case. This is apparently due to the much lower turbulence levels, such that S_L forms a major part of the S_p, and the increase in S_L is reflected in S_p.

Comparing the one- and two-port cases, the drop in in-cylinder velocity due to the increase in effective valve area accounts for the decrease in propagation speed for similar conditions. The data imply a larger decrease in turbulence at 500 rpm and a smaller decrease at 1000 rpm. This may be due to the change in large-scale flow structure between these two speeds.[4]

In-Cylinder Flow Implications

The flame images also lead to a number of conclusions about the in-cylinder flow. For this flow, the boundaries are predetermined, if not fixed, and the decaying turbulence has a particular kind of initiation in terms of energy, duration, and geometry. These constraints lead to repeatable characteristics of the flow, such as mean swirl and turbulence level. The flow in the cylinder that contributes to turbulent combustion is highly three-dimensional and varies over a wide range of scales.

The intensity structure of the flame images presented in this work allows some inferences to be drawn about the flow during the cycle, even though no flow velocity measurements have been made as yet.

Large-Scale Flows

The large-scale flow in this engine apparently consists of swirl that varies from cycle to cycle and a central crossflow for the one-port configuration.

Swirl is inferred from the overall rotation of the flame front, shown in the bottom-view images. It is quite variable, an apparent consequence of a major secondary flow instability in the intake process for both the one- and two-port cases. The average swirl number implied by the data is consistent with independent engine measurements, a swirl number of 1 for one-port flow and 0 for the two-port intake.

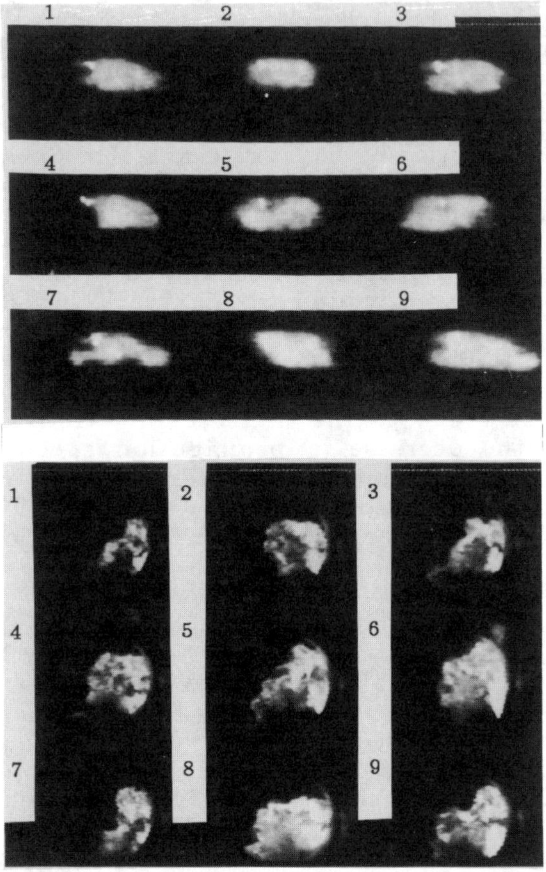

Fig. 25 Flame variation due to flow-flame interaction for nine consecutive cycles, one-port flow, spark at 40 deg-CA BTDC, ϕ_i = 0.75, 1000 rpm: a) side view at 37 deg-CA BTDC and 20 deg-CA BTDC; b) bottom view at 20 deg-CA BTDC and TDC.

In the one-port configuration, there is a crossflow that moves along the pent-roof axis toward the spark plug, deduced from sets of flame-front curves in the previous work.[4] This flow can be seen most prominently in cycle 9 in Fig. 10b. It occurs only in some cycles and seems to disappear at high speed (1000 rpm),[4] as shown by a lack of indentations in the flame fronts in Fig. 25b.

The existence of both the variable swirl and crossflow imply a strongly varying effect of the directed flow caused by the intake ducts. The crossflow does not correlate with the swirl (or lack of it)[4] but may be a related phenomenon.

Although the flow is definitely turbulent, the meaning of turbulence is not clear. Swirl is a mean motion that can

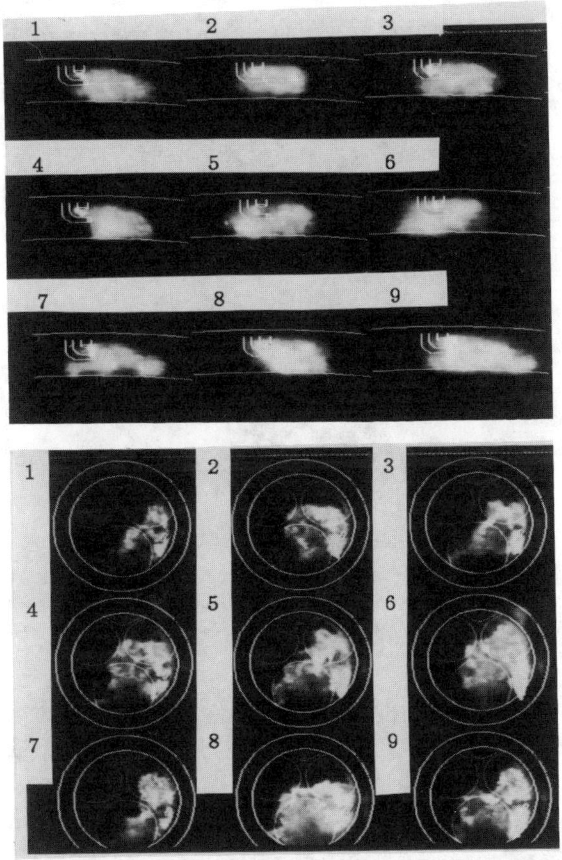

Fig. 26 Flame variation due to flow-flame interaction for nine consecutive cycles with background lines superimposed, one-port flow, spark at 40 deg-CA BTDC, ϕ_i = 0.75, 1000 rpm: a) side view at 37 deg-CA BTDC and 20 deg-CA BTDC; b) bottom view at 20 deg-CA BTDC and TDC.

contribute to turbulence through the presence of squish or because of obstructions such as the spark plug. There is evidence of the presence of various secondary flows in the cylinder. For instance, current experimental work[20,21] has identified secondary flows called "tumble", which seem to make predictable contributions to turbulence characteristics at TDC. These secondary flows are repeatable and thus not a part of turbulence as it is commonly defined. In-cylinder turbulence measurements include these flows, however.

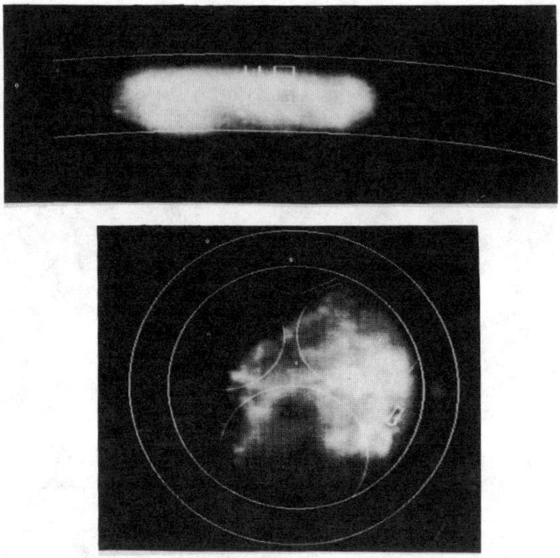

Fig. 27 Example of a cycle with a large flame kernel at 1000 rpm, two-port flow, spark at 40 deg-CA BTDC, ϕ_i = 0.75: a) side view at 37 deg-CA BTDC and 20 deg-CA BTDC; b) bottom view at 20 deg-CA BTDC and TDC.

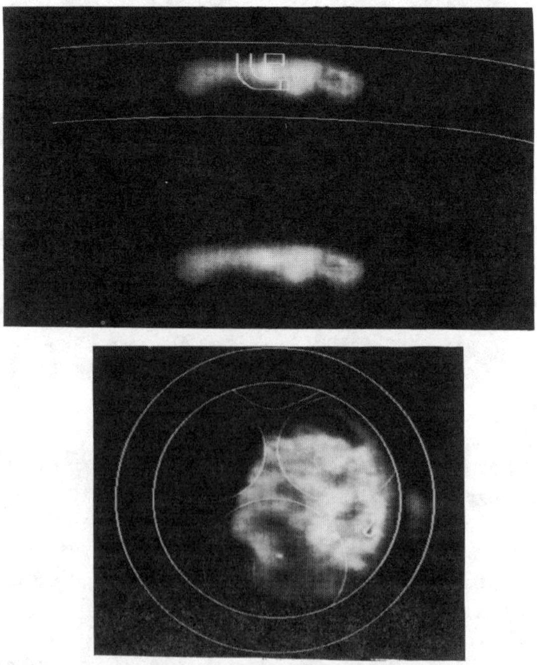

Fig. 28 Flame structures at 1000 rpm, two-port flow, spark at 40 deg-CA BTDC, ϕ_i = 0.75: a) side view at 37 deg-CA BTDC and 20 deg-CA BTDC; b) bottom view at 20 deg-CA BTDC and TDC.

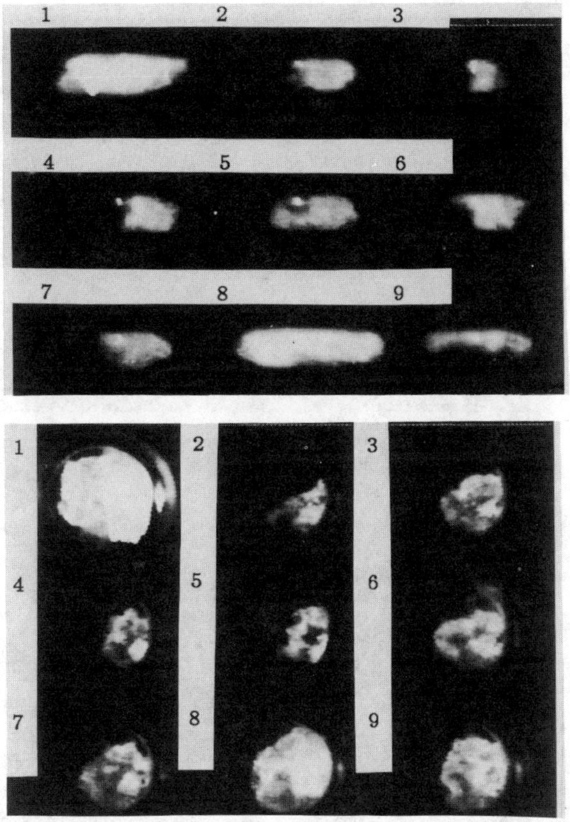

Fig. 29 Flame variation due to flow-flame interaction for nine consecutive cycles, two-port flow, spark at 40 deg-CA BTDC, $\phi_i = 0.75$, 1000 rpm: a) side view at 37 deg-CA BTDC and 20 deg-CA BTDC; b) bottom view at 20 deg-CA BTDC and TDC.

Turbulence

The structure of the brightness variation in the images at the flame front arises from the detailed shape of the flame.[22] The flame is assumed to be thin and convoluted by the flow. These convolutions are in turn a result of the local turbulent flow.

Since the cameras integrate flame radiation along a line of sight, the convolutions of a flame front parallel to the viewing axis will be superimposed, whereas a flame normal to that axis will be more distinct. For the early flame kernel images, the front and rear are superimposed, and any deep convolutions that make the flame front double-valued will also create

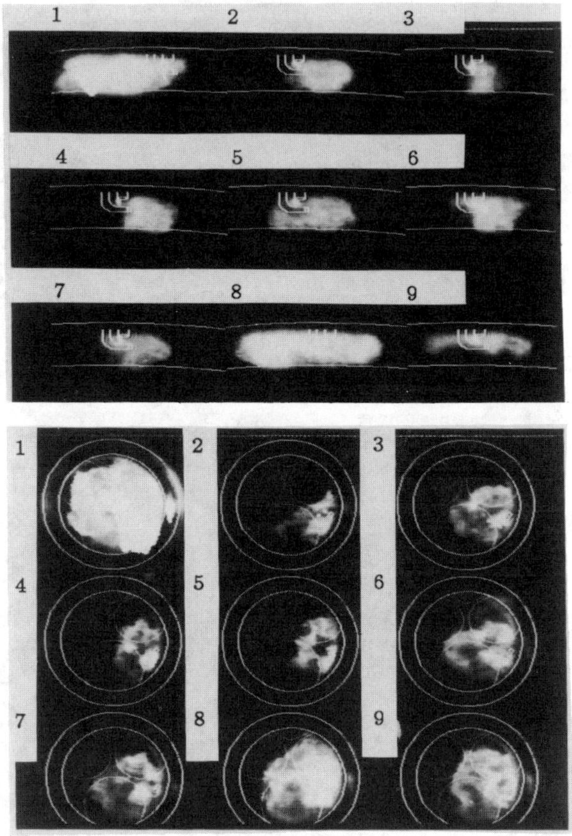

Fig. 30 Flame variation due to flow-flame interaction for nine consecutive cycles with background lines superimposed, two-port flow, spark at 40 deg-CA BTDC, $\phi_i = 0.75$, 1000 rpm: a) side view at 37 deg-CA BTDC and 20 deg-CA BTDC; b) bottom view at 20 deg-CA BTDC and TDC.

confusion in the image. The clearest indication of the true flame convolution is shown in Figs. 31b and 32b since the images capture only a single layer of the flame. In spite of these deficiencies, some conclusions can be reached about the turbulence.

It must be noted that the finest scales of the turbulence, 1 mm and below, cannot be resolved by this imaging system. This explains the fact that the structure of the flames is the same at 1000 rpm as it is at 500 rpm. Although the smallest scale has shrunk by approximately a factor of 2 and the wrinkling of the flame front increased,[22] both these scales are invisible to the cameras used here.

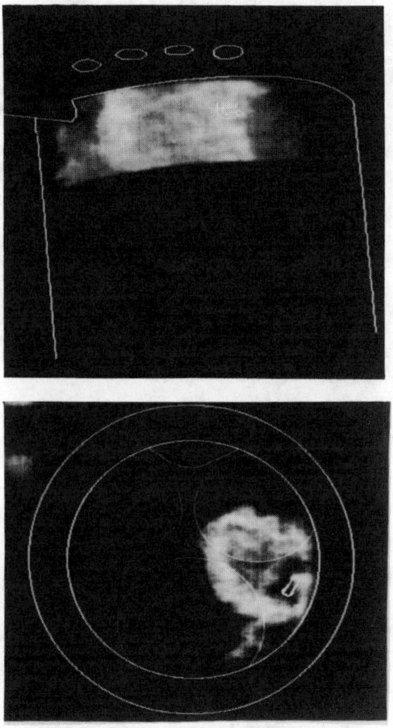

Fig. 31 Combustion completion flame images, one-port flow, spark at 40 deg-CA BTDC, ϕ_i = 0.75, 500 rpm: a) side view at 30 deg-CA BTDC and 50 deg-CA ATDC; b) bottom view at 20 deg-CA BTDC and TDC.

Throughout the image data, both for the early flame kernels and the later flames, shapes of scale of 4-5 mm predominate, based on direct measurements from the images. Clear examples are the convolutions of the early flame kernel in Fig. 24a, and the side-view images of Figs. 31a and 32a. Similar length scales have been found in an engine at comparable conditions with particle imaging velocimetry by Reuss et al.[23] This length scale appears to be the integral length scale of the turbulence, a pervasive scale in turbulence theory.[24] This value for the integral length scale is within the range estimated by Mattavi et al.[11]

It is remarkable that the length scale of the images is constant throughout the data presented here. This is true for a speed change from 500 to 1000 rpm, and a change from a one- to two-port intake flow. Both of these changes alter the turbulence and the flow velocities. Increasing the engine speed increases the turbulence and the turbulent flame speed and

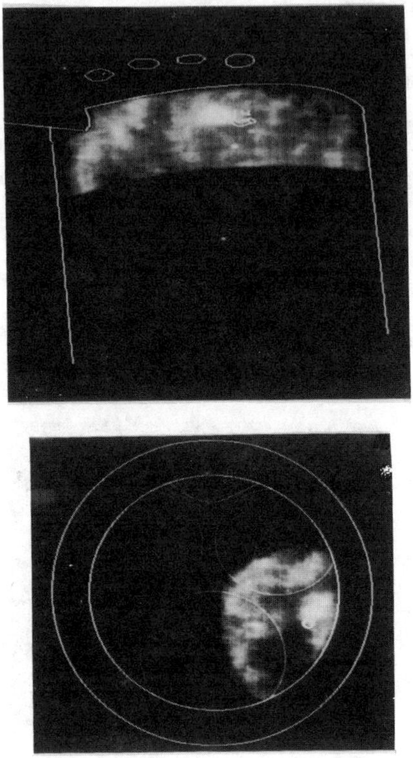

Fig. 32 Combustion completion flame images, two-port flow, spark at 50 deg-CA BTDC, ϕ_i = 0.75, 500 rpm: a) side view at 30 deg-CA BTDC and 50 deg-CA ATDC; b) bottom view at 20 deg-CA BTDC and TDC.

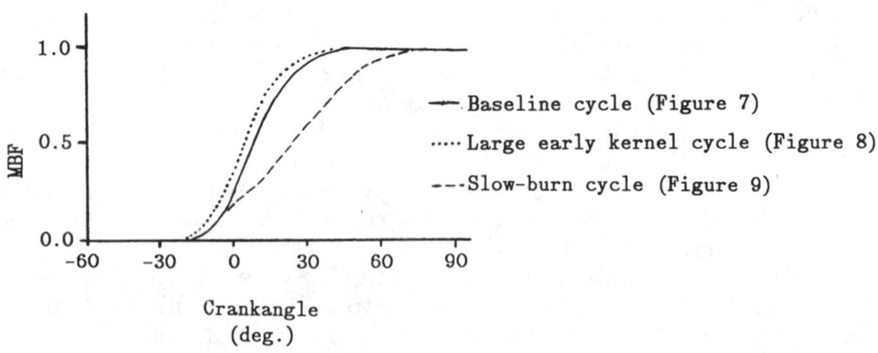

Fig. 33 Relative mass burned fraction curves for different flame shapes and positions.

shrinks the smallest turbulent length scale. The one-port configuration has higher flow velocities and higher flame speeds than the two-port case.

That the visible scales remain similar while the flame speed changes so much implies that the turbulent flame speed is controlled by only the smallest scales of the turbulence.

These conclusions are consistent with the current understanding of the dependence of turbulent flame speed on the rms fluctuating velocity u' given in equation 3 since the two are proportional. A more specific statement is made here about the source of the turbulent flame speed in the finest scale of the turbulence.

Conclusions

The details of flame development in each cycle allow prediction of pressure release and mass burned fraction in the cycle. Flow-flame interactions through the effect of large-scale flows or turbulence result in a variety of flame shapes and speeds that cause observed mass burned rates.

The early flame kernel size and shape usually determine the later flame shape and the time to 1% mass burned. Later burn rate is controlled by the large-scale flame shape that results from an interaction between the early kernel shape, large-scale flow, and the combustion chamber shape. Turbulent flame speed is controlled primarily by the smallest scale of turbulence. These effects form a complete picture of flame development in this engine.

Acknowledgments

Discussions with Rod Rask on the various implications of the flame image data were very helpful. Experimental progress meetings with interested staff members of the Fluid Mechanics Department and the Engine Research Department also provided useful insights.

References

[1]Schipperijn, F. W.,Nagasaka, R., and Sawyer, R. F., "Imaging of Engine Flow and Combustion Processes," SAE Paper 881631, 1988.

[2]Rassweiler, G. M., and Withrow, L., "Motion Pictures of Engine Flames Correlated with Pressure Cards," SAE Transactions, Vol. 42, No. 5, 1938, pp. 185-204.

[3]Bates, S. C., "Flame Imaging Studies in a SI Four-Stroke IC Optical Engine," SAE Paper 890154, 1989.

[4]Bates, S. C., "Cycle-by-Cycle Combustion Variation in a SI Four-Stroke Engine," SAE Paper 892086, 1989.

[5]Bates, S. C., "A Transparent Engine for Flow and Combustion Visualization Studies," SAE Paper 880520, 1988.

[6]Bates, S. C., "A Displaced-Line Velocity Diagnostic and Its Application in a Visualization Engine," Experiments in Fluids, Vol. 7, No. 5, 1989, pp. 335-343.

[7]Kalghatgi, G. T., "Spark Ignition, Early Flame Development and Cyclic Variation in I. C. Engines," SAE Paper 870163, 1987.

[8]Barton, R. K., Lestz, S. S., and Meyer, W. E., "An Empirical Model for Correlating Cycle-by-Cycle Cylinder Gas Motion and Combustion Variations of a Spark Ignition Engine," SAE Paper 710163, 1971.

[9]De Soete, G. C., "Propagation Behavior of Spark Ignited Flames in Early Stages," International Conference on Combustion, Vol. I, Institute of Mechanical Engineers, 1983-3, Paper C59/83, 1983, pp. 93-100.

[10]Herwig, R., Begleris, P., Zettlitz, A., and Ziegler, G. F. W., "Flow Field Effects on Flame Kernel Formation in a Spark-Ignition Engine," SAE Paper 881639, 1988.

[11]Mattavi, J. N., Groff, E. G., and Matekunas, F. A., "Turbulence, Flame Motion and Combustion Chamber Geometry - Their Interactions in a Lean-Combustion Engine," Paper c100/79, Institute of Mechanical Engineers Conference on Fuel Economy and Emissions of Lean Burn Engines, London, June 1979.

[12]Mattavi, J. N., "Effects of Combustion Chamber Design on Combustion in Spark Ignition Engines," SAE Paper 821578, 1982.

[13]Witze, P. O., "The Effect of Spark Location on Combustion in a Variable-Swirl Engine," SAE Paper 820044, 1982.

[14]Groff, E. G., and Matekunas, F. A., "The Nature of Turbulent Flame Propagation in a Homogeneous Spark-Ignited Engine," SAE Paper 800133, 1980.

[15]Matekunas, F. A., "Modes and Measures of Cyclic Combustion Variability," SAE Paper 830337, 1983.

[16]Nakamura, A., Ishii, K., and Sasaki, T., "Application of Image Converter Camera to Measure Flame Propagation in S.I. Engine," SAE Paper 890322, 1989.

[17]Taylor, C. F., The Internal-Combustion Engine in Theory and Practice, Vols. I,II, 2nd Edition, MIT Press, Cambridge, MA, 1985.

[18]Metghalchi, M., and Keck, J. C., "Laminar Burning Velocity of Propane-Air Mixtures at High Temperature and Pressure," Combustion and Flame, Vol. 38, 1980, pp. 143-154.

[19]Metghalchi, M., and Keck, J. C., "Burning Velocities of Mixtures of Air with Methanol, Isooctane, and Indolene at High Temperature and Pressure," Combustion and Flame, Vol. 48, 1982, pp. 191-210.

[20]Vafidas, C., Vorropoulos, G., and Whitelaw, J. H., "Effects of Intake Port and Combustion Chamber Geometry on In-Cylinder Turbulence in a Motored Reciprocating Engine," Fluid Flow and Heat Transfer in Reciprocating Machinery, ASME Publication FED-VOL. 62, 1987, p. 9-16.

[21]Henriot, S., Le Coz, J. F., and Pinchon, P., "Three Dimensional Modeling of Flow and Turbulence in a Four-Valve Spark Ignition Engine - Comparison with LDV Measurements," SAE Paper 890843, 1989.

[22]zur Loye, A. O., and Bracco, F. V., "Two-Dimensional Visualization of Premixed-Charge Flame Structure in an IC Engine," SAE Paper 870454, 1987.

[23]Reuss, D. L., Adrian, R. J., Landreth, C. C., French, D. T., and Fansler, T. D., "Instantaneous Planar Measurements of Velocity and Large-Scale Vorticity and Strain Rate in an Engine Using Particle-Image Velocimetry," SAE Paper 890616, 1989.

[24]Hinze, J. O., Turbulence, McGraw-Hill, New York, 1959.

Flame Propagation Behavior in a Variable Hydrodynamic Constant Volume Combustion Chamber

A. Floch* and M. Trinité†
Université de Rouen, Rouen, France
and
F. Fisson† and T. Kageyama†
Laboratoire d'Energétique et de Détonique, URA 193 au CNRS, ENSMA, Poitiers, France
and
C. H. Kwon‡ and A. Pocheau‡
Université de Provence, Marseille, France

Abstract

During the French Cooperative Research Action (ARC), "Modeling of Combustion Phenomena in Piston Engines" organized by the Centre National de la Recherche Scientifique (CNRS), three identical, variable hydrodynamic condition (vortex, turbulence), constant-volume combustion chambers have been constructed and settled in three CNRS associated laboratories (Poitiers, Rouen and Marseille). These chambers are designed to realize, under the same experimental conditions, different diagnostics: pressure history analysis and schlieren visualization in Poitiers, tomographic visualization and flame structure analysis in Marseille and laser velocimeter analysis in Rouen.

Capability of the device to reproduce the phenomena encountered in real engines is first tested. Influence of an initial turbulence, generated in the combustion chamber by the compression of a propane-air mixture through different perforated blocks, is then studied in detail.

As has been pointed out in engine studies by several authors, the problem is the estimation of the fluctuation intensity. It is shown that consideration of ensemble and cycle-resolved standard deviations, defined

*Complexe de Recherche Interprofessionnel en Aérothermochimie,

† Laboratoire d'Energétique et de Détonique

‡ Laboratoire de Recherches en Combustion

* † ‡ Members of Action de Recherches Concertée sur les Moteurs à pistons (Centre National de la Recherche Scientifique - Programme Interdisciplinaire de Recherches sur les Sciences pour l'Energie et les Matières Premières)

with a cutoff time parameter, leads to different values of turbulence intensity.

Flame propagation parameters are investigated for the same and for different turbulent cases. Flame arrival time, flame speed, and combustion velocity are processed. It is seen that characterization of an initial level of turbulence relies on the combustion processes that are influenced by it. A study of spatial and temporal scales of turbulence is necessary to complete the knowledge of the interaction between combustion and turbulence.

I . Introduction

Interactions between hydrodynamic flowfield (mean flow and turbulence) and combustion have been studied for many years, both in continuous flow[1,2] and in transient flow.[3,4] However, in the piston engine combustion domain, in spite of recent progress largely attributable to the use of large-scale and high-speed computers, there are not yet satisfactory numerical simulation codes. In our view, one reason for this is the lack of comprehension of physical phenomena. In ARC's on the "Modeling of the Combustion Phenomena in Piston Engines" important experimental works are, therefore, included to provide data for model conception and validation. A new type of constant-volume combustion chamber was developed to realize, with a good reproductivity, various hydrodynamic conditions that are encountered in real engines.[5] The purposes of this study are: 1) to show how this chamber works; and 2) to discuss the results obtained in the case of a flame propagating in a turbulent field in which the initial mean velocity is practically zero.

II. Experimental Setup

There are different methods to study turbulent flame propagating in a closed vessel. Bradley and his co-workers[3] have developed a combustion chamber with fans that generate homogeneous and isotropic turbulence fields. They study the influence of the turbulence on burning velocity during the precompression period. Recently Adomeit and his co-workers[4] developed a chamber with a moving grid and a moving piston. With this chamber, which can create compressed homogeneous turbulent fields, they study the flame development in mixture with turbulence but, initially, without mean velocity under evolving pressure conditions. In their chamber, cavity effects on combustion are avoided by careful fitting between the grid holes and the embossing on the piston head.

As Fig. 1 shows, in the device developed for this study various flowfields are generated by means of mixture injection from a variable volume tank T with a moving piston P (120-mm diam, 50-mm stroke) to the combustion chamber V (60 mm x 60 mm section, 150-mm length) through an interchangeable flow generator G (grid, slit and channel) inserted between two parts. The tank and the chamber are initially filled by

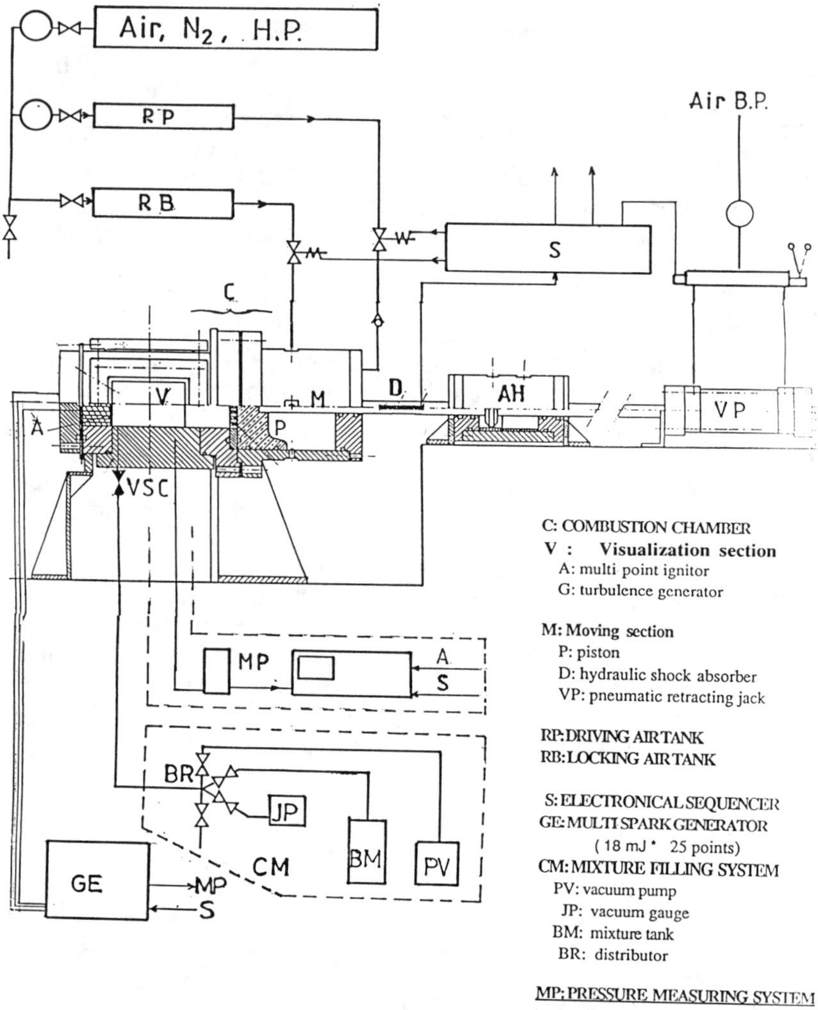

Fig. 1 Schematic of the variable hydrodynamic condition constant volume combustion chamber.

a same mixture with or without oil droplets. The piston is then activated and locked on the inlet surface of the flow generator by means of an electronically controlled S, two-stage compressed air supply (RP, RB) on its rear face. A multipoint parallel ignition system (A on Fig. 1: 25 points, 18 mJ for each spark) is fitted at the other end of the chamber V. Optical accesses 60 mm x 60 mm x 100 mm are available on three surfaces, permitting one to realize schlieren and tomography visualizations and Laser velocimetries from the ignition surface towards the flow generator.

The exposure duration for schlieren pictures on ASA 400 film is less than 5 μs, whereas the tomographic ones are recorded in about 100 μs on a

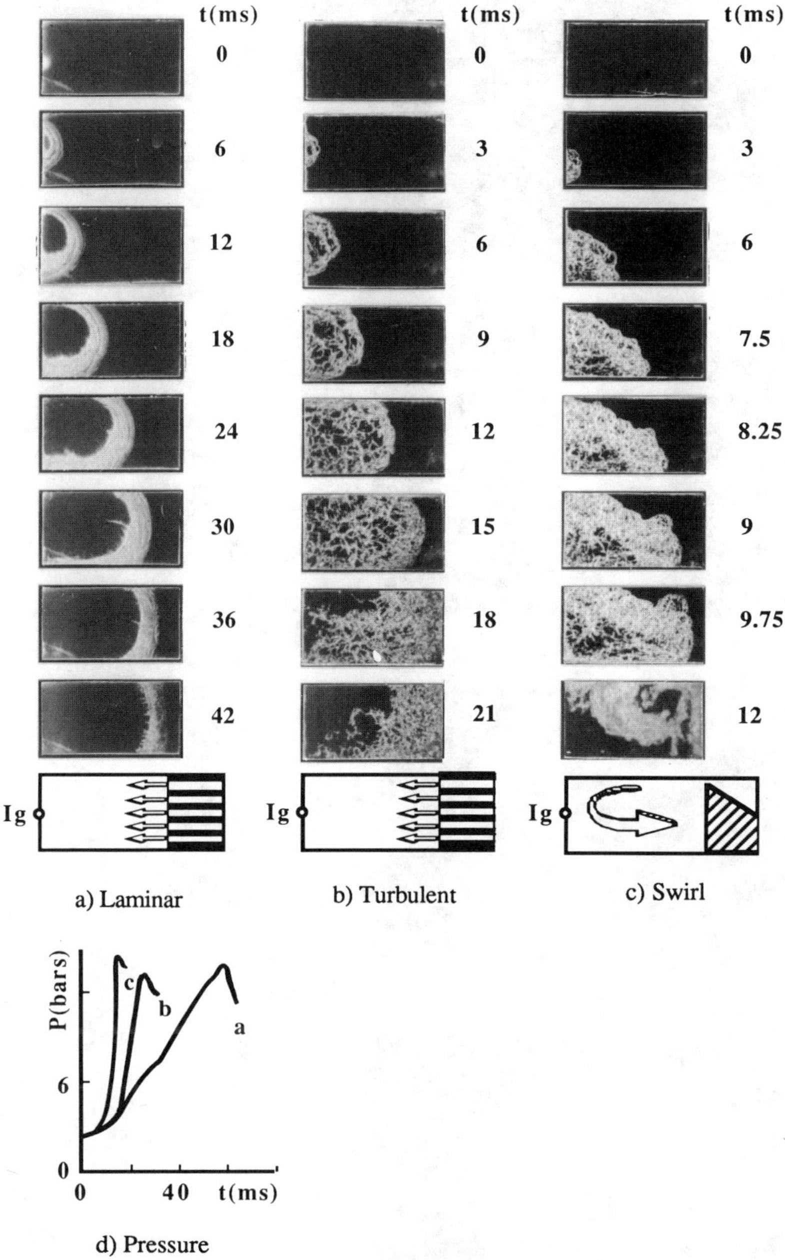

Fig. 2 Influence of the flowfield on a confined premixed flame.

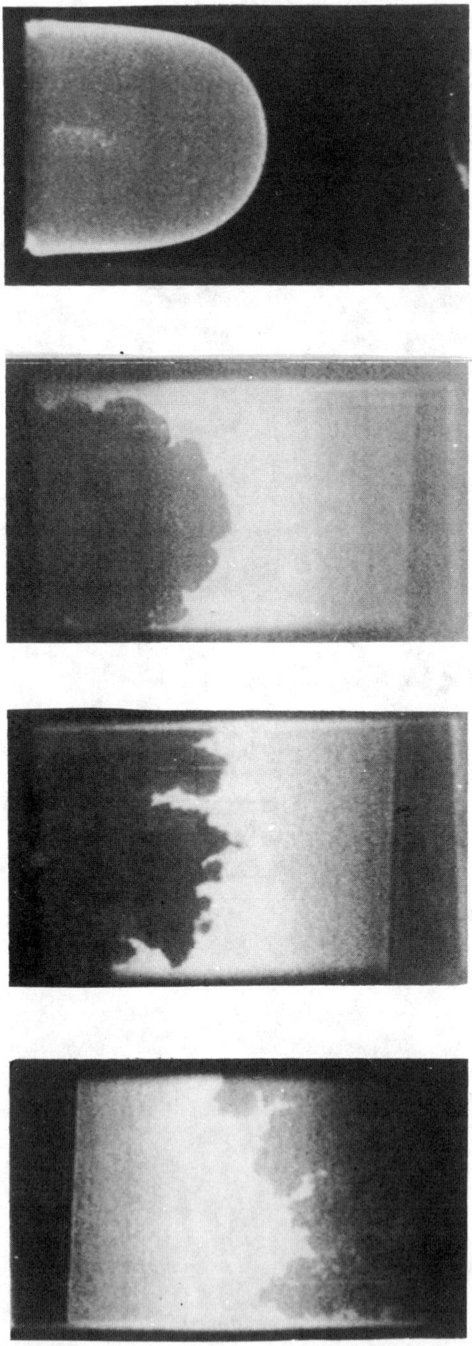

a) Laminar

b) Turbulent (τ_i = 240 ms)

c) Turbulent (τ_i = 110 ms)

d) Turbulent (τ_i = 100 ms)

Fig. 3 Flame structures visualized by laser tomography.

CCD camera equipped with an image-intersifier and associated with a 20-W Argon-ion continous laser.

Pressure measurement is done by means of a Kistler 601 piezo-electrical transducer located on the fourth surface. A propane-air mixture with a 0.9 equivalence ratio and a central single-point ignition have been used exclusively in this work.

III. Chamber Characterization Experiments

In spark-ignited engines, combustion of premixed gas takes place in hydrodynamic fields with intense mean fluid motion (swirl and tumble) and local motion (turbulence). Such fluid motions enable sufficiently rapid combustions to ensure stable and economic running of engines. Consequently, three preliminary tests have been undertaken to verify whether it is possible to reproduce, in this chamber, flowfields encountered in engines:

1) Flame propagating in an initially quiescient mixture.

2) Flame propagating in a distributed turbulent field without initial mean flow.

3) Flame propagating in a turbulent field with initial mean flow.

The flow generator, consisting of a plexiglass block of 60 mm x 60 mm section and 50-mm-length, on which 25 equidistanced, 3-mm-diam channels are drilled, is used for tests 1) and 2), whereas, a generator with a 60-mm-wide and 3-mm-heigh slit has been used to produce a swirl flow in test 3). The filling pressure of the mixture was 1 bar.

A. Laminar Flame

Laminar flame has been studied under static conditions. Following the mixture introduction, the piston has been pushed slowly and locked. Ignition has occurred after several minutes with an initial pressure of about 2.5 bars.

Fig. 2a shows the flame aspect and evolution during flame propagation. From ignition (t=0) to t=18 ms, the flame front develops hemispherically, then elongates until 24 ms with a smooth flame surface and regular pressure increase. At about 25 ms, the rate of pressure increase momentarily lowers. This fact corresponds to the instant at which the flame surface, decreases by touching the surrounding walls. From t=42 ms, slight waving is observed on the flame surface indicating the interaction between flame and acoustic wave. Onset of the phenomenon has been emphasized in this configuration due to the presence of the channels of the flow generator. The laser tomography view on Fig. 3a shows the perfect regularity of the flame front of this laminar flame at the center of the chamber.

Fig. 2d, curve a shows that the maximum pressure has occurred at about t=60 ms, corresponding approximately to overall combustion duration of the laminar flame propagating in this chamber.

B. Turbulent Flames

Turbulent flames have been studied with piston movement. The piston travels the first 45-mm stroke in 50 ms, then slows down in the last 5 mm by hydraulic damper and then locks 90 ms after the beginning of the movement. The reproducibility of the piston motion is analyzed in real time by means of an opto-electronic position transducer installed on the piston rod. Ignition timing has been adjusted at 120 ms after the beginning of the piston motion.

1. Flame propagating in a distributed turbulence field

As can be seen in Fig. 2b, the flame front is perturbed from the iginition to the end of combustion. An elongated hemispherical flame, the surface of which is very perturbed, develops from the very instant of ignition. The flame front does not touch the walls until t=18 ms (uniform black region on the ignitor side). Combustion continues with the development of the flame front, in which the scale of perturbation seems to become smaller and smaller. In this case, the lowering of the pressure increase rate at the instant when the flame contacts the walls is not observed. It is certainly due to the fact that the increase of the flame surface area by turbulence and the decrease of it by wall effects compensate mutually. Overall combustion duration decreases by more than 50% with regard to the laminar case.

Tomographic images in the Fig. 3b, taken at the center of the chamber for three flames that are ignited with different ignition delay τ_i, show that the flame front is much wrinkled and its structure becomes finer and finer with the decrease of ignition delay or with the increase of turbulence intensity. These detailed visualizations are in good agreement with those observed globally by means of the schlieren method.

2. Flame propagating in a flowfield with mean flow and turbulence

A flow generator with a 60-mm-wide and 3-mm-heigh slit has been used to produce a swirl flow with turbulence.

As Fig. 2c and Fig. 2d curve c. show, under swirl and turbulent conditions, the flame core, the surface of which is already very perturbed at the ignition, is convected by the mean flow, while it grows up rapidly. The overall combustion duration becomes less than 10 ms. The schlieren film picture taken at t=12 ms, the instant of the blowback into the chamber of the burned gas in the flow generator cavity, shows clearly the structure of the flow created by means of a slit-type flow generator.

As we have seen previously, the combustion chamber developed for this study is well adapted to reproduce the phenomena encountered in a spark-ignited piston engine. A detailed study has been undertaken first in the case of flame propagation in a mixture with turbulence but without initial mean flow.

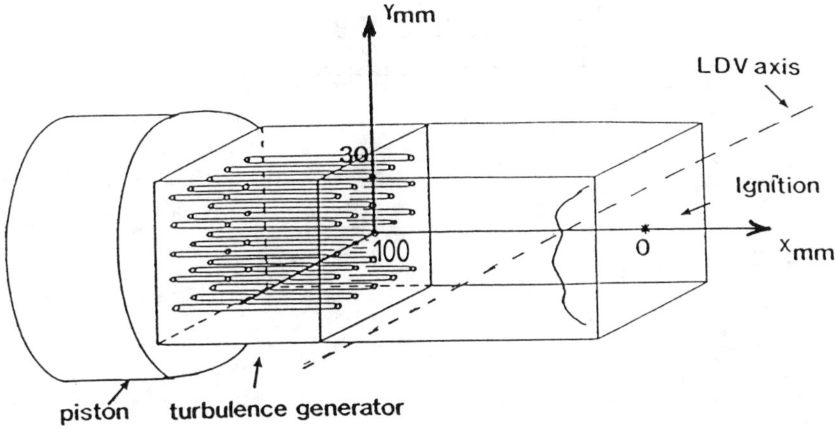

Fig. 4 Diagram and coordinates adopted.

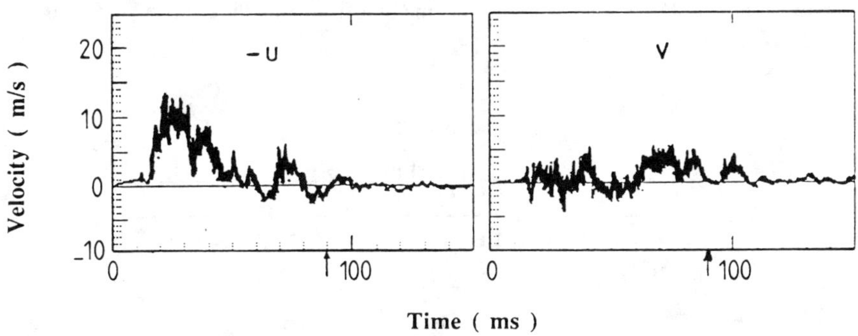

Fig. 5 Raw LDV data of an individual experiment without combustion at x=30 mm and y=0 mm.

IV. Flame-Flow Iinteraction in a Distributed Turbulence Field

In order to generate turbulence of different intensity and scale, four turbulence generators with 25 equidistanced channels of different diameters (2.6 mm, 4.2 mm, 6 mm and 8.5 mm) and spaced 10-mm-apart are examined. Fig. 4 shows the sketch and coordinate system used (x = 0: spark plug side; x = 100: turbulence generator side). In this study, piston velocity is set at a fixed value $\bar{v} = 0.53$ m/s.

A. Problem of Turbulence Definition

Fig. 5 shows an example of instantaneous velocity components measured during one cycle on the center of the chamber for the perforated block with 2.6-mm-diam holes. Mean values of U and V become negligible

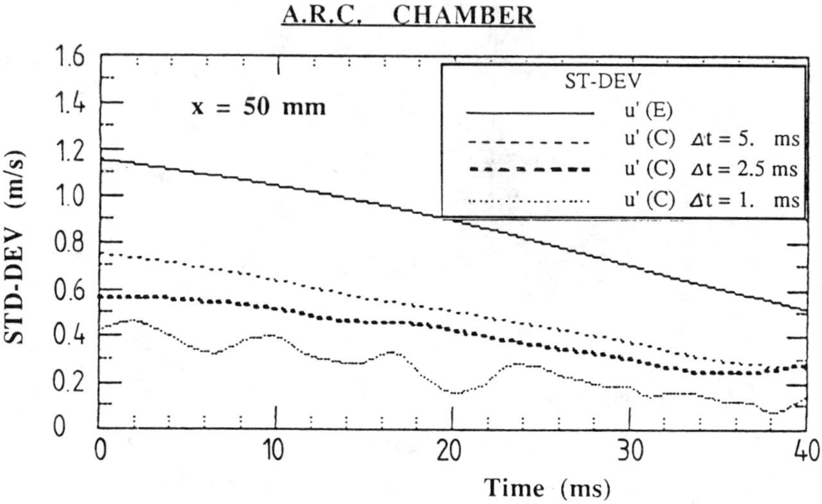

Fig. 6a Effect of the turbulence definition and cutoff time on the deduced turbulence intensity.

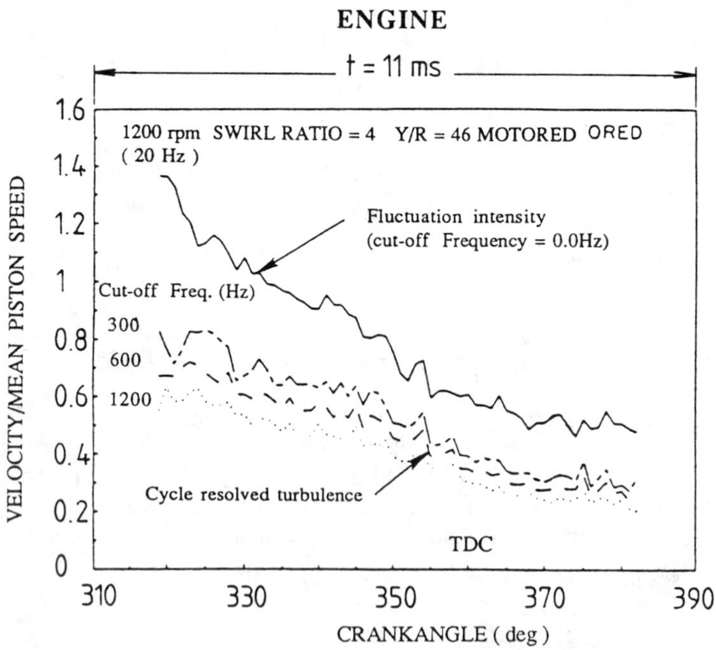

Effect of cutoff frequency on the deduced turbulence intensity (Hall et al. 1986)

Fig. 6b Analagous engine results.

at the end of the stroke, which is about 95 ms. This fact assures that, at the instant of ignition (=95 ms), there is no noticeable mean flow in the chamber.

The study of turbulent flowfield for each turbulence generator has shown that it is possible to assume that combustion is initiated in a homogeneous and isotropic turbulent field. But the problem is the estimation of the fluctuation intensity that must represent turbulence intensity, as has also been pointed out by several authors. As in an engine, the level of turbulence energy depends on the turbulence definition.[5]

The easiest definition of turbulence is based on simple averaging: a quantity is measured in various cycles within a time window around the same time; the average over many cycles is determined; the difference between the value in one cycle and the ensemble average is called the turbulence in that cycle. As mentioned by Bracco, this definition is inadequate because there are, in the chamber, gas motions whose length and time scales are so much larger than those associated with combustion processes that they are perceived as convection by those processes, and yet fluctuations in such gas motions are included in the deduced turbulence intensity when the simple ensemble-average technique is used. This problem about the definition of engine turbulence leads some authors to use some form of filtering technique to separate low-frequency fluctuations.[6,7] These analyses of records obtained in each cycle and based on specific cutoff time or length scales are called cycle-resolved. In this second definition, the turbulence intensity is evaluated as the difference between the value in one cycle and the mean velocity of the cycle.

Fig. 6a gives, at the center of the chamber for the turbulence generator with 2.6-mm-diam holes, a comparative display of turbulence intensity in the case where a simple ensemble averaging is considered and in the cycle-resolved analysis case. The cycle-resolved turbulence intensity

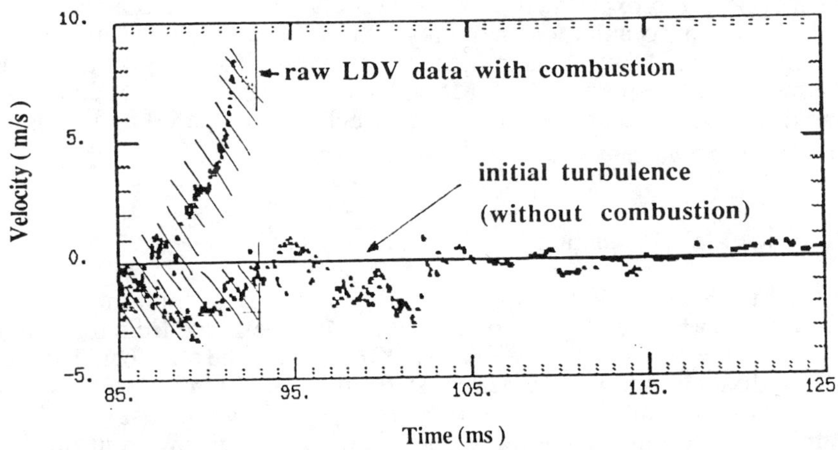

Fig. 7 Modification of the turbulence by the flame (perforated block with 2.6-mm-diam holes).

is roughly a factor of 2 (or a factor of 4 in energy) smaller than the turbulence intensity deduced from simple ensemble averaging. It has also seen the influence of the cutoff time parameter Δt on the cycle-resolved turbulence intensity. When this parameter is changed from 1 ms to 5 ms, the turbulence intensity increases by a factor of 2. Fig. 6b shows some analogous results, obtained by Ref.8 in a ported engine and reported by Ref 5. The same effects of the turbulence definition and of the cutoff frequency on the deduced turbulence intensity are observed too. Furthermore, the turbulence intensity in our combustion chamber, at the time of ignition, is of the same magnitude order as that measured in an engine.

The criteria for the selection of the filter then becomes the basic question. Bracco[5] has adopted the criteria that, in an engine, what is turbulence and what is bulk motion is relative to the process that is influenced by the flow: the combustion in the present study. More specifically, combustion process responds differently to flow fluctuations depending on whether the time and length scales of fluctuations are much larger, of the same order, or much smaller than those of the process. Thus, the characteristic time and length scales of this process becomes the cutoff time and length scale to be used in the analysis of the records. Thus, the time and length scales of the flow must be resolved in each cycle.

In order to illustrate the discussion about fluctuations that must be taken into account in turbulence, raw Laser Doppler Velocimeter data from one cycle were superposed in Fig. 7 in the cases without combustion and with combustion from the time of ignition to the flame arrival time at the center of the chamber. It is seen that the flame propagation time is of the same order of magnitude as the pseudoperiod of low frequency component of the raw LDV data without combustion. Another question to answer is if the increasing velocity due to burned gas expansion modifies the initial turbulence ahead of the flame. Furthermore, time scales are resolved at each cycle in this experiment with a mean data rate of about 50 KHz.

Thus, a complementary analysis is now in progress in order to describe in detail the evaluation of temporal and spatial scales of turbulence by means of two points LDV. Measurements have been also recorded in the different turbulent cases described below and spatial and temporal results are now been processed.[9]

B. Flame-Flow Interaction for Different Levels of Turbulence Intensity

Fig. 8 shows longitudinal mean velocity of the unburned mixture flow generated by expansion of burned gas and measured from the ignition instant at the position X = 70 mm and Y = 0 mm in the chamber. The LDV signal disappears at the moment of flame passage. Up to down are plotted raw LDV data of 25 cycles for each of the following cases: 8a in the laminar case, and for the turbulence generators with different diameter holes: Fig. 8b, 8.5 mm; Fig.8c, 6 mm; Fig. 8d, 4.2 mm; and Fig.8e, 2.6 mm.

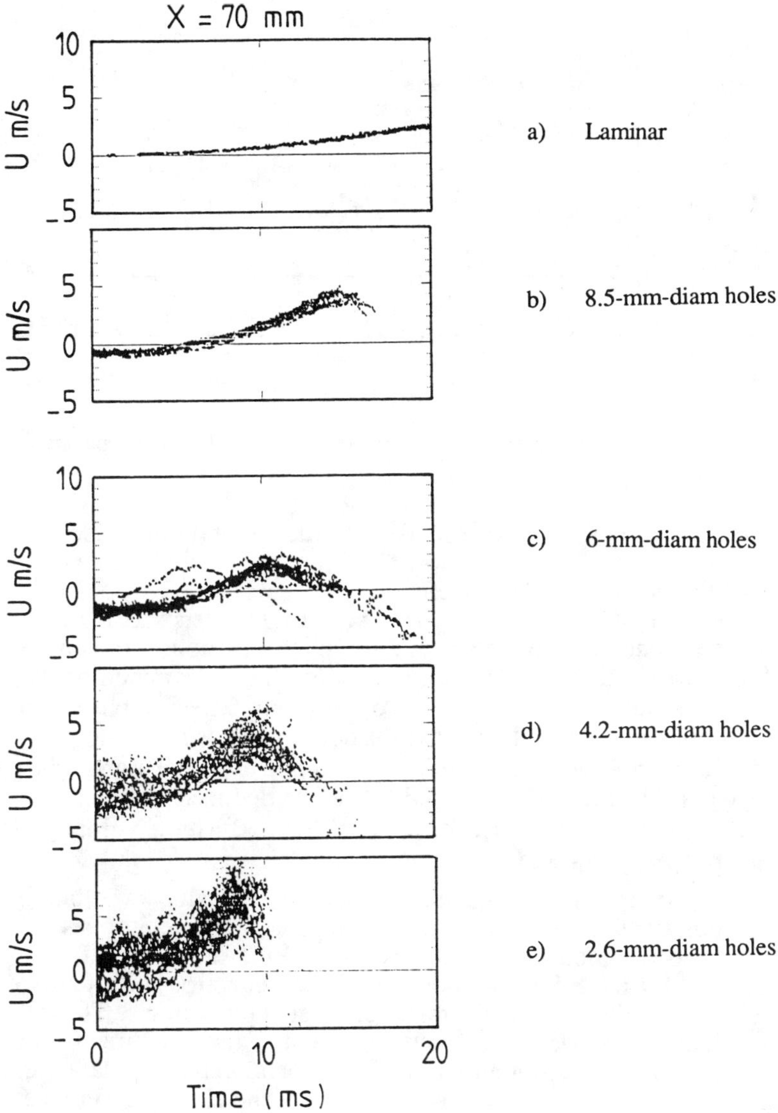

Fig. 8 Evolution of unburned gas longitudinal velocity at X = 70 mm and Y = 0 mm for different levels of initial turbulence intensity.

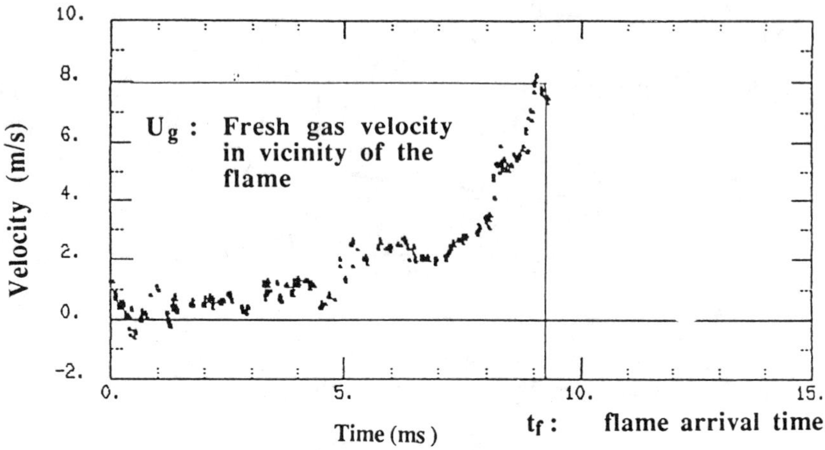

Fig. 9 Information, given by LDV data, about flame propagation.

It is seen that, classically, flame arrival time decreases as the turbulence intensity increases. Furthermore, cyclic fluctuations increase with the initial turbulence intensity.

An LDV method associated with oil droplets seeding has been used to determine burning velocity. At a given position in the chamber, mean flame arrival time t_f over 25 cycles (Fig. 9) is determined by data removal consecutively to evaporation of oil droplets. Mean unburned gas flow x-direction velocity component U_g is determined as the average of the values measured at a same point in individual cycles as the signal disappearance instants. The mean flame speed U_f can be deduced from the knowledge of flame arrival time at different positions. Burning velocity S_u is then determined as $S_u = U_F - U_g$.

Fig. 10a shows initial turbulence intensity vs x position at the ignition time 95 ms for the different turbulence generator cases. The cycle-resolved turbulence intensity increases by a factor of 7 from a value of 0.1 m/s for the 8.5-mm-diam turbulence generator holes to a value of 0.7 m/s for the 2.6-mm-diam turbulence generator holes. These turbulence intensity values were computed with a cutoff time of 5.ms. Flame positions vs time for these different turbulent cases and in the laminar case are shown in Fig. 10b. It is seen that, at the center of the chamber, flame arrival time is about 7 ms for the most turbulent case and about 16 ms for the laminar case. Flame speed curves vs x position for each turbulent case are superposed in Fig. 10c. For the most turbulent case, flame speed is roughly of a factor 4 greater than laminar flame speed. Finally, in Fig. 10d it is observed that turbulent burning velocity S_{UT} is faster than that of the laminar one S_{UL} from the very beginning in the case of high turbulence intensity, and it increases rapidly in the second part of the chamber with flame propagation. In the three most turbulent cases, S_{UT} reaches 6 m/s at the end of propagation, whereas it is only 0.5 m/s in the laminar case. This increase of S_{UT} by a factor of 12, with regards to the laminar burning velocity S_{UL}, is the consequence of the initial turbulence.

V. Conclusion

Our objective to deliver experimental data for model validation has been realized for the case when a flame propagates in an initially

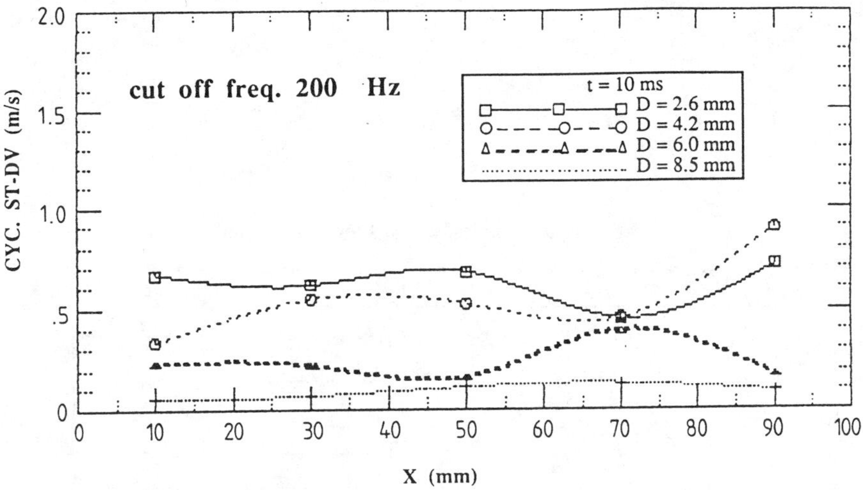

Fig. 10a Initial turbulence intensity in each turbulence case.

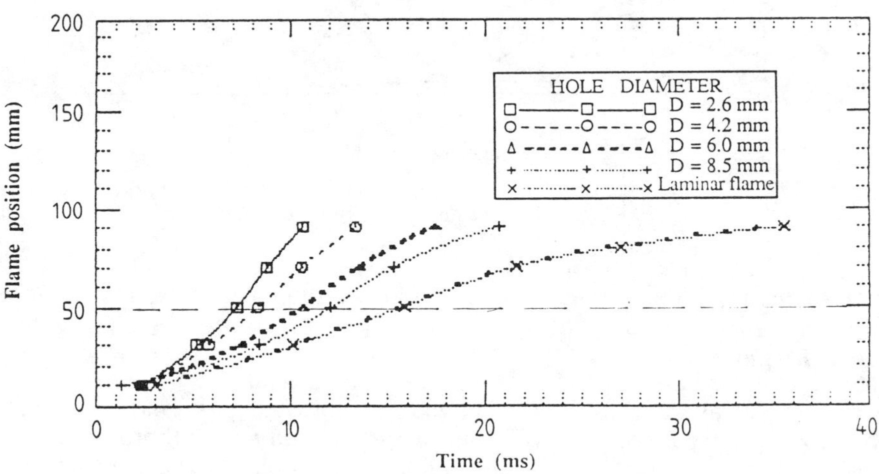

Fig. 10b x flame position vs time.

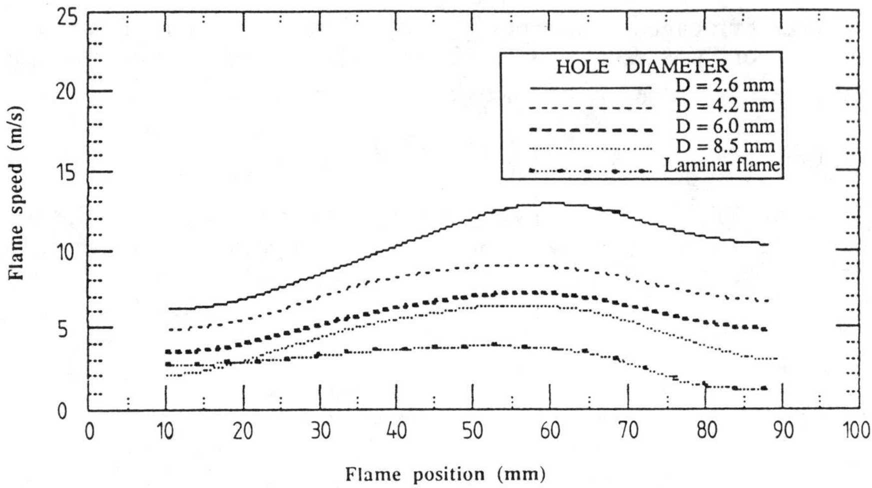

Fig. 10c x direction flame speed vs x position.

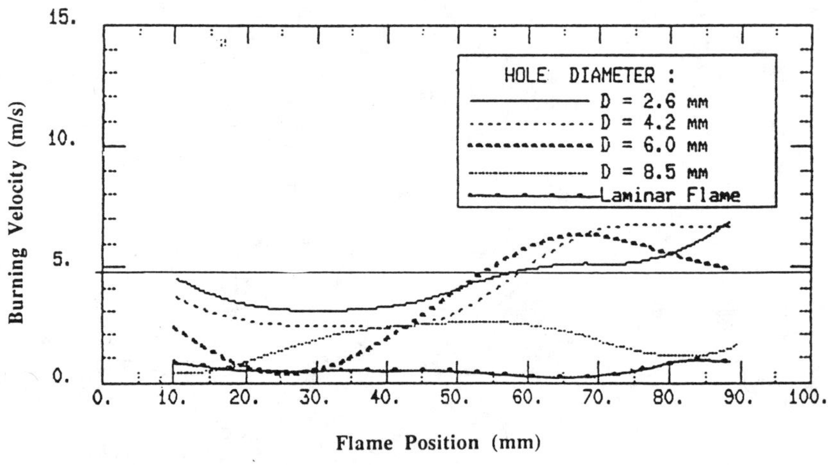

Fig. 10d x direction burning velocity vs x direction.

homogeneous and isotropic turbulent field without mean flow. The device allows a detailed description of turbulent evolution during flame propagation while in an engine; environmental constraints, however, do not permit a full investigation of turbulent caracteristics.

As far as the correlation between turbulent energy and turbulent burning velocity is concerned, it is necessary to study more thoroughly the turbulent energy definition in such a transient situation where cycle-to-cycle variation exists with continuous flame acceleration. In this objective the turbulent length and time scales must rely on those associated with combustion processes. Thus a complementary analysis is now in progress in order to describe, in detail, evolution of temporal and spatial scales of turbulence by means of two points LDV. Measurements have been also recorded in the different cases presented, and a spacial and temporal results are now being processed and under examination in a doctoral work.[9]

References

[1]Strehlow, A., et al., "On the Propagation of Turbulent Flames," Combustion and Flame, Vol. 13, No. 2, 1969, pp. 111-129.

[2]Champion, M., "Etude Théorique et Modélisation d'une Zone de Combustion Turbulente," Docteur ès Sciences Thesis, Univ. of Poitiers, Poitiers, France, 1980.

[3]Bradley, D., et al., "Premixed Turbulent Burning During Explosions," 21 st Symposium (International) on Combustion, The Combustion Institute, Pittsburgh, PE, USA, pp. 497-504, 1986.

[4]Adomeit, G., et al., "Optische Untersuchung der Turbulenten Flammenausbreitung in einem Einhubtriebwerk," SFB Motorische Verbrennung Projekt A5, Aachen,Germany, Oct. 1988.

[5]Bracco, F.V., "Structure of Flames in Premixed-Charge IC Engines," Combustion Sciences and Technology, Vol. 58, No. 1-3, 1988, pp. 209-230.

[6]Rask, R.B., "Comparison of Window, Smoothed-ensemble, and Cycle by Cycle Data Reduction Techniques for Laser Doppler Anemometer Measurements of In-cylinder Velocity", Fluid Mechanics of Combustion Systems, edited by T. Morel, R.P. Lohmann, and J.M. Rackley, American Society of Mechanical Engineering, 1981, p. 11.

[7]Witze, P.O., et al "Direct Measurement of the Turbulent Burning Velocity in a Homogeneous-charge Engine" Sandia Rept. 86 1531, Sandia National Lab., Livermore, CA, USA, 1986.

[8]Hall, M.J., "Cycle Resolved Velocity and Turbulence Measurements near the Cylinder Wall of a Firing S.I. Engine", Society of Automotive Engineers paper No. 860320, Warrendale, PA, USA, 1986.

[9]Floch, A., "Etude de la Turbulence Instationnaire dans une Chambre de Combustion à Volume Constant", Ph. D. Thesis, Univ. of Rouen, Rouen, France, July 1990.

Effect of Turbulent Fluid Motion on Low-Temperature Autoignition of Fuel-Air Mixture Under Piston Compression

Yasuhiko Ohta,* Satoshi Kadowaki,† and Ko Terada‡
Nagoya Institute of Technology, Nagoya, Japan
and
Hitoshi Takahashi§
Meijo University, Nagoya, Japan

Abstract

An experimental investigation of the sensitivity of low-temperature autoignition (cool or blue flame appearance) to in-cylinder mixing process induced by swirl is described. Forced-vortex-like swirls are prepared in the core portion of the simple disc-shaped combustion chamber in a motored engine. Under various swirl intensities, a premixed diethyl-ether / air with a mixture strength 0.8 is compressed up to 578 K, 0.74 MPa at top dead center, and benzene / oxygen / argon mixture with a strength 1.1 is compressed to 1050 K, 1.00 MPa. At equal temperature and pressure environments, intensifying the swirling motion of the mixture retards the low-temperature flame onsets and suppresses their developments up to the succeeding flames. Appearance of low-temperature flames shows no spatial homogeneity under the non-laminar conditions. Turbulent dissipation caused by the swirl is pointed out to play roles as a mixing promoter and reducer of inhomogeneity in the innate heterogeneous in-cylinder autoignition processes. Removal of heat and active species from the local ignition kernels would result in low-temperature flame retardations. Swirl could be one of the effective tools for knock elimination in spark-ignition engines not only through the fast flame propagation but also from a viewpoint of autoignition.

Introduction

The requirement of high efficiency in spark-ignition engines could be achieved by adopting higher compression ratios than those of widely used commercial engines. However, the major problem standing in the way of running engines at higher compression ratios is engine knock.

---

*Associate Professor, Department of Mechanical Engineering.
†Associate Researcher, Department of Mechanical Engineering.
‡Professor, Department of Mechanical Engineering.

It is well known that engine knock is due to an autoignition of the low-temperature charge ahead of the normally propagating flame front. As the propagating flame traverses the cylinder, the pressure and temperature in the end-gas are forced to increase by the combined effect of piston compression and burned gas expansion. It depends on the induction time for the autoignition in an unburned mixture whether the end-gas ignites spontaneously before the propagating flame arrives at the cylinder walls.

The influence of in-cylinder charge turbulence on the flame propagating process initiated by spark discharge is now well recognized. It is seen that the turbulent flame propagation is accelerated almost directly with increase in turbulence intensity. The flow parameters in the fluid in the cylinder should have significance not only for the propagation but also for the autoignition process of end-gas. And it has not been reported in the literature how the turbulence governs the mechanism of the autoignition, contrary to the influence of the turbulence on the flame propagation and the establishment of the spark-initiated flame kernel.

Keck[1] pointed out the delay time elongation in the Shell-Thornton rapid compression machine, compared with the delays in the MIT machine. The Thornton machine has long-stroke double pistons and very high clearance height, equivalent to the cylinder bore; but the MIT machine has a short-stroke single piston and thin pancake-shaped combustion chamber. Rollup vortices caused by piston movement are thought to explain the difference between both machines, though it is mere conjecture.

Tizard and Pye [2] made an experimental comparison of the compression ignition delays of mixtures in a rapid compression machine having a rotat-ing vane with those of an ordinary quiet machine. The delays of agitated mixtures were almost double those in nonagitated mixtures. Heat losses from the walls, increased three times by vane rotations, were considered responsible for these retardations in their paper.

Dumas et al.[3] obtained the temperature and peroxides profiles in a turbulent low-temperature burner flame under atmospheric conditions. In the turbulent cool flame, peroxide concentration did not follow the temperature rise immediately and increased late. This suggests that the cool flame appearance could be retarded by the turbulence.

The roles of cool flames as precursors of engine knock was, at one time, seriously questioned. Smith et al.[4] and Warnatz [5] proposed high-temperature reaction mechanisms based upon a single experimental result. However, popular low-speed engine knock is nowadays recognized again as the one associated with low-temperature oxidation reactions under 1000 K (for example, Sandia CFR News [6]). The high-temperature concept is now considered valid only for the high-speed and high-temperature knocking. Ohta et al.[7] have recently shown clear evidence which emphasizes the important effect of low-temperature flame reactions occurring in compression stroke on the hot flame onset.

This paper focuses on preliminary evaluation of the effect of fluid motion on the compression ignitions of the fuel / air mixtures in relation to engine knock. Of primary interest is whether the fluid motion or turbulence will retard the low-temperature flame appearances in the cylinder

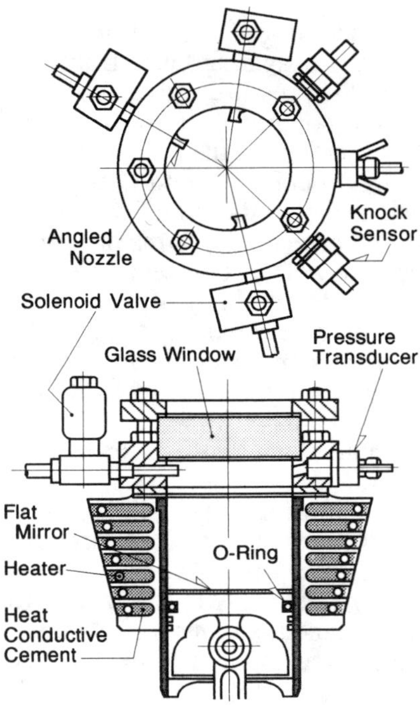

Fig. 1 Simplified longitudinal section and top view of optically-accessed engine.

under engine-like conditions, concerning not only the cool flame, the first low-temperature flame, but also the blue flame, the second low-temperature flame.

Experimental Technique

A single-cycle visualized engine, modified to have extensive optical access and simple geometries, from an original Yamaha 2J2 single cylinder motorcycle engine with 87 mm bore and 84 mm stroke, was motored. Details are described elsewhere.[8] Figure 1 shows a schematic diagram of the piston-cylinder system. The modifications include an electric heater around the cylinder for improved temperature control, and addition of an O-ring for a piston ring to reduce leakage of charge. The combustion chamber was a simple disc type, perfectly symmetric right circular cylinder. Valves have been replaced with three solenoid valves located in the side wall of the combustion chamber. Three angled nozzles (outer diam. 7mm, inner diam. 4mm, bent angle 45°), shown in Fig. 2, were attached on each radial inlet port, projected 7 mm from the wall into the chamber, to introduce the mixture into the combustion chamber in the prescribed direction. Let screw angle θ be the angle between the tangent to the cylinder and the opening surface of the angled nozzles as shown in Fig. 2; swirling motion of the intake mixture could be adjusted by reducing θ from 90°. This configuration established a combined vortex which showed a

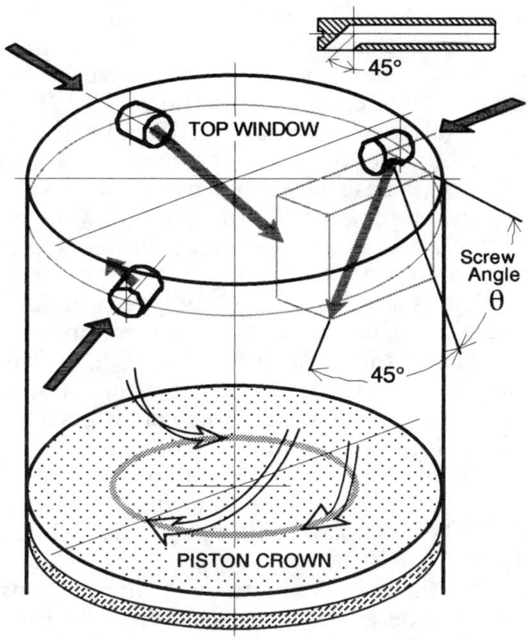

Fig. 2 Angled nozzle attached to each inlet port to establish swirl in combustion chamber, and set up of the nozzle openings and swirl established in combustion chamber before compression.

forced vortex in the center core portion of about half the chamber diameter and a free vortex closer to the wall.

Cylinder pressure was measured using a strain-gauge type transducer (Kyowa, PE30KF). Knock sensors (NGK, non-resonant type) were used to detect cylinder vibrations. The engine was driven by an electric motor at a very low engine speed of 240 rev/min (4 s^{-1}) to provide the time needed for the induction time required for the autoignition of the mixture, and in anticipation of turbulence relaxation to a state of near isotropy and homogeneity prior to autoignition.

The mixture, well mixed beforehand, was prepared in the storage tank, which is connected to the cylinder through valves left open. A single cycle piston compression of the charge would start by a transfer to a regular motoring through closing valves at bottom dead center (BDC) from a continuous noncompression motoring. The small torque of the driving motor resulted in an angular velocity difference between compression and expansion strokes.

The first half of the tests were run on an 0.8 of stoichiometric mixture of diethyl-ether and air, to bring this work into relation with the previous ones.[8, 9] The mixture was compressed from an initial mixture temperature of 328 K with nearly the same wall temperature. The time-averaged compressed temperature was 578 ±2 K at top dead center (TDC). The compression ratio was 8.7 with clearance height of about 11 mm.

The second half of the tests were run on a benzene / oxygen / argon = 3 / 21 / 79 mixture with an equivalence ratio of 1.1. It was compressed from an initial temperature of 543 K up to 1050 ±4 K at TDC. The compression ratio was 19.0. Benzene and methane are known as the fuels which produce no cool flames and start from blue flame when they ignite.

Temperature measurements of the charge were made using an uncoated type-K fine wire (25 μm) thermocouple associated with thermal inertia compensation, at the position of 12 mm away radially from the chamber center. Details are described elsewhere (see Ref. 9). For these, compressions were done under nonreacting conditions by using air. Compensations added on were less than 12 % of the thermocouple e.m.f. itself. Intensifying the swirl by reducing screw angle θ from 90° scarcely affected the compressed temperature in the center portion of the charge. The increased heat loss from the walls, though it was small, was compensated by a slight rise of cylinder wall temperature to secure a gas temperature of 578 K at TDC. The required wall temperature was, for example, 331 K at θ=60° compared with the original 328 K at θ=90° in the first half of the tests.

Cool, blue, or hot flame is identified by the implications of HCO band emission or red coloration as described elsewhere.[10] Blue light emitted from the compressed charge was monitored here through the glass cylinder head of the engine by a photomultiplier (Hamamatsu Photonics, 931A) associated with a blue glass filter (Toshiba, V-42, 320-510 nm) to detect the appearance of the low-temperature flames and their intensities, on the basis of the previous knowledge. Monitored light source was restricted only from the center portion of about 25 mm diam. in the combustion chamber.

A hot-wire anemometer was used to measure the fluid motion and turbulence characteristics, such as intensity and scales, under flameless condition with air as a fluid, only for the first half of the tests. A platinum-iridium wire of 11 μm diam. was heated up to the wire temperature of 723 K, located parallel to and 14 mm away from the cylinder axis. For the fluid temperature changes and conductive heat losses to wire supports, the outputs were compensated as described in Hamamoto et al.[11] Turbulence characteristics were obtained using the relations that are effective in stable pipe flows, treating the swirl as a mean velocity. In-cylinder swirl allows the reduction of uncertainties on flow direction when hot-wire method is applied to a high relative turbulence intensity flow: i.e., the measurement was difficult under weak swirl where θ was larger than 80°. The mean swirl velocities measured by the hot-wire were confirmed by the high-speed schlieren images of the in-cylinder flows (for optical set up, see Ref. 8).

High-speed color pictures were made with a rotating prism camera (Hitachi, HIMAC 16HS) at 4000 frames/s for the latter half of the tests to observe behaviors of blue flame appearances. The film used was Fuji Color RT500 (ISO 500), and was developed through push processing.

Results and Discussion

Light emission monitored was only from the center core about 12.5 mm in radius. Fluid flows in this center portion were measured, though only in a few limited cases of the first half of tests. The swirl ratio was 4 when θ=80°, and 11 when θ=60° at TDC, i.e., the swirl velocity was 35 and 100 cm/s, respectively, at a position 14 mm away from the cylinder axis. Turbulence intensity u' associated with the swirls ranged from 21 to 29 cm/s, and integral length scale was about 4 mm; no clear differences were shown among various swirls.

Figure 3 shows the series of pressure and blue light emission records of the compression ignition processes of the first half of tests, with the screw angle θ as a parameter. These traces were obtained in different swirl environments at equal temperature and pressure. The pressure at the TDC was set 0.74 MPa if flameless at TDC, and the time-averaged compressed temperature 578 K as mentioned in the previous chapter. As the flows were turbulent in the cylinder, temperature fluctuation accompanied about an amplitude of ±6 K p-p. Blue light emission traces are plotted downwards. It will be easily seen that autoignition does not occur under the intense swirl when the screw angle θ is less than 60°. In the case of screw

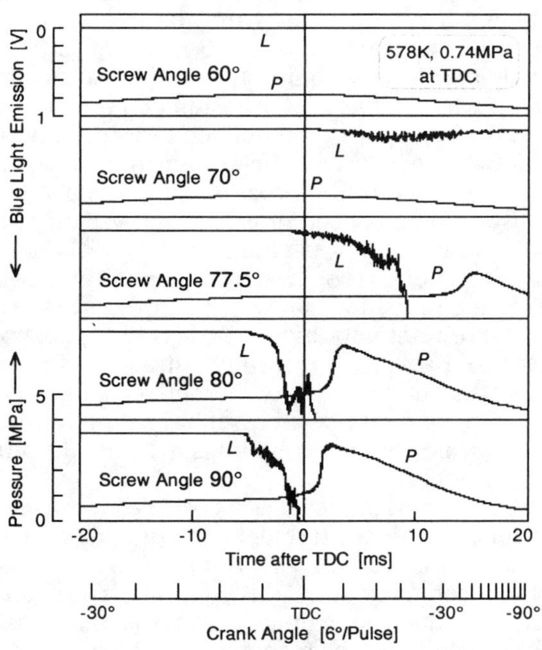

Fig. 3 Series of pressure and light emission records of the cool-, blue- and hot-flame ignition processes with various swirls. Diethyl-ether / air ϕ=0.8, engine speed 240 rev/min, 578 K / 0.74 MPa at TDC.

angle of 90°, 80°, or 77.5° each light emission trace swings off outside of the figure, which shows the low-temperature flames have developed into hot flames. This fact is confirmed by pressure rises that correspond with the light emission swings. On each blue light emission trace, two temporary hesitations are observed, which are recognized as the typical degenerate chain-branching reactions in cool flames. These autoignitions start from cool flame onsets, followed by blue and hot flame; the beginnings of the light emission represent cool flame appearances though the cool flame peaks prior to the hesitations might somewhat overlap on the blue flame onset.

It can be seen that the cool flame appearances are shifted till later or delayed as the swirl velocity increases, and cool flame developments become slow. The appearance and development of cool flame, the first stage of the low-temperature flames, are exactly suppressed by the swirl; both precool (in the so-called τ_1 region) and cool flame reactions (in the so-called τ_2 region) should be impeded in their progress under piston compression.

Ohta and Takahashi[8] have shown using schlieren pictures that the inflammation of cool flame occurs in the mixture originating from local flame kernels counted by the hundred, and then followed by only a few blue-flame sources in a motored engine in a case of no swirl. Cool flames do not appear in a completely homogeneous manner in a space where mixture is charged, but they occur from the kernels distributed in the charge. However, its appearance is much more uniform compared with that of blue flame onsets. Autoignition would always occur heterogeneously to a certain extent in the case of piston compression, even if compression starts from a homogeneity in temperature and concentrations. The same conclusion was given through the examination of heat release rates in the low-temperature autoignition processes, though a visualization technique was not used.[9] High-speed direct movie was difficult to obtain in the first half of the tests due to weak luminosity in cool flames.

Lewis and von Elbe[12], recognizing the nonuniformity, stated that the kernels of ignition are formed within which temperature and concentrations of active species are relatively high. Dumas et al.[3] expressed that the turbulent burner cool flame has a distributed structure, and called it "packet nature". Though the mechanism of kernel formation is still unclear, nonuniform appearance might be an essential nature of the low-temperature flames under the nonlaminar conditions, whether the mixture is compressed by piston or not.

Flame strain concept is accepted as a possible mechanism for the flame quenching of popular premixed turbulent propagating flame (for example, Abdel-Gayed et al.[13]), but for the retardation or the restriction of autoignited low-temperature flames, this concept may be called in question.

Turbulence is the fluctuation or deviation from the average value in the strict sense. However, it must be recognized that the swirl with turbulence is the mixing promoter and the reducing effect of nonuniformity in the innate heterogeneous field. The species and heat produced by low-temperature preflame reactions in the very neighborhood of the kernels act autocatalitically on their own chemical reactions, because the low-temperature oxidation reactions are always "chain-thermal reactions", in which the chain reactions release some amount of heat and are accelerated through the temperature rises induced by their own heat liberations. Several portions locally surpass the adjacent layers in temperature and active species, i.e., kernels will be forced to transfer and diffuse heat and species to close vicinity, when mixed with its surroundings by the turbulence. This might be a reasonable explanation for cool flame retardation caused by swirl. Since ignition is a chain-thermal explosion phenomenon, it is determined by the ratio between the rate of heat and active species liberations in the kernel and the rate of removals from it. In spite of the decrease of the cool flame emission peak as the swirl was strengthened, ignition still developed up to blue and hot flames in 90°, 80°, or 77.5° case as shown in Fig. 3. The mixing induced by swirls could not yet have completely eliminated the nonuniformity in the reaction volume in the center core of the chamber; the extent of the heterogeneity is reduced successively. The mixing effects by larger scale motions than the Kolmogorov level, such as of integrated length scale and Taylor micro scale, are considered to have little effects in these cases. Cool flames appear at hundreds of kernels in the chamber if swirl does not exist.[8] From each local kernel cool flame propagates, and envelops the adjacent mixture within a distance of about 1 mm. Both integrated scale and micro scale are larger than this distance. In these tests turbulence intensity u' seems to be almost identical, so that smaller scale mixing effects around the distributed kernels have to be considered rather than strain rate, u'/λ (u': turbulence intensity, λ: Taylor micro scale) on the continuous surface of popular propagating flame front.

Dent[14] has used the Kolmogorov time scale (τ) and dissipative mixing rate ($1/\tau$) to explain that the soot oxidation rate is characterized by turbulent dissipation at the Kolmogorov level. Following this concept, the dissipative mixing through the turbulence in the charges is estimated. Figure 4 shows the comparison of the mixing rates in the smallest scale of turbulence between the two cases established with different screw angles, 80° and 60°, under nonreacting, flow measurement condition. It can be seen that intensifying the swirl results in the promotion of small scale mixing in a compression period (-18° to TDC) corresponding to the essential period up to the autoignitions.

Lewis and von Elbe[12] have mentioned that there is no reason to believe a preparatory reaction is essential for the propagation of the cool flame. Though it is not known whether cool flame could propagate or not, if the propagation of the cool flame has the same characteristics as the normal propagating flame to the fluid motion, its propagation velocity should be increased by turbulence. The experimental result obtained here suggests that fluid motion does not accelerate the cool flame, and the precool flame reactions (τ_1 reactions) are the precursors for cool flame propagation, if one regards the spatial development of cool flame as the cool flame propagation.

The effects of the swirl on the blue flame appearances are hardly extracted from these first half of tests shown in Fig. 3, because blue flame

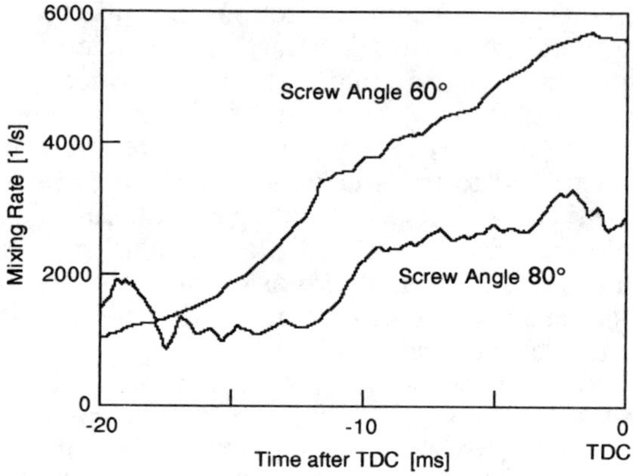

Fig. 4 Comparison of mixing rate of Kolmogorov level between cases of different swirl intensities in the period prior to autoignition.

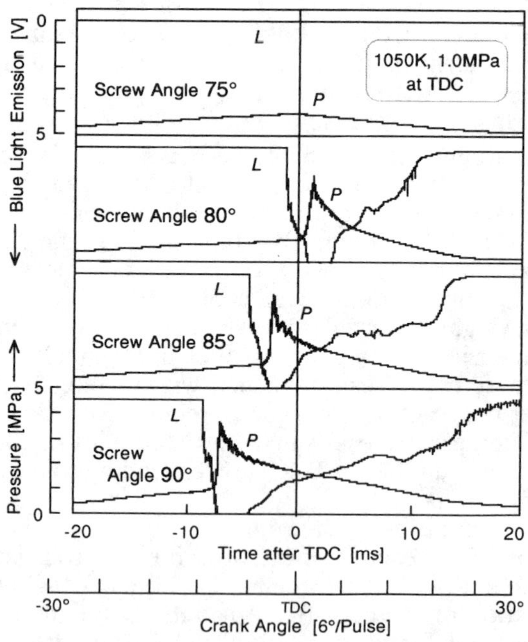

Fig. 5 Series of pressure and light emission records of the blue- and hot-flame ignition processes with various swirls. Benzene / O_2 / Ar ϕ=1.1, engine speed 240 rev/min, 1050 K / 1.00 MPa at TDC.

always appears while the cool flame is degenerating, not after it disappears. Figure 5 is the result of the latter half of tests using benzene as the fuel, with the screw angle θ as a parameter. The compressed temperature was 1050 K at TDC with a fluctuation of about ±30 K p-p. Hot-wire method was no longer applicable for these cases because the mixture temperature was too high. There was not much difference in mean swirl velocity with a certain screw angle θ between the first and second half of tests, within a limited accuracy given by the high-speed schlieren images of the in-cylinder flows. Autoignition breaks out from a blue flame, not proceeded by cool flames. The blue flame onset is retarded with an intensified swirl in the similar manner as the cool flame appearance was done by a swirl. It can be seen that the swirl also lays restraint on the development of preblue flame reactions (τ_2 reactions).

Blue light emission at the blue flame onset is intense from the very beginning of the onset, even when the onset is retarded. It is distinctly separated by swirl intensity whether the blue flame could occur or not. Blue flame never occurs without a succession of hot flame appearances, in contrast to the tenacious behavior of the cool flame onset. Besides, cool flame sometimes degenerates without development to the succeeding flames.

It is not clear whether the development of blue flame itself is suppressed or not after it appears, because of the possibility of hot-flame coexistence among the blue flame as mentioned below.

Cylinder pressure correlated blue flame appearances in the cases of screw angles θ=90° and 80° are shown in Fig. 6. Every two successive contour images, at intervals of 250 μs, are shown in each expression, digitized from high-speed direct color photographs. Inside the center small circle is the optical access area of photomultiplier. Under the former no swirl condition, θ=90° corresponding to the bottom figure in Fig. 5, the flames appear at 9° before TDC as several pieces of cloud in pale blue, white, or sometimes pale pink colors. It could be considered that many kernels of hot-flame arise soon after the initiation of blue flames. The latter, the screw angle θ=80° corresponding to the second figure in Fig. 5, is a case under the influence of swirl. A disk-shaped deep blue "blue flame" kernel appears at 2° before TDC, first from almost a single kernel and develops in the surrounding area of the chamber, where swirl induced mixing effects are relatively weak. Slower spatial development is also noticed compared with no swirl case. If swirl is provided, the hot-flame kernels might be kept from appearing and more pure blue flame would come in sight, emitting intense blue light which essentially characterizes blue flame itself by its coloration.

Directly initiated blue flame is also spatially heterogeneous in the geometric sense as the one preceded by cool flames unless an intense swirl is provided. The mixing process induced by swirl is also effective for the blue flame appearance. Mixing induced by swirl in the charge would make the same type of reactions occur in every space at every instant.

It is a well known fact that the addition of lead alkyls to the fuel retards the blue flame appearance but has little effect on the cool flame. Of

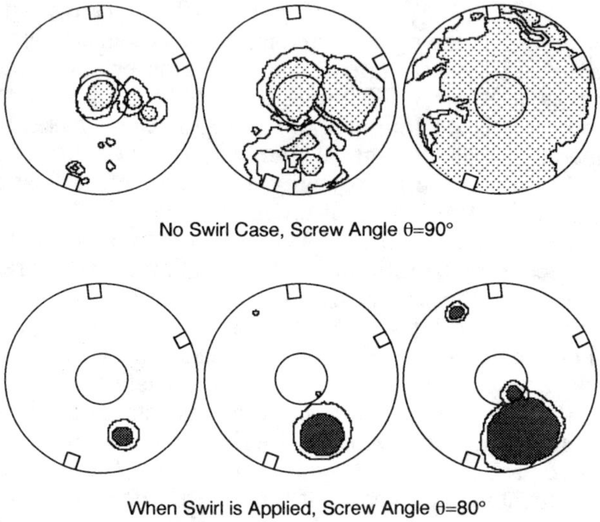

Fig. 6 Different behaviors of blue flame appearances (digital-processed image contours). Benzene / O_2 / Ar ϕ=1.1, Engine speed 240 rev/min, 1050 K / 1.00 MPa at TDC, Inside the small center solid circle is the optical access area of photomultiplier. Upper: without swirl, θ=90°, blue flame appears at 9° bTDC. Lower: swirl is applied, θ=80°, blue flame appears at 2° bTDC.

particular interest, increasing the octane rating of the engine cycle could be more easily achieved by controlling the blue flame onset through fluid motion.

Concluding Remarks

Forced-vortex-like swirls are prepared in the core portion in a motored engine cylinder. A premixed diethyl-ether / air with an equivalence ratio 0.8 is compressed up to 578 K, 0.74 MPa at TDC under the different swirl intensities.

Swirls in the fuel / air mixture retard the compression-ignited cool flame appearances and suppress their developments. Turbulent dissipation at the Kolmogorov level induced by swirl might play roles as a mixing promoter and reducer of inhomogeneity in the innate heterogeneous autoignition processes in piston engine cylinder.

Benzene / oxygen / argon mixture with an equivalence ratio 1.1 is adopted to independently examine the effect of swirl on the blue flame onset. It is compressed to 1050 K, 1.00 MPa at TDC. The blue flame onset is retarded with intensified swirl in the similar manner as the cool flame was retarded by the swirl.

In-cylinder fluid motion could not accelerate preflame reactions. Nonuniformity of the onsets might be an essential nature of the appearances of low-temperature flames. Removal of heat and active species from the ignition kernels through the mixing motion of the swirl would lead to low-temperature flame retardations. Swirl could be one of the effective tools for knock elimination in spark-ignition engines not only through the fast flame propagation but also through the retardation of autoignitions.

Acknowledgment

The first author would like to thank the Nitto Foundation, Aichi, Japan for its support and aid.

References

[1]Keck, J. C., and Hu, H., "Explosions of Adiabatically Compressed Combustible Gas Mixtures," paper IVA1-4 presented in 10th ICDERS, Berkeley, CA, 1985, or "Explosions of Adiabatically Compressed Gas in A Constant Volume Bomb," 21th Symposium (International) on Combustion, The Combustion Institute, Pittsburgh, PA, 1988, pp. 528-529.

[2]Tizard, H. T., and Pye, D. R., "Experiments on the Ignition of Gases by Sudden Compression," Philosophical Magazine, Series 6, Vol. 44, No. 259, 1922, pp. 79-121.

[3]Dumas, G. M. L., Barbarin, V., Ben-Aim, R. I., and Al-Andari, J., "Peroxides Analysis in Turbulent Cool Flames of n-Heptane," 20th Symposium (International) on Combustion, Poster Session 120, Ann Arbor, MI, 1984.

[4]Smith, J. R., Green, R. M., Westbrook. C. K., and Pitz, W. J., "An Experimental and Modeling Study of Engine Knock," 20th Symposium (International) on Combustion, The Combustion Institute, Pittsburgh, PA, 1984, pp. 91-100.

[5]Warnatz, J., "Chemistry of Flame Propagation and Auto-Ignition in Hydrocarbon-Air Mixtures up to Octane," paper IC4-1 presented in 10th ICDERS, Berkeley, CA, 1985.

[6]Sandia National Laboratories, Combustion Research Facility News, 1987, 9-1, p. 2, and 9-4, p. 4.

[7]Ohta, Y., Hayashi, A. K., Takahashi, H., and Fujiwara, T., "Consequence of Temperature and Pressure-Time History for Autoignition," Progress in Astronautics and Aeronautics, Vol. 105; Dynamics of Reactive Systems; Flames and Configurations, edited by J. R. Bowen, J. -C. Leyer, and R. I. Soloukhin, AIAA, New York, 1986, pp. 93-103.

[8]Ohta, Y., and Takahashi, H., "Homogeneity and Propagation of Autoignited Cool and Blue Flames," Progress in Astronautics and Aeronautics, Vol. 95; Dynamics of Flames and Reactive Systems, edited by J. R. Bowen, N. Manson, A. K. Oppenheim, and R. I. Soloukhin, AIAA, New York, 1984, pp. 236-246.

[9]Ohta, Y., and Takahashi, H., "Temperature and Pressure Effects in Cool and Blue Flames," Progress in Astronautics and Aeronautics, Vol. 88; Flames, Lasers and Reactive Systems, edited by J. R. Bowen, N. Manson, A. K. Oppenheim, and R. I. Soloukhin, AIAA, New York, 1983, pp. 38-56.

[10]Ohta, Y., and Furutani, M., "Identification of Cool and Blue Flames in Compression Ignition," paper En1 presented in 11th International Symposium on Combustion Processes, Miedzyzdroje, Poland, Sept., 1989, and to appear in Archivum Combustionis, Polish Academy of Sciences, 1990.

[11]Hamamoto, Y., Tomita, E., Tanaka, Y., and Katayama, T., "The Effect of Swirl on Spark-Ignition Engine Combustion," Proceedings of International Symposium on Diagnostics and Modeling of Combustion in Reciprocating Engines (COMODIA), Tokyo, 1985, pp. 413-422.

[12]Lewis, B. and von Elbe, G., Combustion, Flames, and Explosions of Gases, Academic, New York, 1961, pp. 161-162.

[13]Abdel-Gayed, R. G., and Bradley, D., "Criteria for Turbulent Propagation Limits of Premixed Flames," Combustion and Flame, Vol. 62, 1985, pp. 61-68.

[14]Dent, J. C., "Turbulent Mixing Rate - Its Effect on Smoke and Hydrocarbon Emissions from Diesel Engines," Society of Automotive Engineers, Paper 800092, 1980.

Experimental Study of N-Butane Autoignition in a Rapid Compression Machine

J. D. Gabano, T. Kageyama, and F. Fisson
Laboratoire d'Energétique et de Détonique, URA 193 au CNRS, ENSMA, Poitiers, France

Abstract

Measurements of autoignition delays for n-butane and oxidizer (21% of oxygen and 79% of argon in volume) with an equivalence ratio of 0.9 are performed in an original rapid compression machine apparatus developed and designed in our laboratory. The mixture is compressed under the pressure condition of about 10.7 bar in a range of temperature from 770 K to 900 K. Pressure histories and blue light emission records clearly show the behavior characteristic of a two-stage ignition process for temperatures above 800 K during the postcompression interval. Indeed, cool flames are observed to produce a slight pressure buildup because of enough heat release and also to sensitize the mixture with intermediate chemical species responsible for the recorded blue chemiluminescence before complete hot ignition. Finally, the variations of induction delays vs temperature are examined and discussed. The results are then compared with some obtained with isobutane in near experimental conditions at pressure around 12.6 bar within the same range of temperature.

Introduction

Economic and ecological imperatives have led engine designers within recent years, to improve the efficiency of spark ignition engines and to reduce exhaust pollutants.

Although engine power and efficiency both increase with compression ratio, these parameters cannot be optimized for many contemporary engines due to knock limitations, strengthened by the use of lead-free fuels. It is now widely accepted that knock is due to autoignition of the end-gas, which is the last part of the charge to be consumed by the normal flame propagation which is and responsible for strong pressure vibrations. The pressure oscillations at frequencies that are characteristic of sound waves in the medium contained in the chamber and determined by the chamber geometry are due to the relaxation of a pressure gradient within the combustion chamber as autoignition of end-gas occurs.[1,2,3]

Autoignition is preceded by low-temperature oxidation characterized by slow reactions (cool and blue flames), which could be established on laboratory time scales by using rapid compression machines[4,5] or motored engines.[6,7] In recent works on end-gas chemistry, there appears to be some question as to the importance of the preknock reactions occurring in the end-gas. Recent studies performed by Green and co-workers[8,9] seem to indicate a significant role for low-temperature end-gas chemistry, in the sense that it provides enough heat release so that the end-gas temperature increases to a level where hot ignition can occur. In any event it is clear at the present time that it is necessary to take low-temperature chemistry ($T < 1000$ K) into account in properly describing the autoignition process in knock models, especially when the end-gas compression duration is long, as outlined by Ohta et al.[10] This point of view has been enhanced by a recent experimental study made by the authors[1] where engine knock was simulated in a static preheated spark-ignited combustion chamber. It was found that chemical activity was important enough to be detected using the high-speed laser schlieren technique at temperatures below 900 K for n-butane.

The aim of this study is to examine experimentally in detail the onset and the development of low-temperature reactions that lead to autoignition, using a mixture consisting of $n\text{-}C_4H_{10} / O_2 / Ar = 2.91 / 21.00 / 79.00$, equivalence ratio $\Phi = 0.9$ under pressure conditions around 10.7 bar. Argon has been used to replace nitrogen in mixtures with air in order to increase the ratio of heat capacities of the mixture to the value $\gamma = 1.59$, which also augments the compressed gas temperature.

In order to observe the influence of the molecular structure, some experiments have also been made by replacing n-butane by isobutane, the mixture composition being then $i\text{-}C_4H_{10} / O_2 / Ar = 2.91 / 21.00 / 79.00$, equivalence ratio $\Phi = 0.9$.

Experimental Measurements

The experimental measurements were carried out in a rapid compression machine, which was previously designed to study the combustion initiated by a spark ignition so as to better approximate combustion in real engine conditions rather than with a classical constant volume bomb.

The modular structure of the apparatus, which facilitates the adjustment of arrangements to different cases of use, is the original aspect of that rapid compression machine. A sketch of the machine is shown in Fig. 1. It is made of three parts, which are the actuater, the binding cam,

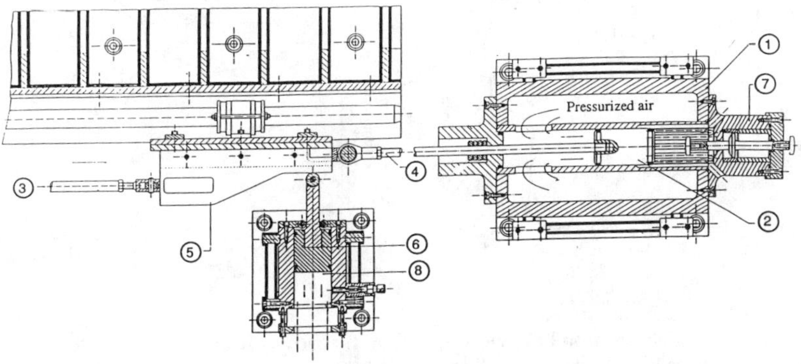

Fig. 1 Sketch of the rapid compression machine.

and the cylindrical combustion chamber. The actuater works as an air gun, whose driving force is ensured by pressurized air stored in a cylindrical tank (1) that surrounds the driving cylinder (2). From an initial movement promoted by a pneumatic jack (3), the driving piston, used as a valve, is connected with the high pressure holder, and then run rapidly. As it moves, the driving piston carries, through the connecting rod (4), the translation of the wagon supporting the binding cam (5). The latter causes the compression of the mixture initially introduced in the cylindrical combustion chamber by pushing on the piston rod (6) perpendicular to its displacement. At full stroke a gradual oil shock absorber (7) stops the actuater.

The rapid compression machine has a combustion chamber (8) of 60-mm bore and 50-mm stroke. The compression ratio was 5.76 for experiments with n-butane and 6.49 for tests with isobutane, the height of the chamber at top dead center (tdc) being 10.5 mm and 9.1 mm, respectively, because of the use of different cams.

The whole of the cylindrical combustion chamber, initially filled at atmospheric pressure, was heated by a 440-W electrical resistance heater wrapped around the cylinder and operated by one on/off controller. The initial temperature was varied from 300 K to 360 K to obtain different temperature values at the end of compression.

The cylinder head was fitted with a 10-mm-diam. silicon dioxide window, placed at the center, in order to collect emission light from the compressed charge.

A piezo-electric pressure transducer (Kistler 601) and a photomultiplier (Hamamatsu R106) combined with a blue color filter (Kodak Wratten No. 47) were used for pressure and blue light emission records. Signals were recorded by a 12 bits digital oscilloscope (Nicolet 3091) and then transferred to a microcomputer (CPU Intel 8088).

The rapid compression machine was driven so that the compression was performed within about 20 ms. Assuming isentropic compression of an adiabatic core gas, which is expected to control the ignition delays,[11] temperatures of the compressed charges at the end of compression T_{tdc}

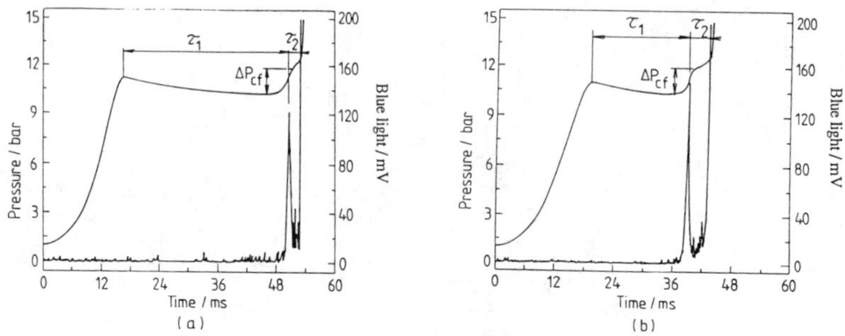

Fig. 2 Two stage ignition behaviour of n-butane for 2 different values of temperature T_{tdc}: a) T_{tdc} = 820 K; b) T_{tdc} = 836 K.

were calculated from the measured initial values of pressure and temperature and the final pressure at top dead center P_{tdc} . Thus, it is clear that this calculation does not consider the effect of in-cylinder heat transfer[12] accountable to the existence of a cooled layer at cylinder walls. Therefore, calculated temperatures represent the maximum possible temperature experienced by the contents of the combustion chamber.

In ignition delay measurements the pressure at the end of compression was set to remain constant around 10.7 bar or 12.6 bar, according to the value of the compression ratio. The temperature at the end of compression was varied, ranging from 770 K to 900 K.

Experimental Results and Discussion

N-Butane Autognition

Figure 2 shows the typical profiles in pressure and blue light emission vs time in the rapid compression machine for two values of the temperature, 820 K (Fig. 2a) and 836 K (Fig. 2b). In the range of experiment of temperatures above 800 K, the evolution of a two-stage autoignition phenomenon could be observed both on pressure and blue light emission records.

As clearly shown in Fig. 2b , after a first induction period τ_1 from the end of compression time, a pressure buildup ΔP_{cf} greater than 1 bar is observed, whereas a blue luminescence peak is recorded at the time fitting with the inflection point of the pressure increase ΔP_{cf} , corresponding to a maximum in reaction rate. The first stage flame, the so-called cool flame in the literature,[13,14] is then followed by a drop of reaction until the final hot ignition after a second induction time τ_2 , as revealed by the intense burst in pressure and blue light emission. Note that the maximum explosion pressure, followed by pressure oscillations that are characteristic of sound waves in the chamber at tdc,[15] has deliberately not been recorded so as to measure and determine with more accuracy the pressure buildups ΔP_{cf} and the values of τ_1 and τ_2 delays .

For temperatures below 800 K only the first stage could be recorded. Indeed, neither pressure raise ΔP_{cf} nor blue light emission peak could be detected prior to the explosion after an induction period τ , as shown in Fig. 3, where the temperature is only 800 K.

This tends to show a different chemical behavior when the temperature exceeds 800 K. Indeed, low-temperature reactions are then responsible for heat release, seen through the pressure augmentation ΔP_{cf}, and produce intermediate chemical species, such as excited formaldehyde,[16] accountable to the blue chemiluminescence before the occurrence of hot ignition.

Figure 4 shows the evolution of pressure histories for five different temperatures T_{tdc} ranging from 790 K to 869 K. The major features of pressure histories through the postcompression interval and the way that

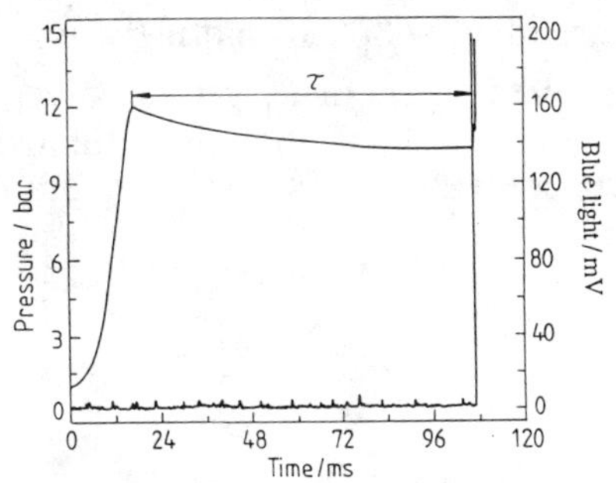

Fig. 3 One-stage ignition behaviour of n-butane at the temperature T_{tdc} = 800 K.

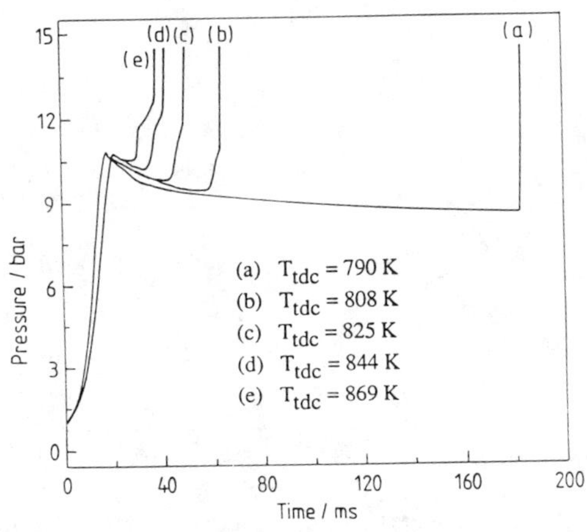

Fig. 4 Evolution of ignition behavior of n-butane for temperatures T_{tdc} ranging from 790 K to 869 K.

these records relate to temperature change in our experiments are qualitatively in good agreement with the phenomena observed by different authors.[4, 12, 17]

In Fig.5 the Arrhenius plots of the experimental ignition delay measurements, performed at the pressure P_{tdc} = 10.7 bar, illustrate their evolution vs the temperature established at the end of compression, which was varied between 770-900 K. The inverted and regular triangle symbols are the first induction period τ_1 and the second induction period τ_2 respectively, which approximately lay out on the straight dotted lines. The black dots linked by the solid line are representative of the evolution of the global ignition delay $\tau = \tau_1 + \tau_2$, which is shortened when the temperature is raised.

It is interesting to note that the extension of τ_2 as τ_1 is shortened when the temperature is raised. Furthermore, the change of behavior as the temperature exceeds 800 K can be observed on Arrhenius curves, assuming global expressions for delays related to the temperature in the form $\tau_1 = A_1 \exp(E_1/RT)$ and $\tau_2 = A_2 \exp(-E_2/RT)$.[17] Indeed, as τ_2 could not be measured below 800 K and if the experimental global delay dots τ may be considered as the measurement of τ_1, it appears that the τ_1 delay curve bends, in the sense that the global activation energy E_1 decreases as the temperature exceeds 800 K. In return, the global activation

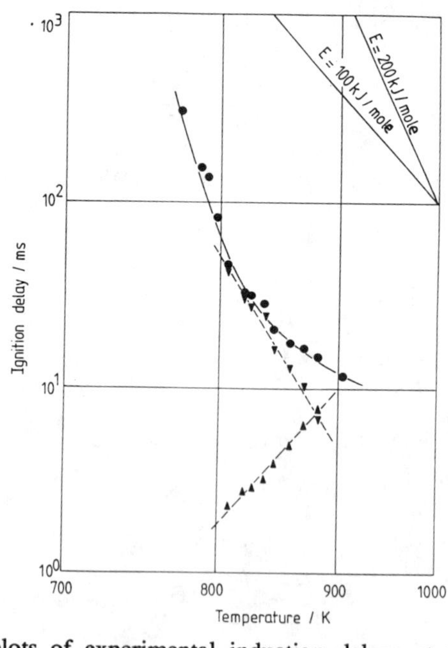

Fig. 5 Arrhenius plots of experimental induction delays at the pressure around 10.7 bar for n-$C_4H_{10}/O_2/Ar$ = 2.91 / 21.00 / 79.00, Φ = 0.9.
● $\tau = \tau_1 + \tau_2$; ▼ τ_1; ▲ τ_2.

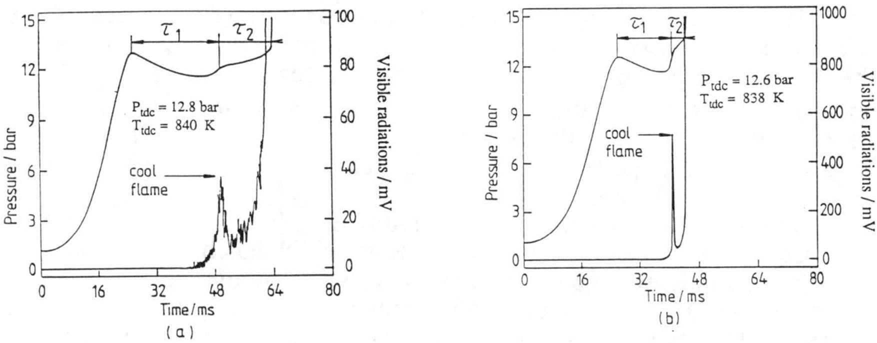

Fig. 6 Comparison of isobutane (a) and n-butane (b) autoignition behavior.

energies E_1 and E_2 are observed to remain constant (E_1 = 145 kJ/mole, and E_2 = 100 kJ/mole) in the temperature range of 800-900 K. Thus, the fact that the global delay τ curve is bent as the temperature is increased towards 900 K is well explained by the evolution of τ_1 and τ_2 .

It must be stated that for temperatures above 900 K the measurement of induction delays τ_1 and τ_2 are quite critical because their durations become shorter than the time of compression, which is about 20 ms. In these conditions it is difficult to determine the real start of chemical activity along the compression interval, if we take into account the influence of the compression duration.[10] Furthermore, the temperature value of 900 K was the upper limit of good operation of the machine, i.e. providing fast enough compression. Nevertheless, it would be interesting to observe the behavior for temperatures ranging from 900 K to 1000 K so as to determine if the "negative temperature coefficient" domain can be perceived, where the global delay τ is extended as the temperature is raised before it is shortened again.[4]

Isobutane Autoignition

In the range of experiment temperatures for isobutane, varying from 790 K to 890 K, the autoignition phenomenon is always a two-stage process, having nevertheless different characteristics than those of n-butane.

In Fig. 6 are compared pressure histories and visible radiation records related to isobutane autoignition at P_{tdc} = 12.8 bar, T_{tdc} = 840 K (Fig. 6a) and to n-butane autoignition at P_{tdc} = 12.6 bar, T_{tdc} = 838 K (Fig. 6b). These records allow one to outline the typical differences of behavior between the two fuels:

1) First, for very similar experimental conditions, the global autoignition delay value is much longer in the case of isobutane.

2) Second, the pressure augmentation due to the development of a cool flame in the isobutane/oxidizer mixture is more than twice as less as the butane/oxidizer mixture. This confirms the results obtained by Green and co-workers[8,9] indicating that under similar operating conditions less heat is generated during the development of the low-temperature flame, producing a more moderate pressure increase in the compressed gas of isobutane than in the case of n-butane.

3) Finally, note that the cool flame luminescence is extremely weak in the case of isobutane. The visible radiation, which has been collected by the photomultiplier tube on both records of Fig. 6 without the use of a blue filter, shows that the chemiluminescence peak level due to excited intermediate species derived from the oxidation of isobutane is 15 times lower than that of light emitted by oxidized products of n-butane.

Figure 7 shows the Arrhenius plots of τ and τ_1 and τ_2 isobutane ignition delays measurements performed under a pressure condition of 12.6 bar. These curves can be compared with those of n-butane illustrated in Fig. 5. In the range of experiment temperatures for isobutane, as contrasted to n-butane, no one-stage behavior could be detected. The τ_1 data fit fairly well a straight line within the whole range of experiment temperatures, whose slope allows one to determine the global activation energy E_1 = 240 kJ/mole. In return, the variation of τ_2 vs the temperature shows a more complex dependency as it first decreases as the temperature

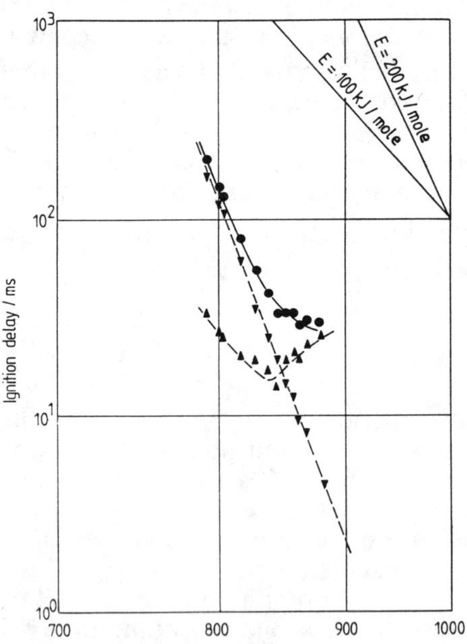

Fig. 7 Arrhenius plots of experimental induction delays at the pressure around 12.6 bar for $i\text{-}C_4H_{10}/O_2/Ar$ = 2.91 / 21.00 / 79.00, Φ = 0.9.
● $\tau = \tau_1 + \tau_2$; ▼ τ_1; ▲ τ_2.

is raised from 790 K before it increases, showing a minimum around 840 K. This, of course, induces a bending of the global delay curve as the temperature is raised over 840 K. Furthermore, the τ_2 delay curve remains much higher than that of n-butane, as compared to τ_1 straight line.

Conclusions and Perspectives

This study characterizes the autoignition process of n-butane and isobutane in the range of temperature between 800-900 K, which are typical values of end-gas temperature prior to knock occurrence in spark ignition engines.[18,19] It showed the capability of the rapid compression machine designed and built in our laboratory of achieving this task and providing data for autoignition delays for pressures of 10.7 bar and 12.6 bar and temperatures ranging from 770 K to 900 K, which corresponds to the field of intermediate temperature chemistry, which corresponds to the field of intermediate temperature governing engine knock onset and where very little data have yet been collected.[4]

This study showed that the autoignition of the mixture consisting of $n\text{-}C_4H_{10} / O_2 / Ar = 2.91/ 21.00/ 79.00$, with an equivalence ratio of 0.9, is a one-stage process below 800 K and becomes a two-stage process as the temperature exceeds 800 K, indicating that it is not possible to describe with proper accuracy the evolution of global autoignition delay by a one-step mechanism solely. The existence of the second stage t_2 has been evidenced for temperatures above 800 K with a time dependence related to the temperature, showing a negative temperature coefficient.

The two-stage process has been evidenced by both pressure increases and blue light emission records, indicating the development of cool flames[16,20] responsible for heat release and producing excited chemical species.

In addition, experiments made with isobutane showed some major differences from n-butane in the development of the two-stage process of autoignition:

1) The heat released by cool flame is less important at the end of the first induction delay τ_1.

2) The luminescence related to the intermediate products of cool flame is far weaker.

3) The second induction delay τ_2 always remains much higher, as compared to the first induction delay τ_1, which decreases much more with increasing temperature. The global activation energy E_1 of isobutane is 2.4 times that of n-butane.

This tends to prove the important role of the oxidation mechanisms during the τ_2 lag leading to hot ignition, which are initiated with stable intermediate species produced at the end of τ_1 and, thus, depending on the reactive nature of the τ_1 oxidation mechanisms, result in a more or less significant heating of the altered fuel/ oxidizer mixture.

In future studies, we intend to perform laser shadow or schlieren high-speed cinematographies in order to visualize the development of autoignition and the structure of the change in the compressed charge, especially during the transition from the first to the second stage and to observe accurately where in the chamber the reactions are initiated, so as to check the validity of the assumption of the existence of an adiabatic core gas. Furthermore, the influence of equivalence ratio on the two-stage ignition behavior of the tow isomeric structures of butane in fuel/oxidizer mixtures will be studied.

References

[1]Gabano, J.D., Kageyama, T., Fisson, F., and Leyer, J.C., "Experimental Simulation of Engine Knock by Means of a Preheated Static Combustion Chamber," 22 nd Symposium (International) on Combustion, The Combustion Institute, Pittsburgh, USA, 1989, pp. 447-454.

[2]Pitz, W.J., Westbrook, C.K., "Interactions Between Laminar Flame and End-gas Autoignition," Progress in Aeronautics and Astronautics: Dynamics of Reactive Systems, Pt.1, Vol. 105, AIAA, New York, USA, 1986, pp.69-77.

[3]Kono, M., Shiga, S., Kumagai, S., and Iinuma, K., "Thermodynamic and Experimental Determinations of Knock Intensity by Using a Spark-Ignited Rapid Compression Machine," Combustion and Flame, Vol. 54, New York, USA, 1983, pp. 33-47.

[4]Ohta, Y., and Takahashi, H. "N-Butane Ignition in a Wide Range of Temperature," Progress in Aeronautics and Astronautics, Vol. 113, AIAA, New York, USA, 1988, pp.225-237.

[5]Shiga, S., Kono, M., Iinuma, K., Karasawa, T., and Kurabayashi, T., "Further Investigation of Knock Intensity by a Thermodynamic Model and Experiments Using a Rapid Compression Machine," Paper presented at International Symposium on Diagnostics and Modeling of Combustion in Reciprocationg Engines (COMODIA 85), Tokyo, 1985.

[6]Leppard, W.R.; "The Autoignition Chemistry of N-Butane: An Experimental Study," Society of Automotive Engineers, TP No. 872150, Warrendale, PA., USA, 1987.

[7]Ohta, Y., and Takahashi, H., "Homogeneity and Propagation of Autoignited Cool and Blue Flames," Progress in Aeronautics and Astronautics, Vol.95, AIAA, New York, USA, 1984, pp.236-247.

[8]Green, R.M., Cernansky,N.P., Pitz, W.J. and Westbrook, C.K., "The Role of Low Temperature Chemistry in the Autoignition of n-Butane," Society of Automotive Engineers, TP No. 872108, Warrendale, PA, USA, 1987.

[9]Green, R.M., Parker, C.D., Pitz, W.J. and Westbrook,C.K., "The Autoignition of Isobutane in a Knocking Spark Ignition Engine," Society of Automotive Engineers, TP No. 870169, Warrendale, PA, USA, 1987.

[10]Ohta, Y., Hayashi, A.K., Takahashi, H., and Fujiwara, T., "Consequence of Temperature-pressure-time History for Autoignition," Progress in Aeronautics and Astronautics: Dynamics of Reactive Systems, Pt.1, Vol. 105, AIAA, New York, USA, 1984, pp.93-103.

[11]Keck, J., and Hu, H., "Autoignition of Adiabatically Compressed Combustible Gas Mixtures," Society of Automotive Engineers, TP No. 872110, Warrendale, PA, USA, 1987.

[12]Franck, J., Griffiths, J.F., and Nimmo. W., "The Control of Spontaneous Ignition Under Rapid Compression," 21st Symposium (International) on Combustion, The Combustion Institute, Pittsburgh, USA, 1986, p. 447.

[13]Levedahl, W.J., "Multistage Autoignition of Engine Fuels," 5th Symposium (International) on Combustion, The Combustion Institute, Pittsburgh, USA, 1955, pp.372-385.

[14]Jost, W., "Knock Reaction," 9th Symposium (International) on Combustion, The Combustion Institute, Pittsburgh, USA, 1963, pp.1013-1022.

[15]Griffiths, J.F., and Nimmo, W., "Spontaneous Ignition and Engine Knock under Rapid Compression," Combustion and Flame, Vol. 60, New York, USA, 1985, pp.215-218.

[16]Sokolik, A.S., "Self Ignition and Detonation in Gases," Israël Program for Scientific Translations, Jerusalem, ISRAEL, 1963.

[17]Rögener, H., Zeitschrift für Electrochemie, Vol. 53, 1949, p.389.

[18]Gluckstein, M.E., and Walcutt, C., "End-gas Temperature-pressure Histories and their Relation to Knock," Society of Automotive Engineers Transactions, Vol. 69, Warrendale, PA, USA, 1961, pp.529-533.

[19]Marie, J.J., and Cottereau, M.J., "Single-shot Temperature Measurement by CARS in an I.C. Engine for Normal and Knocking Conditions," Society of Automotive Engineers, TP No. 870458, Warrendale, PA, USA, 1987.

[20]Malmberg, E.W., Smith, M.L., Bigler, J.E., and Bobbit, J.A., "A Study of Cool Flames and Associated Reactions in an Engine," 5th Symposium (International) on Combustion, The Combustion Institute, Pittsburgh, USA, 1955, pp.385-392.

Author Index

PROGRESS IN ASTRONAUTICS AND AERONAUTICS SERIES VOLUMES

*1. **Solid Propellant Rocket Research** (1960)
Martin Summerfield
Princeton University

*2. **Liquid Rockets and Propellants** (1960)
Loren E. Bollinger
Ohio State University
Martin Goldsmith
The Rand Corp.
Alexis W. Lemmon Jr.
Battelle Memorial Institute

*3. **Energy Conversion for Space Power** (1961)
Nathan W. Snyder
Institute for Defense Analyses

*4. **Space Power Systems** (1961)
Nathan W. Snyder
Institute for Defense Analyses

*5. **Electrostatic Propulsion** (1961)
David B. Langmuir
Space Technology Laboratories, Inc.
Ernst Stuhlinger
NASA George C. Marshall Space Flight Center
J.M. Sellen Jr.
Space Technology Laboratories, Inc.

*6. **Detonation and Two-Phase Flow** (1962)
S.S. Penner
California Institute of Technology
F.A. Williams
Harvard University

*7. **Hypersonic Flow Research** (1962)
Frederick R. Riddell
AVCO Corp.

*8. **Guidance and Control** (1962)
Robert E. Roberson, Consultant
James S. Farrior
Lockheed Missiles and Space Co.

*9. **Electric Propulsion Development** (1963)
Ernst Stuhlinger
NASA George C. Marshall Space Flight Center

*10. **Technology of Lunar Exploration** (1963)
Clifford I. Cummings
Harold R. Lawrence
Jet Propulsion Laboratory

*11. **Power Systems for Space Flight** (1963)
Morris A. Zipkin
Russell N. Edwards
General Electric Co.

*12. **Ionization in High-Temperature Gases** (1963)
Kurt E. Shuler, Editor
National Bureau of Standards
John B. Fenn, Associate Editor
Princeton University

*13. **Guidance and Control – II** (1964)
Robert C. Langford
General Precision Inc.
Charles J. Mundo
Institute of Naval Studies

*14. **Celestial Mechanics and Astrodynamics** (1964)
Victor G. Szebehely
Yale University Observatory

*15. **Heterogeneous Combustion** (1964)
Hans G. Wolfhard
Institute for Defense Analyses
Irvin Glassman
Princeton University
Leon Green Jr.
Air Force Systems Command

16. **Space Power Systems Engineering** (1966)
George C. Szego
Institute for Defense Analyses
J. Edward Taylor
TRW Inc.

17. **Methods in Astrodynamics and Celestial Mechanics** (1966)
Raynor L. Duncombe
U.S. Naval Observatory
Victor G. Szebehely
Yale University Observatory

18. **Thermophysics and Temperature Control of Spacecraft and Entry Vehicles** (1966)
Gerhard B. Heller
NASA George C. Marshall Space Flight Center

---
*Out of print.

*19. **Communication Satellite Systems Technology** (1966)
Richard B. Marsten
Radio Corporation of America

*20. **Thermophysics of Spacecraft and Planetary Bodies: Radiation Properties of Solids and the Electromagnetic Radiation Environment in Space** (1967)
Gerhard B. Heller
NASA George C. Marshall Space Flight Center

21. **Thermal Design Principles of Spacecraft and Entry Bodies** (1969)
Jerry T. Bevans
TRW Systems

22. **Stratospheric Circulation** (1969)
Willis L. Webb
Atmospheric Sciences Laboratory, White Sands, and University of Texas at El Paso

23. **Thermophysics: Applications to Thermal Design of Spacecraft** (1970)
Jerry T. Bevans
TRW Systems

24. **Heat Transfer and Spacecraft Thermal Control** (1971)
John W. Lucas
Jet Propulsion Laboratory

25. **Communication Satellites for the 70's: Technology** (1971)
Nathaniel E. Feldman
The Rand Corp.
Charles M. Kelly
The Aerospace Corp.

26. **Communication Satellites for the 70's: Systems** (1971)
Nathaniel E. Feldman
The Rand Corp.
Charles M. Kelly
The Aerospace Corp.

27. **Thermospheric Circulation** (1972)
Willis L. Webb
Atmospheric Sciences Laboratory, White Sands, and University of Texas at El Paso

28. **Thermal Characteristics of the Moon** (1972)
John W. Lucas
Jet Propulsion Laboratory

29. **Fundamentals of Spacecraft Thermal Design** (1972)
John W. Lucas
Jet Propulsion Laboratory

30. **Solar Activity Observations and Predictions** (1972)
Patrick S. McIntosh
Murray Dryer
Environmental Research Laboratories, National Oceanic and Atmospheric Administration

31. **Thermal Control and Radiation** (1973)
Chang-Lin Tien
University of California at Berkeley

32. **Communications Satellite Systems** (1974)
P.L. Bargellini
COMSAT Laboratories

33. **Communications Satellite Technology** (1974)
P.L. Bargellini
COMSAT Laboratories

34. **Instrumentation for Airbreathing Propulsion** (1974)
Allen E. Fuhs
Naval Postgraduate School
Marshall Kingery
Arnold Engineering Development Center

35. **Thermophysics and Spacecraft Thermal Control** (1974)
Robert G. Hering
University of Iowa

36. **Thermal Pollution Analysis** (1975)
Joseph A. Schetz
Virginia Polytechnic Institute
ISBN 0-915928-00-0

37. **Aeroacoustics: Jet and Combustion Noise; Duct Acoustics** (1975)
Henry T. Nagamatsu, Editor
General Electric Research and Development Center
Jack V. O'Keefe, Associate Editor
The Boeing Co.
Ira R. Schwartz, Associate Editor
NASA Ames Research Center
ISBN 0-915928-01-9

38. **Aeroacoustics: Fan, STOL, and Boundary Layer Noise; Sonic Boom; Aeroacoustics Instrumentation** (1975)
Henry T. Nagamatsu, Editor
General Electric Research and Development Center
Jack V. O'Keefe, Associate Editor
The Boeing Co.
Ira R. Schwartz, Associate Editor
NASA Ames Research Center
ISBN 0-915928-02-7

39. Heat Transfer with Thermal Control Applications (1975)
M. Michael Yovanovich
University of Waterloo
ISBN 0-915928-03-5

40. Aerodynamics of Base Combustion (1976)
S.N.B. Murthy, Editor
J.R. Osborn,
Associate Editor
Purdue University
A.W. Barrows
J.R. Ward,
Associate Editors
Ballistics Research Laboratories
ISBN 0-915928-04-3

41. Communications Satellite Developments: Systems (1976)
Gilbert E. LaVean
Defense Communications Agency
William G. Schmidt
CML Satellite Corp.
ISBN 0-915928-05-1

42. Communications Satellite Developments: Technology (1976)
William G. Schmidt
CML Satellite Corp.
Gilbert E. LaVean
Defense Communications Agency
ISBN 0-915928-06-X

43. Aeroacoustics: Jet Noise, Combustion and Core Engine Noise (1976)
Ira R. Schwartz, Editor
NASA Ames Research Center
Henry T. Nagamatsu,
Associate Editor
General Electric Research and Development Center
Warren C. Strahle,
Associate Editor
Georgia Institute of Technology
ISBN 0-915928-07-8

44. Aeroacoustics: Fan Noise and Control; Duct Acoustics; Rotor Noise (1976)
Ira R. Schwartz, Editor
NASA Ames Research Center
Henry T. Nagamatsu,
Associate Editor
General Electric Research and Development Center
Warren C. Strahle,
Associate Editor
Georgia Institute of Technology
ISBN 0-915928-08-6

45. Aeroacoustics: STOL Noise; Airframe and Airfoil Noise (1976)
Ira R. Schwartz, Editor
NASA Ames Research Center
Henry T. Nagamatsu,
Associate Editor
General Electric Research and Development Center
Warren C. Strahle,
Associate Editor
Georgia Institute of Technology
ISBN 0-915928-09-4

46. Aeroacoustics: Acoustic Wave Propagation; Aircraft Noise Prediction; Aeroacoustic Instrumentation (1976)
Ira R. Schwartz, Editor
NASA Ames Research Center
Henry T. Nagamatsu,
Associate Editor
General Electric Research and Development Center
Warren C. Strahle,
Associate Editor
Georgia Institute of Technology
ISBN 0-915928-10-8

47. Spacecraft Charging by Magnetospheric Plasmas (1976)
Alan Rosen
TRW Inc.
ISBN 0-915928-11-6

48. Scientific Investigations on the Skylab Satellite (1976)
Marion I. Kent
Ernst Stuhlinger
NASA George C. Marshall Space Flight Center
Shi-Tsan Wu
University of Alabama
ISBN 0-915928-12-4

49. Radiative Transfer and Thermal Control (1976)
Allie M. Smith
ARO Inc.
ISBN 0-915928-13-2

50. Exploration of the Outer Solar System (1976)
Eugene W. Greenstadt
TRW Inc.
Murray Dryer
National Oceanic and Atmospheric Administration
Devrie S. Intriligator
University of Southern California
ISBN 0-915928-14-0

51. Rarefied Gas Dynamics, Parts I and II (two volumes) (1977)
J. Leith Potter
ARO Inc.
ISBN 0-915928-15-9

52. Materials Sciences in Space with Application to Space Processing (1977)
Leo Steg
General Electric Co.
ISBN 0-915928-16-7

53. Experimental Diagnostics in Gas Phase Combustion Systems (1977)
Ben T. Zinn, Editor
Georgia Institute of Technology
Craig T. Bowman, Associate Editor
Stanford University
Daniel L. Hartley, Associate Editor
Sandia Laboratories
Edward W. Price, Associate Editor
Georgia Institute of Technology
James G. Skifstad, Associate Editor
Purdue University
ISBN 0-015928-18-3

54. Satellite Communications: Future Systems (1977)
David Jarett
TRW Inc.
ISBN 0-915928-18-3

55. Satellite Communications: Advanced Technologies (1977)
David Jarett
TRW Inc.
ISBN 0-915928-19-1

56. Thermophysics of Spacecraft and Outer Planet Entry Probes (1977)
Allie M. Smith
ARO Inc.
ISBN 0-915928-20-5

57. Space-Based Manufacturing from Nonterrestrial Materials (1977)
Gerard K. O'Neill, Editor
Brian O'Leary, Assistant Editor
Princeton University
ISBN 0-915928-21-3

58. Turbulent Combustion (1978)
Lawrence A. Kennedy
State University of New York at Buffalo
ISBN 0-915928-22-1

59. Aerodynamic Heating and Thermal Protection Systems (1978)
Leroy S. Fletcher
University of Virginia
ISBN 0-915928-23-X

60. Heat Transfer and Thermal Control Systems (1978)
Leroy S. Fletcher
University of Virginia
ISBN 0-915928-24-8

61. Radiation Energy Conversion in Space (1978)
Kenneth W. Billman
NASA Ames Research Center
ISBN 0-915928-26-4

62. Alternative Hydrocarbon Fuels: Combustion and Chemical Kinetics (1978)
Craig T. Bowman
Stanford University
Jorgen Birkeland
Department of Energy
ISBN 0-915928-25-6

63. Experimental Diagnostics in Combustion of Solids (1978)
Thomas L. Boggs
Naval Weapons Center
Ben T. Zinn
Georgia Institute of Technology
ISBN 0-915928-28-0

64. Outer Planet Entry Heating and Thermal Protection (1979)
Raymond Viskanta
Purdue University
ISBN 0-915928-29-9

65. Thermophysics and Thermal Control (1979)
Raymond Viskanta
Purdue University
ISBN 0-915928-30-2

66. Interior Ballistics of Guns (1979)
Herman Krier
University of Illinois at Urbana-Champaign
Martin Summerfield
New York University
ISBN 0-915928-32-9

***67. Remote Sensing of Earth from Space: Role of "Smart Sensors"**(1979)
Roger A. Breckenridge
NASA Langley Research Center
ISBN 0-915928-33-7

68. Injection and Mixing in Turbulent Flow (1980)
Joseph A. Schetz
Virginia Polytechnic Institute and State University
ISBN 0-915928-35-3

69. Entry Heating and Thermal Protection (1980)
Walter B. Olstad
NASA Headquarters
ISBN 0-915928-38-8

70. Heat Transfer, Thermal Control, and Heat Pipes (1980)
Walter B. Olstad
NASA Headquarters
ISBN 0-915928-39-6

71. Space Systems and Their Interactions with Earth's Space Environment (1980)
Henry B. Garrett
Charles P. Pike
Hanscom Air Force Base
ISBN 0-915928-41-8

72. Viscous Flow Drag Reduction (1980)
Gary R. Hough
Vought Advanced Technology Center
ISBN 0-915928-44-2

73. Combustion Experiments in a Zero-Gravity Laboratory (1981)
Thomas H. Cochran
NASA Lewis Research Center
ISBN 0-915928-48-5

74. Rarefied Gas Dynamics, Parts I and II (two volumes) (1981)
Sam S. Fisher
University of Virginia
ISBN 0-915928-51-5

75. Gasdynamics of Detonations and Explosions (1981)
J.R. Bowen
University of Wisconsin at Madison
N. Manson
Université de Poitiers
A.K. Oppenheim
University of California at Berkeley
R.I. Soloukhin
Institute of Heat and Mass Transfer, BSSR Academy of Sciences
ISBN 0-915928-46-9

76. Combustion in Reactive Systems (1981)
J.R. Bowen
University of Wisconsin at Madison
N. Manson
Université de Poitiers
A.K. Oppenheim
University of California at Berkeley
R.I. Soloukhin
Institute of Heat and Mass Transfer, BSSR Academy of Sciences
ISBN 0-915928-47-7

77. Aerothermodynamics and Planetary Entry (1981)
A.L. Crosbie
University of Missouri-Rolla
ISBN 0-915928-52-3

78. Heat Transfer and Thermal Control (1981)
A.L. Crosbie
University of Missouri-Rolla
ISBN 0-915928-53-1

79. Electric Propulsion and Its Applications to Space Missions (1981)
Robert C. Finke
NASA Lewis Research Center
ISBN 0-915928-55-8

80. Aero-Optical Phenomena (1982)
Keith G. Gilbert
Leonard J. Otten
Air Force Weapons Laboratory
ISBN 0-915928-60-4

81. Transonic Aerodynamics (1982)
David Nixon
Nielsen Engineering & Research, Inc.
ISBN 0-915928-65-5

82. Thermophysics of Atmospheric Entry (1982)
T.E. Horton
University of Mississippi
ISBN 0-915928-66-3

83. Spacecraft Radiative Transfer and Temperature Control (1982)
T.E. Horton
University of Mississippi
ISBN 0-915928-67-1

84. Liquid-Metal Flows and Magnetohydrodynamics (1983)
H. Branover
Ben-Gurion University of the Negev
P.S. Lykoudis
Purdue University
A. Yakhot
Ben-Gurion University of the Negev
ISBN 0-915928-70-1

85. Entry Vehicle Heating and Thermal Protection Systems: Space Shuttle, Solar Starprobe, Jupiter Galileo Probe (1983)
Paul E. Bauer
McDonnell Douglas Astronautics Co.
Howard E. Collicott
The Boeing Co.
ISBN 0-915928-74-4

86. Spacecraft Thermal Control, Design, and Operation (1983)
Howard E. Collicott
The Boeing Co.
Paul E. Bauer
McDonnell Douglas Astronautics Co.
ISBN 0-915928-75-2

87. Shock Waves, Explosions, and Detonations (1983)
J.R. Bowen
University of Washington
N. Manson
Université de Poitiers
A.K. Oppenheim
University of California at Berkeley
R.I. Soloukhin
Institute of Heat and Mass Transfer, BSSR Academy of Sciences
ISBN 0-915928-76-0

88. Flames, Lasers, and Reactive Systems (1983)
J.R. Bowen
University of Washington
N. Manson
Université de Poitiers
A.K. Oppenheim
University of California at Berkeley
R.I. Soloukhin
Institute of Heat and Mass Transfer, BSSR Academy of Sciences
ISBN 0-915928-77-9

89. Orbit-Raising and Maneuvering Propulsion: Research Status and Needs (1984)
Leonard H. Caveny
Air Force Office of Scientific Research
ISBN 0-915928-82-5

90. Fundamentals of Solid-Propellant Combustion (1984)
Kenneth K. Kuo
Pennsylvania State University
Martin Summerfield
Princeton Combustion Research Laboratories, Inc.
ISBN 0-915928-84-1

91. Spacecraft Contamination: Sources and Prevention (1984)
J.A. Roux
University of Mississippi
T.D. McCay
NASA Marshall Space Flight Center
ISBN 0-915928-85-X

92. Combustion Diagnostics by Nonintrusive Methods (1984)
T.D. McCay
NASA Marshall Space Flight Center
J.A. Roux
University of Mississippi
ISBN 0-915928-86-8

93. The INTELSAT Global Satellite System (1984)
Joel Alper
COMSAT Corp.
Joseph Pelton
INTELSAT
ISBN 0-915928-90-6

94. Dynamics of Shock Waves, Explosions, and Detonations (1984)
J.R. Bowen
University of Washington
N. Manson
Université de Poitiers
A.K. Oppenheim
University of California at Berkely
R.I. Soloukhin
Institute of Heat and Mass Transfer, BSSR Academy of Sciences
ISBN 0-915928-91-4

95. Dynamics of Flames and Reactive Systems (1984)
J.R. Bowen
University of Washington
N. Manson
Université de Poitiers
A.K. Oppenheim
University of California at Bereley
R.I. Soloukhin
Institute of Heat and Mass Transfer, BSSR Academy of Sciences
ISBN 0-915928-92-2

96. Thermal Design of Aeroassisted Orbital Transfer Vehicles (1985)
H.F. Nelson
University of Missouri-Rolla
ISBN 0-915928-94-9

97. Monitoring Earth's Ocean, Land, and Atmosphere from Space – Sensors, Systems, and Applications (1985)
Abraham Schnapf
Aerospace Systems Engineering
ISBN 0-915928-98-1

98. Thrust and Drag: Its Prediction and Verification (1985)
Eugene E. Covert
Massachusetts Institute of Technology
C.R. James
Vought Corp.
William F. Kimzey
Sverdrup Technology AEDC Group
George K. Richey
U.S. Air Force
Eugene C. Rooney
U.S. Navy Department of Defense
ISBN 0-930403-00-2

99. Space Stations and Space Platforms – Concepts, Design, Infrastructure, and Uses (1985)
Ivan Bekey
Daniel Herman
NASA Headquarters
ISBN 0-930403-01-0

100. Single- and Multi-Phase Flows in an Electromagnetic Field: Energy, Metallurgical, and Solar Applications (1985)
Herman Branover
Ben-Gurion University of the Negev
Paul S. Lykoudis
Purdue University
Michael Mond
Ben-Gurion University of the Negev
ISBN 0-930403-04-5

101. MHD Energy Conversion: Physiotechnical Problems (1986)
V.A. Kirillin
A.E. Sheyndlin
Soviet Academy of Sciences
ISBN 0-930403-05-3

102. Numerical Methods for Engine-Airframe Integration (1986)
S.N.B. Murthy
Purdue University
Gerald C. Paynter
Boeing Airplane Co.
ISBN 0-930403-09-6

103. Thermophysical Aspects of Re-Entry Flows (1986)
James N. Moss
NASA Langley Research Center
Carl D. Scott
NASA Johnson Space Center
ISBN 0-930403-10-X

104. Tactical Missile Aerodynamics (1986)
M.J. Hemsch
PRC Kentron, Inc.
J.N. Nielsen
NASA Ames Research Center
ISBN 0-930403-13-4

105. Dynamics of Reactive Systems Part I: Flames and Configurations; Part II: Modeling and Heterogeneous Combustion (1986)
J.R. Bowen
University of Washington
J.-C. Leyer
Université de Poitiers
R.I. Soloukhin
Institute of Heat and Mass Transfer, BSSR Academy of Sciences
ISBN 0-930403-14-2

106. Dynamics of Explosions (1986)
J.R. Bowen
University of Washington
J.-C. Leyer
Université de Poitiers
R.I. Soloukhin
Institute of Heat and Mass Transfer, BSSR Academy of Sciences
ISBN 0-930403-15-0

107. Spacecraft Dielectric Material Properties and Spacecraft Charging (1986)
A.R. Frederickson
U.S. Air Force Rome Air Development Center
D.B. Cotts
SRI International
J.A. Wall
U.S. Air Force Rome Air Development Center
F.L. Bouquet
Jet Propulsion Laboratory, California Institute of Technology
ISBN 0-930403-17-7

108. Opportunities for Academic Research in a Low-Gravity Environment (1986)
George A. Hazelrigg
National Science Foundation
Joseph M. Reynolds
Louisiana State University
ISBN 0-930403-18-5

109. Gun Propulsion Technology (1988)
Ludwig Stiefel
U.S. Army Armament Research, Development and Engineering Center
ISBN 0-930403-20-7

110. Commercial Opportunities in Space (1988)
F. Shahrokhi
K.E. Harwell
University of Tennessee Space Institute
C.C. Chao
National Cheng Kung University
ISBN 0-930403-39-8

111. Liquid-Metal Flows: Magnetohydrodynamics and Applications (1988)
Herman Branover, Michael Mond, and Yeshajahu Unger
Ben-Gurion University of the Negev
ISBN 0-930403-43-6

112. Current Trends in Turbulence Research (1988)
Herman Branover, Michael Mond, and Yeshajahu Unger
Ben-Gurion University of the Negev
ISBN 0-930403-44-4

113. Dynamics of Reactive Systems Part I: Flames; Part II: Heterogeneous Combustion and Applications (1988)
A.L. Kuhl
R & D Associates
J.R. Bowen
University of Washington
J.-C. Leyer
Université de Poitiers
A. Borisov
USSR Academy of Sciences
ISBN 0-930403-46-0

114. Dynamics of Explosions (1988)
A.L. Kuhl
R & D Associates
J.R. Bowen
University of Washington
J.-C. Leyer
Université de Poitiers
A. Borisov
USSR Academy of Sciences
ISBN 0-930403-47-9

115. Machine Intelligence and Autonomy for Aerospace (1988)
E. Heer
Heer Associates, Inc.
H. Lum
NASA Ames Research Center
ISBN 0-930403-48-7

116. Rarefied Gas Dynamics: Space-Related Studies (1989)
E.P. Muntz
University of Southern California
D.P. Weaver
U.S. Air Force Astronautics Laboratory (AFSC)
D.H. Campbell
University of Dayton Research Institute
ISBN 0-930403-53-3

117. Rarefied Gas Dynamics: Physical Phenomena (1989)
E.P. Muntz
University of Southern California
D.P. Weaver
U.S. Air Force Astronautics Laboratory (AFSC)
D. Campbell
University of Dayton Research Institute
ISBN 0-930403-54-1

118. Rarefied Gas Dynamics: Theoretical and Computational Techniques (1989)
E.P. Muntz
University of Southern California
D.P. Weaver
U.S. Air Force Astronautics Laboratory (AFSC)
D.H. Campbell
University of Dayton Research Institute
ISBN 0-930403-55-X

119. Test and Evaluation of the Tactical Missile (1989)
Emil J. Eichblatt Jr.
Pacific Missile Test Center
ISBN 0-930403-56-8

120. Unsteady Transonic Aerodynamics (1989)
David Nixon
Nielsen Engineering & Research, Inc.
ISBN 0-930403-52-5

121. Orbital Debris from Upper-Stage Breakup (1989)
Joseph P. Loftus Jr.
NASA Johnson Space Center
ISBN 0-930403-58-4

122. Thermal-Hydraulics for Space Power, Propulsion and Thermal Management System Design (1989)
William J. Krotiuk
General Electric Co.
ISBN 0-930403-64-9

123. Viscous Drag Reduction in Boundary Layers (1990)
Dennis M. Bushnell
Jerry N. Hefner
NASA Langley Research Center
ISBN 0-930403-66-5

124. Tactical and Strategic Missile Guidance (1990)
Paul Zarchan
Charles Stark Draper Laboratory, Inc.
ISBN 0-930403-68-1

125. Applied Computational Aerodynamics (1990)
P.A. Henne
Douglas Aircraft Company
ISBN 0-930403-69-X

126. Space Commercialization: Launch Vehicles and Programs (1990)
F. Shahrokhi
University of Tennessee Space Institute
J.S. Greenberg
Princeton Synergetics Inc.
T. Al-Saud
Ministry of Defense and Aviation Kingdom of Saudi Arabia
ISBN 0-930403-75-4

127. Space Commercialization: Platforms and Processing (1990)
F. Shahrokhi
University of Tennessee Space Institute
G. Hazelrigg
National Science Foundation
R. Bayuzick
Vanderbilt University
ISBN 0-930403-76-2

128. Space Commercialization: Satellite Technology (1990)
F. Shahrokhi
University of Tennessee Space Institute
N. Jasentuliyana
United Nations
N. Tarabzouni
King Abulaziz City for Science and Technology
ISBN 0-930403-77-0

129. Mechanics and Control of Large Flexible Structures (1990)
John L. Junkins
Texas A&M University
ISBN 0-930403-73-8

130. Low-Gravity Fluid Dynamics and Transport Phenomena (1990)
Jean N. Koster
Robert L. Sani
University of Colorado at Boulder
ISBN 0-930403-74-6

131. Dynamics of Deflagrations and Reactive Systems: Flames (1991)
A. L. Kuhl
Lawrence Livermore National Laboratory
J.-C. Leyer
Université de Poitiers
A. A. Borisov
USSR Academy of Sciences
W. A. Sirignano
University of California
ISBN 0-930403-95-9

132. Dynamics of Deflagrations and Reactive Systems: Heterogeneous Combustion (1991)
A. L. Kuhl
Lawrence Livermore National Laboratory
J.-C. Leyer
Université de Poitiers
A. A. Borisov
USSR Academy of Sciences
W. A. Sirignano
University of California
ISBN 0-930403-96-7

133. Dynamics of Detonations and Explosions: Detonations (1991)
A. L. Kuhl
Lawrence Livermore National Laboratory
J.-C. Leyer
Université de Poitiers
A. A. Borisov
USSR Academy of Sciences
W. A. Sirignano
University of California
ISBN 0-930403-97-5

134. Dynamics of Detonations and Explosions: Explosion Phenomena (1991)
A. L. Kuhl
Lawrence Livermore National Laboratory
J.-C. Leyer
Université de Poitiers
A. A. Borisov
USSR Academy of Sciences
W. A. Sirignano
University of California
ISBN 0-930403-98-3

(Other Volumes are planned.)